# Landscape and Power in Geographical Space as a Social-Aesthetic Construct

Olaf Kühne

# Landscape and Power in Geographical Space as a Social-Aesthetic Construct

Olaf Kühne
Eberhard Karls University
Tuebingen
Germany

ISBN 978-3-319-72901-5       ISBN 978-3-319-72902-2   (eBook)
https://doi.org/10.1007/978-3-319-72902-2

Library of Congress Control Number: 2017962030

© Springer International Publishing AG 2018
This work is subject to copyright. All rights are reserved by the Publisher, whether the whole or part of the material is concerned, specifically the rights of translation, reprinting, reuse of illustrations, recitation, broadcasting, reproduction on microfilms or in any other physical way, and transmission or information storage and retrieval, electronic adaptation, computer software, or by similar or dissimilar methodology now known or hereafter developed.
The use of general descriptive names, registered names, trademarks, service marks, etc. in this publication does not imply, even in the absence of a specific statement, that such names are exempt from the relevant protective laws and regulations and therefore free for general use.
The publisher, the authors and the editors are safe to assume that the advice and information in this book are believed to be true and accurate at the date of publication. Neither the publisher nor the authors or the editors give a warranty, express or implied, with respect to the material contained herein or for any errors or omissions that may have been made. The publisher remains neutral with regard to jurisdictional claims in published maps and institutional affiliations.

Printed on acid-free paper

This Springer imprint is published by Springer Nature
The registered company is Springer International Publishing AG
The registered company address is: Gewerbestrasse 11, 6330 Cham, Switzerland

# Foreword

This book is the product of almost two decades of continuous occupation with issues of landscape and power. It represents a condensed cross section of my publications on that theme, most of which have appeared in German: in particular the four monographs *Landschaft in der Postmoderne. Das Beispiel des Saarlandes* (2006—my Ph.D. thesis in the Faculty of Sociology), *Distinktion Macht Landschaft. Zur sozialen Definition von Landschaft* (2008), *Stadt – Landschaft – Hybridität. Ästhetische Bezüge im postmodernen Los Angeles mit seinen modernen Persistenzen* (2012) and *Landschaftstheorie und Landschaftspraxis* (2013).[1] This book aims, however, not just to make my own research up to the present point in time available to an international readership, but also to document the ongoing discussion about landscape in the German-speaking countries of Central Europe—a discussion undoubtedly rooted in language and culture. Hence, my coverage of this debate will focus especially on the literature in German.

Broad studies of this kind are scarcely imaginable without the critical stimulus of one's fellow researchers in this and kindred areas which, in my case, extend into practice and art. Among my fellow academics I would like to thank in particular Vera Denzer, Ludger Gailing, Markus Leibenath, Heidi Megerle, Rainer Kazig, Sebastian Kinder, Jochen Kubiniok, Barbara Neumann, Christoph Moning, Markus Reinke, Olaf Schnur, Annette Spellerberg, Boris Stemmer, Uta Stock-Gruber and Klaus Sachs. And special thanks must go to Diedrich Bruns, with whom I have for more than 10 years delved into the consequences of a constructivist perspective on landscape. For their observations on landscape and art my thanks are due to Harald Hullmann and Karl-Heinz Einberger; and, for stimulating conversations and constructive remarks from the point of view of professional practice, especially to Holger Zeck, Detlef Reinhard, Hanns Albert Letter, Reinhard Guth, Volker Wild, Bertold Huwig, Kurt Kniebe, Jörn Wallacher and Ulrich Franke. Finally, I would

---

[1]These titles could be rendered in English as: 'Landscape in Postmodernity: the Saarland' (2006); 'Distinction, Power, Landscape: on the Social Definition of Landscape' (2008); 'City – Landscape – Hybridity: Aesthetic Dimensions in Postmodern Los Angeles' (2012); and 'Landscape in Theory and Practice' (2013).

like to thank my present and former colleagues, especially Corinna Jenal, Albert Rossmeier, Antje Schönwald and Florian Weber, who have worked together with me in this field. For critical support on language matters, I thank Walter Strauss and especially for translation Joseph Swann.

My greatest debt of thanks, however, is to my wife, Sibylle Berger, who has continuously supported me in my work and provided many valuable and significant impulses.

Tuebingen  
June 2017

Olaf Kühne

# Contents

| | | |
|---|---|---|
| **1** | **Introduction** | 1 |
| | References | 6 |
| **2** | **Theoretical Foundations** | 11 |
| | 2.1 Social Constructivism | 11 |
| | 2.2 Landscape | 16 |
| | 2.3 Power | 22 |
| | 2.4 Aesthetics | 31 |
| | References | 38 |
| **3** | **The Genesis of Social Landscapes and Their Physical Manifestations** | 49 |
| | 3.1 The Genesis of Social Landscapes in Co-evolution with the Development of Physical Spaces in the German Language Area | 49 |
| |     3.1.1 Etymological Origins of the German Concept of *Landschaft* ('Landscape') | 49 |
| |     3.1.2 Landscape Painting from the Renaissance to the German Romantics | 51 |
| |     3.1.3 The Sociocritical Dimension of 'Landscape'—Life Beyond the City Streets | 55 |
| |     3.1.4 Landscape as Cultural Heritage—the Concept of *Heimat* (Home Environment) | 56 |
| |     3.1.5 A Specifically German Tale—The 'Wild Woods' | 57 |
| |     3.1.6 The Rejection of Romantic Concepts of Cultural Landscape and Home Environment Under the National Socialists, and the Post-war Rise of a Scientific Approach to Landscape | 58 |

|  |  | 3.1.7 | Post-industrial Landscapes and the Contemporary Understanding of Landscape in Germany | 60 |
|---|---|---|---|---|
|  | 3.2 | | The Genesis of Social Landscapes Outside the German Language Area | 64 |
|  | References | | | 70 |
| 4 | **The Social Genesis of the Definition of Landscape** | | | 79 |
|  | 4.1 | | Socialization of Landscape Constructs | 79 |
|  |  | 4.1.1 | General Socialization of Landscape in Childhood, Youth, and Early Adulthood | 81 |
|  |  | 4.1.2 | Influences on Landscape Awareness: Gender, Mobility, Mass Media | 84 |
|  |  | 4.1.3 | Socialization of the Concept of Landscape Among Specialists | 85 |
|  |  | 4.1.4 | Relation Between Lay and Expert Views of Landscape | 88 |
|  | 4.2 | | Social Distinction and Landscape | 90 |
|  |  | 4.2.1 | Social Distinction and the Aesthetics of Landscape | 90 |
|  |  | 4.2.2 | Social Distinction and the Aesthetics of Planning | 97 |
|  |  | 4.2.3 | Social Distinction and Landscape Experts—The Aesthetics of the Urban-Rural Hybrid | 98 |
|  |  | 4.2.4 | Appropriated Physical Landscape as an Embodiment of Social Distinction | 100 |
|  |  | 4.2.5 | Contingent Paradigms: The Conservation of 'Historical Cultural Landscape' and Its Alternatives | 105 |
|  | 4.3 | | Landscape and Power | 112 |
|  |  | 4.3.1 | Landscape, Power, and Economics | 113 |
|  |  | 4.3.2 | Landscape as a Medium of Symbolic Communication at the Interface of Science, Politics, Administration, and Civil Society | 125 |
|  |  | 4.3.3 | Landscape, Social Capital, and Power | 143 |
|  |  | 4.3.4 | The Concept of Landscape in Schoolbooks and 'Fact' Books for Children and Young People | 152 |
|  |  | 4.3.5 | The New Governance Paradigm—Perpetuating or Overthrowing Power Structures? | 158 |
|  | 4.4 | | Interim Summary with Further Reflections on the Interrelations of Socialization, Distinction, Power, and Landscape | 162 |
|  | References | | | 168 |
| 5 | **Case Studies** | | | 195 |
|  | 5.1 | | Landscape Between Modernization and Mystification: The American Grid and the Frontier | 195 |
|  | 5.2 | | Motorized Space—The Development of Los Angeles as an Urban-Rural Hybrid | 201 |

|   |     | 5.2.1 | Historical Aspects of the Co-evolution of Settlement and Transportation Infrastructure in LA | 202 |
|---|-----|-------|----------------------------------------------------------------------------------------------|-----|
|   |     | 5.2.2 | Impact of the Private Automobile on the Life and Environment of Los Angeles | 206 |
|   |     | 5.2.3 | Interim Summary | 210 |
|   | 5.3 | Landscape and Power in the Development of Eastern Europe—The Example of Warsaw | | 211 |
|   |     | 5.3.1 | Warsaw—City of Socialist Modernism | 212 |
|   |     | 5.3.2 | Warsaw—City of Post-socialist Postmodernism | 215 |
|   |     | 5.3.3 | Interim Summary | 221 |
|   | 5.4 | Power Conflicts—Landscape and the Impact of Renewable Energies | | 223 |
|   | 5.5 | Model Railroad Landscapes—Power Versus Contingency | | 228 |
|   | 5.6 | Landscape and Power in the Media—The Reproduction of Social Landscape Stereotypes in Internet Videos of Southern California | | 230 |
|   | References | | | 239 |
| 6 | **Conclusion** | | | 251 |
|   | References | | | 255 |

# Chapter 1
# Introduction

Landscape is a complex construct with a wide 'semantic train' (Hard 1969) of "associations, emotions, and evocations" (Hard 2002 [1983], p. 178) that allow the concept to be applied to any spatial context whose internal ordering "somehow suggests (or can be taken to suggest) a harmonious experiential whole" (Schrage 2004, p. 63). The topic has recently gained increasing attention in political, public, and scientific debate (see Stobbelaar and Pedroli 2011). Climate change, the growing importance of renewably sourced energy, the demographic shift, and the transformation of industrial into service-based societies are some of the many developments connected with change in our concepts of landscape. Added relevance has come from the insights of the 'new cultural geography' into the interactions of society and what it constitutes as 'space' (see Cosgrove and Jackson 1987; Gregory and Ley 1988; Cosgrove and Domosh 1997). For landscape is, as Freyer has argued, a good deal more than "an assemblage of natural facts": it is "a stretch of land with an inherently human dimension, and as such a reflexive construct" (Freyer 1996 [1966], p. 70).

That this construct is largely a matter of aesthetic interpretation was noted by Georg Simmel as far back as 1913. Simmel wrote (1996, p. 191): "Where we really see landscape rather than just an agglomeration of natural objects, we have a work of art *in statu nascendi*." Consistently with Simmel's dictum, Hauser and Kamleithner (2006, p. 74) argue that landscape is among the most "central, frequently used and, for that very reason, unclear concepts of the past thousand years of European political and intellectual history." Not only among specialists, but also in public and political discourse, the concept of landscape reveals a flexibility in its contextual application that leads Gailing and Leibenath (2012) to propose the following characterization:

- The term 'landscape' is comparatively open in meaning and, overall, has positive connotations; hence, it offers wide scope for identification.
- It brackets together a range of different, spatially oriented research issues.

- It gives rise to opposing and, indeed, contradictory interpretations, and hence stimulates "reflection and controversy [...] about different concepts of landscape and research ideas, and above all about how to design the spaces in which we live" (Gailing and Leibenath 2012, p. 96; for similar comments see Jones and Daugstad 1997; Schenk 2013).

Central aspects of current debate and reflection about landscape (escpecially in the German language area) are (a) whether it should be understood as a physical entity independent of the observer (as proposed in positivist/essentialist perspectives) or simply as a social and/or individual construct; and (b) whether—and if so to what extent—the city, too, can be seen as 'landscape'. The extension of the term 'landscape' to the city has in recent decades become increasingly common in sociological and urban planning contexts. The usage is associated not only with declared 'wild' areas in urban settings (e.g. parks, ruderal vegetation etc.), but also with the historical development of the notion of landscape as an essentially aesthetic way of looking at nature (Fischer 2011; Kühne 2012b), and with the use of natural materials for cultural purposes (e.g. house-building with stone)—a point to be further discussed later.

To ask what can be classified as landscape, and when and by whom it can be so classified, is to enter into a process of negotiation with diverse cultural and social aspects. Both the question of the social interpretation of landscape and that of the articulation of individual and personal interests inherent in that concept point to the broader issue of the origins of landscape discourse in societal power structures. Moreover, power is invariably linked with at least relative powerlessness (Paris 2005). This is also true of landscape. Where, for example, an area has already been taken over by industry, it is no longer possible for alternative uses (e.g. forestry) to gain a foothold. Certain ways of looking at and interpreting landscape will, in fact, always dominate over others; this is evident even in European funding policies, which support some uses of space and suppress others (see e.g. Kost and Schönwald 2015a). Gerhard Hard puts it like this: "Today any halfway competent, methodologically reflected, socio-geographical fieldwork will—hopefully as a matter of course—ask not only what social, economic, ecological and historical evidence can be detected in a specific area or landscape, but also what cannot, for that very reason, be detected (even though its implications for that specific environment might well have been more important). It will go further, too, and ask what we perhaps think we are seeing simply because we project our (possibly false) prior knowledge onto something whose meaning is, in fact, entirely different" (Hard 2008, p. 268). This way of reading landscape implies continuous reflection on power structures: What does a particular physical context *not* show? What does a particular society *not* interpret and accept as landscape? Moreover, what power structures lie behind such absences?

These questions frame the constructivist view of landscape that forms a theoretical background to this book. A fundamental tenet of social constructivism is that knowledge of the world in which we live comes only through interaction with those with whom we live. It is impossible, then, to know the world 'as it is'; we can

# 1 Introduction

access it only in pre-interpreted form. This is not to deny the existence of physical things, nor to deny their meaning for society. Social constructivism is concerned, rather, with how these meanings come about and how they are communicated. Wolfgang Welsch illustrates the relation between the two levels—that of things and that of the social construction of their meanings—with the example of architecture: "Architecture always has a twofold effect, real and symbolic. At the level of 'reality', it defines our living spaces and the scope of our actions; at the symbolic level, it shapes our ideas of urbanity, the environment and society. Doing so, it affects not only our practical lives but also our imaginations, desires and aspirations" (Welsch 1993, p. 13)

The social constructivist perspective has a long tradition in the social and cultural sciences, calling on such names as Husserl (1973 [1929]), Merleau-Ponty (1945), Berger and Luckmann (1966), Mead and Morris (1967), Blumer (1973), Schütz (1971 [1962], 1971, 2004 [1932]), and Schutz and Luckmann 1973). Specifically in the area of landscape research it has also shaped discussion for some decades—see e.g. Cosgrove (1984), Greider and Garkovich (1994), Duncan (1995), Makhzoumi (2002), Kühne (2006a, b, 2008, 2013), Jones (2007), DeLue (2008), Gailing (2008, 2012), Paasi (2008) Stakelbeck and Weber (2012), Bruns and Kühne (2013), Gailing and Leibenath (2015), Kost and Schönwald (2015b), Kühne and Weber (2015). A question closely connected with the constructivist perspective is: Who has power to define the terms of the debate, and how is that authority distributed? Applied to landscape this could be phrased as the question of the relation between what we call landscape and the societal power structures and processes that underlie this concept—a question that has come to the fore in the sciences concerned with the shaping and interpretation of geographical space since the late 1990s (see e.g. Olwig 1984, 1995, 2002; Mitchell 2001, 2002a, b; Schein 1997; Kühne 2008a, b; Mitchell 2008; Wescoat 2008; Kost and Schönwald 2015a; Leibenath 2015). Most of these studies, however, are more concerned with the impact of power on physical space than with the processes by which our ideas of landscape, and the preferences, norms and values with which it is connected, are socially constructed. It is these, nevertheless, that underlie such physical incursions. The present book aims to compensate for this bias (see also in this respect Kühne 2001, 2005, 2006a, 2008a, b, 2012, 2013; Kühne and Schönwald 2015); after all, as Schein (1997, p. 663) has succinctly put it: "When the action results in a tangible landscape element, or total ensemble, the cultural landscape becomes the discourse materialized."

The focus on the social foundations of the definition of landscape demands a widening of the perspective from classical geographical concerns to sociological (especially here the sociology of knowledge) and philosophical (especially aesthetic) points of view. Nevertheless, although the discussion of globalization and 'neoliberalism' in the contemporary socio-spatial sciences takes account of economic and political considerations and their implications with regard to the exercise and impact of power, it generally ignores administrative workings and civil, as well as scientific involvement in these processes. This, too, is a gap the present book aspires to fill.

The turn to constructivist positions in sociological research implies an intensified interest in language and allied sign systems, something Berger and Luckmann (1966, p. 68) already clearly expressed half a century ago: "Normally, of course, the decisive system is linguistic. Language objectivates the shared experiences and makes them available to all within the linguistic community" (see also Burr 2005; and in more specifically spatial contexts Daniels and Cosgrove 1988; Schein 1993; Greider and Garkovich 1994; Myers 1996; Cosgrove and Domosh 1997; Weber 2013). The quotation underlines the need for a high level of cultural sensibility in landscape research, given the firm rooting of the concept of landscape in cultural strata. Attention in this book will focus primarily on German-language cultures, but also to some extent on those of the English-speaking world. An overview of similar concepts in other cultures will be given in Chap. 3.

The social constructivist perspective taken in this book informs several different fields of sociologically oriented landscape research, including geography, psychology and philosophy, as well as architecture, urban and landscape planning, and sociology itself. A multi-perspective approach of this kind promises to do more justice to the complexity of the topic of landscape than any approach from a single research angle such as environmental psychology or geography. Given the wide range of disciplines in which landscape research is conducted, it cannot be conceived in linear terms: it consists, after all, of many parallel strands whose origins are in some cases independent of each other, and which may, indeed, cultivate mutual ignorance or even opposition, while other strands may be closely interwoven. Research initiatives may in some cases have been abandoned—and perhaps taken up again later in other contexts. The result is that there is today no consensus about what is and is not 'landscape', how it is constituted, or how it should in future be regarded. One of the tasks of the present book, then, is to provide an overview of the central strands of landscape research and to examine the implications of landscape in the various contexts in which it is intimately connected with the exercise of power.

To do this is to reflect on the meaning of power in social relationships—for, as Bourdieu (1991, p. 30) remarks, "power over space is power in one of its most privileged forms". Moreover, it is to reflect on the spatial manifestations of power at various levels: that of physical space, that of landscape (including the individual perception of physical space) as a social construct, and that of the social dimension of change occurring in that space (Kühne 2008a). On the one hand "landscape today [...] is determined by an aesthetic of 'nature' that reflects in the most subtle manner structures of ownership and power" (Franzen 2003, p. 124); on the other, power over nature means power over man (Althusser and Balibar 1968/69; Hickel 1984)—even if nature and landscape are not identical (see e.g. Haber 2006).

Against the background of an increasing sensibility to the natural environment in wide segments of the population, the last half century has witnessed vociferous complaints about changes to the landscape. Stigmatized as "loss, faulty planning, sell-out, exploitation, pollution, disfigurement and destruction" (Kaufmann 2005, p. 10), these can be interpreted as the result of an unequal distribution of power, both on the level of physical space and on that of the discourses in which landscape and change are defined (see also Krebs 2005). The aesthetic dimension of landscape also determines its symbolic relevance for processes of social distinction (see Bourdieu 1979; Zukin 1993; Duncan and Duncan 2004), whether these have to do with how it is perceived and valued, or with how it is physically distributed—e.g. which social group is to live where and under what terms. In this respect, the argument of this book is directed primarily to scholars and practitioners of the many disciplines and sub-disciplines concerned with physical space and landscape, as well as to those concerned with its perception and articulation. These include not only sociologists and philosophers but also geographers, landscape architects and planners.

Academic discourse on power is often seen as necessarily 'critical' (Leibenath 2015), and this is frequently taken (or even required—e.g. by Narr 2010) to mean critical in neo-Marxist terms. I have no quarrel with critical thinking in the philosophical sense of serious reflection, of non-affirmative inquiry into current structures of power and landscape, or of social-constructivist approaches to the contingency of language. But the perspective adopted here will not be neo-Marxist; rather it will take its cue from the theory of equity and fairness proposed by Rawls (1971), Nussbaum (2006), Tremmel (2009), Fainstein (2010) and Sen (2011)—which does not mean, however, that neo-Marxist diagnoses of the contemporary world should be rejected simply because they lead to different conclusions. The underlying concept of equity proposed here is based on the maximizing of vital opportunities, which, as Dahrendorf (2007, p. 44) observes, means "first and foremost opportunities to choose—i.e. options". Moreover, Dahrendorf goes on to say: "These require two things: the right to participate and the offer of a range of goods and activities from which to choose"; one might add that the choices in question must be meaningful (Kühne 2011a). The equitable distribution of opportunities can be derived from the second principle of John Rawls' theory of justice, namely that "Social and economic inequalities are to be arranged so that they are both (a) to the greatest benefit of the least advantaged, and (b) attached to offices and positions open to all under conditions of fair equality and opportunity" (Rawls 1971, p. 302). Fair chances for all means, in this context, that people of similar ability should have similar opportunities in life. Systematic inequality of access to offices and positions in society (e.g. on grounds of age, gender, sexual orientation, origins and ancestry is thereby ruled out (see also Kühne 2011b). It is not enough, however, to reduce this to legal terms (e.g. equality of statutory rights to social positions), as this takes no account of individual starting points. Given the practical difficulty of modifying a person's primary social status, Rawls demands intervention by the state to compensate for such inequities—a basically liberal position, which he sees as entailing taxation, and with it a restriction of the liberal tenet of untrammeled property rights. Dahrendorf (2007, p. 86) argues that secondary inequities are acceptable "so long as they do not put the winners in a position to hinder the full social participation of others or to prevent them on grounds of poverty from exercising their civil rights". While admitting that the formal regulation of social life contravenes the "hope of finally attaining an ideal human society" (2008, p. 67), Dahrendorf points out that such hope invariably contains a seed of inequality, of the division of society into friend and foe, and "hence [of] intolerance and the abuse of power" (ibid. 76; for a more detailed account of this position see Kühne 2011a and Kühne and Meyer 2015).

The present book is the result of some two decades of continuous involvement with theoretical and practical issues of landscape and planning. Precisely this latter aspect, with its methodology of participatory observation, has provided insights into the inner workings of power with relation to space and spatial planning (see especially Sects. 4.3.2 and 4.3.3. I have also made wide use of my own research findings, whether based on quantitative questionnaires (e.g. Kühne 2006a; Kühne and Spellerberg 2010, 2011), qualitative and quantitative media analysis (e.g. Kühne 2012a), qualitative interviews (as in Kühne and Schönwald 2015), or (socio) theoretical reflection (e.g. Kühne 2008a, 2014, 2015).

Apart from this introduction and a concluding summary, the book has four main chapters. Chapter 2 (Theoretical Foundations) is concerned largely with the perspectives and concepts of social constructivism, landscape, power, and aesthetics that underlie the entire work. Chapter 3 (The Genesis of Social Landscapes and their Physical Manifestations) considers the co-evolution of social constructs of landscape and human impacts on physical space. This already broaches issues of power and the development of concepts of landscape that form the matter of Chap. 4 (The Social Genesis of the Definition of Landscape). Here various processes of socialization are described in which landscape is constituted not only in itself but also as a medium of social distinctions. Chapter 5 (Case Studies) considers examples of the impact of power on and in landscape, from physical organization through the 'American grid' or through the imposition of socialist doctrine, to the struggle for interpretive authority over the values and ideals involved in particular landscape concepts. One such context cited here is the expansion of renewable energy sources; another is the communicative structures of model railway enthusiasts.

## References

Althusser, L., & Balibar, E. (1968/69). Lire ‚le Capital. Two Volumes. Paris: Maspero.
Berger, P., & Luckmann, T. (1966). *The social construction of reality*. New York: Penguin Books.
Blumer, H. (1973). Der methodologische Standort des symbolischen Interaktionismus. In Arbeitsgruppe Bielefelder Soziologen (Ed.), *Alltagswissen, Interaktion und gesellschaftliche Wirklichkeit* (Vol. 1, pp. 80–146). Reinbek bei Hamburg: Rowohlt.
Bourdieu, P. (1979). *La distinction: critique sociale du jugement*. Paris: Les Editions de Minuit.
Bourdieu, P. (1991). Physischer, sozialer und angeeigneter physischer Raum. In M. Wentz (Ed.), *Stadt-Räume. Die Zukunft des Städtischen* (pp. 25–34). Frankfurt a. M.: Suhrkamp.
Bruns, D., & Kühne, O. (2013). Landschaft im Diskurs. Konstruktivistische Landschaftstheorie als Perspektive für künftigen Umgang mit Landschaft. *Naturschutz und Landschaftsplanung, 45* (3), 83–88.
Burr, V. (2005). *Social constructivism*. London, New York: Busch-Lüty.
Cosgrove, D. E. (1984). *Social formation and symbolic landscape*. London, Sydney: University of Wisconsin Press.

# References

Cosgrove, D., & Domosh, M. (1997). Author and authority. Writing the new cultural geography. In J. Duncan & D. Ley (Eds.), *Place, culture, representation* (pp. 25–38). London: Routledge.

Cosgrove, D., & Jackson, P. (1987). New directions in cultural geography. *Area, 19*, 95–101.

Dahrendorf, R. (2007). *Auf der Suche nach einer neuen Ordnung. Vorlesungen zur Politik der Freiheit im 21. Jahrhundert.* München: C. H. Beck.

Dahrendorf, R. (2008). *Die Versuchungen der Unfreiheit. Die Intellektuellen in Zeiten der Prüfung.* München: C.H. Beck.

Daniels, S., & Cosgrove, D. (1988). Introduction: Iconography and landscape. In D. Cosgrove & S. Daniels (Eds.), *The iconography of landscape. Essays on the symbolic representation, design and use of environments* (pp. 1–10). Cambridge et al.: Cambridge University Press.

DeLue, R. (2008). Elusive landscapes and shifting grounds. In R. DeLue & J. Elkins (Eds.), *Landscape theory* (pp. 3–14). New York, London: Cambridge University Press.

Duncan, J. (1995). Landscape geography, 1993-94. *Progress in Human Geography, 19*(3), 414–422.

Duncan, J., & Duncan, N. (2004). *Landscapes of privilege. The politics of the aesthetic in an American Suburb.* New York, London: Routledge.

Fainstein, S. (2010). *The just city.* Ithacta: Cornell University Press.

Fischer, L. (2011). Landschaft und der doppelte Effekt von Arbeit. Vortrag auf der Konferenz: Konstituierung von Kulturlandschaft: Wie wird Landschaft gemacht? Given lecture on 12th May 2011 in Hannover.

Franzen, B. (2003). Driftende Landschaften. Einige Überlegungen zur zeitgenössichen Landschaftswahrnehmung. In I. Flagge (Ed.), *Architektur und Wahrnehmung. Jahrbuch Licht und Architektur 2003* (pp. 122–125). Darmstadt: Rudolf Müller.

Freyer, H. (1996 [1966]). Landschaft und Geschichte. In G. Gröning & U. Herlyn (Eds.), *Landschaftswahrnehmung und Landschaftserfahrung* (pp. 69–90). Münster: lit.

Gailing, L. (2008). Kulturlandschaft—Begriff und Debatte. In D. Fürst, L. Gailing, K. Pollermann & A. Röhring (Eds.), *Kulturlandschaft als Handlungsraum. Institutionen und Governance im Umgang mit dem regionalen Gemeinschaftsgut Kulturlandschaft* (pp. 21–34). Dortmund: Rohn-Verlag.

Gailing, L. (2012). Sektorale Institutionensysteme und die Governance kulturlandschaftlicher Handlungsräume. Eine institutionen- und steuerungstheoretische Perspektive auf die Konstruktion von Kulturlandschaft. *Raumforschung und Raumordnung, 70*(2), 147–160.

Gailing, L., & Leibenath, M. (2012). Von der Schwierigkeit, Landschaft' oder, Kulturlandschaft' allgemeingültig zu definieren. *Raumforschung und Raumordnung, 70*(2), 95–106.

Gailing, L., & Leibenath, M. (2015). The social construction of landscapes: Two theoretical lenses and their empirical applications. *Landscape Research, 40*(2), 123–138.

Gregory, D., & Ley, D. (1988). Culture's geographies. *Environment and Planning D: Society and Space, 6*(2), 115–227.

Greider, T., & Garkovich, L. (1994). Landscapes: The social construction of nature and the environment. *Rural Sociology, 59*(1), 1–24.

Haber, W. (2006). Kulturlandschaften und die Paradigmen des Naturschutzes. Stadt+Grün, 55, 20–25.

Hard, G. (1969). Das Wort Landschaft und sein semantischer Hof. Zur Methode und Ergebnis eines linguistischen Tests. *Wirkendes Wort, 9*, 3–14.

Hard, G. (2002 [1983]). Zu Begriff und Geschichte von "Natur" und "Landschaft" in der Geographie des 19. und 20. Jahrhunderts. In G. Hard (Ed.), *Landschaft und Raum. Aufsätze zur Theorie der Geographie* (pp. 171–210). Osnabrück: University Press Rasch.

Hard, G. (2008). Der Spatial Turn, von der Geographie her beobachtet. In J. Döring & T. Thielmann (Eds.), *Spatial turn. Das Raumparadigma in den Kultur- und Sozialwissenschaften* (pp. 263–316). Bielefeld: transkript.

Hauser, S., & Kamleithner, C. (2006). *Ästhetik der Agglomeration.* Wuppertal: Müller und Busmann.

Hickel, E. (1984). Tod der Natur—Frauen, Ökologie und wissenschaftliche Revolution. *Wechselwirkung, 23*, 34–37.

Husserl, E. (1973 [1929]). *Cartesianische Meditationen und Pariser Vorträge*. Husserliana I. Den Haag: Martinus Nijhoff.
Jones, M. (2007). The European landscape convention and the question of public participation. *Landscape Research, 32*(5), 613–633.
Jones, J., & Daugstad, K. (1997). Usages of the 'cultural landscape' concept in Norwegian and Nordic landscape administration. *Landscape Research, 22*(3), 267–281.
Kaufmann, S. (2005). *Soziologie der Landschaft*. Wiesbaden: Springer.
Kost, S., & Schönwald, A. (Eds.). (2015a). *Landschaftswandel—Wandel von Machtstrukturen*. Wiesbaden: Springer.
Kost, S., & Schönwald, A. (2015b). Einleitung. In S. Kost & A. Schönwald (Eds.), *Landschaftswandel—Wandel von Machtstrukturen* (pp. 9–16). Wiesbaden: Springer.
Krebs, S. (2005). Was heißt hier Landschaft? Eine transatlantische Beziehungsgeschichte. In B. Franzen & St. Krebs (Eds.), *Landschaftstheorie. Texte der Cultural Landscape Studies* (pp. 304–324). Köln: Walther König.
Kühne, O. (2001). The interaction of industry and town in Central Eastern Europe—An intertemporary comparison based on systems theory and exemplified by Poland. *Die Erde, 132*(2), 161–185.
Kühne, O. (2005). *Landschaft als Konstrukt und die Fragwürdigkeit der Grundlagen der konservierenden Landschaftserhaltung—eine konstruktivistisch-systemtheoretische Betrachtung. Beiträge zur Kritischen Geographie 4*. Wien: Verein kritische Geographie.
Kühne, O. (2006a). *Landschaft in der Postmoderne. Das Beispiel des Saarlandes*. Wiesbaden: Deutscher Universitätsverlag.
Kühne, O. (2006b). Landschaft und ihre Konstruktion—theoretische Überlegungen und empirische Befunde. *Naturschutz und Landschaftsplanung, 38*, 146–152.
Kühne, O. (2008a). *Distinktion, Macht, Landschaft. Zur sozialen Definition von Landschaft*. Wiesbaden: Springer.
Kühne, O. (2008b). Kritische Geographie der Machtbeziehungen—konzeptionelle Überlegungen auf der Grundlage der Soziologie Pierre Bourdieus. *Geographische Revue, 10*(2), 40–50.
Kühne, O. (2011a). Die Konstruktion von Landschaft aus Perspektive des politischen Liberalismus. Zusammenhänge zwischen politischen Theorien und Umgang mit Landschaft. *Naturschutz und Landschaftsplanung, 43*(6), 171–176.
Kühne, O. (2011b). Akzeptanz von regenerativen Energien—Überlegungen zur sozialen Definition von Landschaft und Ästhetik. *Stadt+Grün, 60*(8), 9–13.
Kühne, O. (2012a). Urban nature between modern and postmodern aesthetics: Reflections based on the social constructivist approach. *Quaestiones Geographicae, 31*(2), 61–70.
Kühne, O. (2012b). *Stadt—Landschaft—Hybridität. Ästhetische Bezüge im postmodernen Los Angeles mit seinen modernen Persistenzen*. Wiesbaden: Springer.
Kühne, O. (2013). *Landschaftstheorie und Landschaftspraxis. Eine Einführung aus sozialkonstruktivistischer Perspektive*. Wiesbaden: Springer.
Kühne, O. (2014). Das Konzept der Ökosystemdienstleistungen als Ausdruck ökologischer Kommunikation. Betrachtungen aus der Perspektive Luhmannscher Systemtheorie. *Naturschutz und Landschaftsplanung, 46*(1), 17–22.
Kühne, O. (2015). Weltanschauungen in regionalentwickelndem Handeln—die Beispiele liberaler und konservativer Ideensysteme. In O. Kühne & F. Weber (Eds.), *Bausteine der Regionalentwicklung* (pp. 55–69). Wiesbaden: Springer.
Kühne, O., & Meyer, W. (2015). Gerechte Grenzen? Zur territorialen Steuerung von Nachhaltigkeit. In O. Kühne & F. Weber, F. (Eds.), *Bausteine der Regionalentwicklung* (pp. 25–40). Wiesbaden: Springer.
Kühne, O., & Schönwald, A. (2015). *San Diego—Eigenlogiken, Widersprüche und Entwicklungen in und von, America's finest city*. Wiesbaden: Springer.
Kühne, O., & Spellerberg, A. (2010). *Heimat und Heimatbewusstsein in Zeiten erhöhter Flexibilitätsanforderungen. Empirische Untersuchungen im Saarland*. Wiesbaden: Springer.
Kühne, O., & Spellerberg, A. (2011). Heimat in ihrer sozialen Bedeutung—das Beispiel des Saarlandes. *Rheinische Heimatpflege, 48*(4), 295–302.

# References

Kühne, O., & Weber, F. (2015). Der Energienetzausbau in Internetvideos—eine quantitativ ausgerichtete diskurstheoretisch orientierte Analyse. In S. Kost & A. Schönwald (Eds.), *Landschaftswandel—Wandel von Machtstrukturen* (pp. 113–126). Wiesbaden: Springer VS.

Leibenath, M. (2015). Landschaften und Macht. In S. Kost & A. Schönwald (Eds.), *Landschaftswandel—Wandel von Machtstrukturen* (pp. 17–26). Wiesbaden: Springer.

Makhzoumi, J. M. (2002). Landscape in the Middle East. An inquiry. *Landscape Research, 27*(3), 213–228.

Mead, G., & Morris, C. (1967). *Mind, self & society from the standpoint of a social behaviorist.* Chicago: University of Chicago Press.

Merleau-Ponty, M. (1945). *Phénoménologie de la perception.* Paris: Gallimard.

Mitchell, D. (2001). *Cultural geography. A critical introduction.* Oxford: Blackwell.

Mitchell, W. (2002a). Introduction. In W. Mitchell (Ed.), *Landscape and power* (pp. 1–4). Chicago, London: University of Chicago Press.

Mitchell, W. (2002b). Imperial landscape. In W. J. T. Mitchell (Ed.), *Landscape and power* (pp. 5–34). Chicago, London: University of Chicago Press.

Mitchell, D. (2008). New axioms for reading the landscape: Paying attention to political economy and social justice. In J. Wescoat & D. Johnston (Eds.), *Political economies of landscape change: Places of integrative power* (Vol. 89, pp. 29–50). Dordrecht: Springer.

Myers, G. (1996). Naming and placing the other: Power and the urban landscape in Zanzibar. *Tijdschrift voor economische en sociale geografie, 87*(3), 237–246.

Narr, W.-D. (2010). Wie kommt's zum Raum-Echo? http://www.raumnachrichten.de/ressource/buecher/1187-raum. Accessed April 21, 2016.

Nussbaum, M. (2006). *Frontiers of justice: Disability, nationality, species membership.* Cambridge, London: Harvard University Press.

Olwig, K. (1984). *Nature's ideological landscape.* London: New Left Books.

Olwig, K. (1995). Reinventing common nature: Yosemite and Mt. Rushmore—A meaning tale of a double nature. In W. Cronon (Ed.), *Uncommon ground: Towards reinventing nature* (pp. 379–408). New York: W.W. Norton and Company.

Olwig, K. (2002). *Landscape, nature, and the body politic. From Britain's renaissance to America's new world.* London: University of Wisconsin Press.

Paasi, A. (2008). Finnish landscape as social practice. Mapping identity and scale. In M. Jones & K. Olwig (Eds.), *Nordic Landscapes. Region and Belonging on the Northern Edge of Europe* (pp. 511–539). Minneapolis, London: University of Minnesota Press.

Paris, R. (2005). *Normale Macht. Soziologische essays.* Konstanz: UVK Verlagsgesellschaft mbH.

Rawls, J. (1971). *A theory of justice.* Cambridge, Massachusetts: Harvard University Press.

Schein, R. (1993). Representing urban America: 19th-century views of landscape, space, and power. *Environment and Planning. D: Society and Space, 11*(1), 7–21.

Schein, R. (1997). The place of landscape: A conceptual framework for interpreting an American scene. *Annals of the Association of American Geographers, 87*(4), 660–680.

Schenk, W. (2013). Landschaft als zweifache sekundäre Bildung—historische Aspekte im aktuellen Gebrauch von Landschaft im deutschsprachigen Raum, namentlich in der Geographie. In D. Bruns & O. Kühne (Eds.), *Landschaften: Theorie, Praxis und internationale Bezüge* (pp. 23–34). Schwerin: Oceano.

Schrage, D. (2004). Abstraktion und Verlandschaftlichung. Moderne Räume und Artifizialität. In W. Eßbach, S. Kaufmann, D. Verdicchio, W. Lutterer, S. Bellanger & G. Uerz (Eds.), *Landschaft, Geschlecht, Artefakte. Zur Soziologie naturaler und artifizieller Alteritäten* (pp. 63–78). Würzburg: Ergon.

Schütz, A. (1971 [1962]). *Gesammelte Aufsätze 1. Das Problem der sozialen Wirklichkeit.* Den Haag: Martinus Nijhoff.

Schütz, A. (1971b). *Gesammelte Aufsätze 3. Studien zur phänomenologischen Philosophie.* Den Haag: Martinus Nijhoff.

Schütz, A. (2004 [1932]). *Der sinnhafte Aufbau der sozialen Welt. Eine Einleitung der sozialen Welt.* Konstanz: UVK Verlagsgesellschaft mbH.

Schütz, A., & Luckmann, T. (1973). *The structures of the life-world* (Vol. 1). Evanston, Illinois: Northwestern University Press.

Sen, A. (2011). *The idea of justice*. Cambridge, Massachusetts: Belknap Press/Harvard University Press.

Simmel, G. (1996 [1913]). Philosophie der Landschaft. In G. Gröning & U. Herlyn (Eds.), *Landschaftswahrnehmung und Landschaftserfahrung. Texte zur Konstitution und Rezeption von Natur als Landschaft* (pp. 91–105). München: Minerva Publication.

Stakelbeck, F., & Weber, F. (2012). Almen als alpine Sehnsuchtslandschaften: Aktuelle Landschaftskonstruktionen im Tourismusmarketing am Beispiel des Salzburger Landes. In D. Bruns & O. Kühne (Eds.), *Landschaften: Theorie, Praxis und internationale Bezüge* (pp. 235–252). Schwerin: Oceano.

Stobbelaar, D., & Pedroli, B. (2011). Perspectives on landscape identity. A conceptual challenge. *Landscape Research, 36*(3), 321–339.

Tremmel, J. (2009). *A theory of intergenerational justice*. London: Earthscan.

Weber, F. (2013). *Soziale Stadt—Politique de la Ville—Politische Logiken. (Re-)Produktion kultureller Differenzierungen in quartiersbezogenen Stadtpolitiken in Deutschland und Frankreich*. Wiesbaden: Springer VS.

Welsch, W. (1993). Städte der Zukunft. Philosophische Überlegungen. In Kulturkreis der Deutschen Wirtschaft im Bundesverband der Deutschen Industrie (Ed.), *Wohnen und Arbeiten. Städtebauliches Modellprojekt Schwerin-Lankow* (pp. 12–18). Heidelberg: Spektrum.

Wescoat, J. (2008). Introduction. Three faces of power in landscape change. In J. L. Wescoat & D. M. Johnston (Eds.), *Political economies of landscape change: Places of integrative power*. GeoJournal library Volume 89 (pp. 1–27). Dordrecht: Springer.

Zukin, S. (1993). *Landscapes of power. From Detroit to Disney World*. Berkeley, Los Angeles, London: University of California Press.

# Chapter 2
# Theoretical Foundations

This chapter provides the theoretical and conceptual foundation of the present work. These comprise on the one hand a survey of the main theoretical perspectives of a social constructivist approach to landscape and its aesthetics, and on the other a brief outline of the concept of power.

## 2.1 Social Constructivism

A notable gap has opened in recent decades between the largely constructivist positions of the social sciences and humanities and the (for the most part) realist perspective of the natural sciences (Egner 2010). The tension can be felt within individual sciences like psychology, sociology and geography, whose roots lie in both cultures, and which make an essential contribution to landscape research. Trepl (2012, p. 29) illustrates the different approaches in terms of their attitude to concepts: "Concepts for the natural scientist are on the whole means to an end; for the human and social scientist the understanding of concepts is itself the end"—so long, I would add, as we are not talking about those aspects of the human sciences, like quantitative sociological research, that predominantly use the methods of natural science.

The realism of natural science springs from an acceptance of the possibility of objective knowledge and of its attainment through appropriate empirical procedures. This, in turn, presumes the existence of a reality structured independently of human knowing processes but accessible through those processes (Gergen 1985; Bailer-Jones 2005; Burr 2005; Gergen and Gergen 2009). The extreme case of realism, known as 'naïve realism', amounts to an unconditional belief in the reality of the physically perceived world (Wetherell and Still 1998). Burr (1998, 2005) distinguishes three aspects of what we call 'reality': truth as opposed to falsehood, materiality as opposed to illusion, and essence (what a thing is in itself) as opposed to construct (what human/social knowledge makes of it). In contrast to such realism, constructivism sees reality as the product of everyday social practice: the result not just of thought but of action (Berger and Luckmann 1966). This immediately involves a critical dimension: "Social constructivism insists that we take a critical stance toward our taken-for-granted ways of understanding the world, including ourselves" (Burr 2005, pp. 2–3; see also Schütz and Luckmann 1973). From the thesis that "reality is socially constructed" Berger and Luckmann (1966, p. 1) derive the conclusion that "the sociology of knowledge must analyze the processes in which this occurs".

Hence, far from being anti-empirical, social constructivist research is devoted to empirical investigation of "which specific interpretations of reality do, and which do not, achieve normative status" (Kneer 2009, p. 5). The focus here is on the contextual anchoring of empirical knowledge—indeed, the constructivist question itself arises in the context of specific scientific concerns. Since the initial appearance of the movement in the 1960s, many research fields—from gender and cultural studies, through critical and discursive psychology, to discourse analysis, deconstruction, post-structuralism, and postmodern approaches in general—have adopted social constructivist ideas (Burr 2005; Kühne 2006a, 2008a; Gergen and Gergen 2009). In contrast to radical constructivism (see e.g. Luhmann 1984; Maturana and Varela 1987; Glasersfeld 1995), social constructivism, however, is not rooted in the biological sciences, nor does it pursue a full-scale epistemological program. It is more concerned with pre-scientific life practices and their social dimensions of change (Hacking 1999; Miggelbrink 2002; Egner 2010; Kühne 2013a; Lynch 2016). Reusswig (2010, p. 79) cites the example of climate change to illustrate the latent conflict in these approaches from a constructivist point of view: "That natural scientists so often reject the basic thesis […] of the social 'construction' of climate is due as a rule to an erroneously derived, abstract perception of social actors and systems as 'physical objects' bound up in some sort of causal chain with natural processes."

Social constructivist approaches are particularly indebted to the phenomenological sociology of Alfred Schütz, which in turn takes its cue on the one hand from Max Weber's sociology of *Verstehen* (understanding informed by empathy), with its key concept of 'meaningfully derived action', and on the other from the 'phenomenological philosophy' of Edmund Husserl. This, according to Zahavi (2007, p. 13), entails "the philosophical analysis of the various forms in which objects appear". For Husserl (1973 [1929]), intersubjective experience is constitutive for the development of the individual subject, who can only experience 'the world' as a member of society. Linking Weber's sociological approach with Husserl's philosophy, Schütz (2004 [1932]) extended this by introducing the concept of 'social meaning': i.e. the meanings men and women accord to their actions. This shifted the focus of sociological research onto the individual subject and the meanings that may be associated with their actions—a way of thinking fundamentally different from that of natural science, whose objects develop no self-awareness, no understanding of the world, and no interests, motives or 'meanings' (see Schütz 1971 [1962]).

## 2.1 Social Constructivism

In line with these reflections, interpretations of the (socially generated) world by social scientists can be thought of as "second level constructs: constructs of constructs formed in and by the social activities of men and women whose behavior scientists observe and seek to explain in accordance with the procedures of their science" (Schütz 1971 [1962], p. 7). 'Construct' here "does not imply an intentional act, but a preconscious cultural process" (Kloock and Spahr 2007, p. 56). For every act of perception is informed by prior knowledge of the world in the form of abstractions (Schütz 1971)—as in the present instance with respect to 'landscape': "nowhere is there such a thing as a pure and simple fact" (Schütz 1971 [1962], p. 5; Burr 2005). The social construction of 'world' takes place in the fusion of individual sense percepts into an overall picture; nor should perception in this sense be thought of as an isolated event: it is the result of "a highly complex process of interpretation in which present percepts are related to past ones" (Schütz 1971 [1962], pp. 123–124) and referential structures thereby actualized.

Of central importance in the experience of 'world' as the social reality in which we live, learn, grow and suffer (Schütz and Luckmann 1973) is a process of typification that gains in importance with increasing social distance: "The typifications of social interaction become progressively anonymous the farther away they are from the face-to-face situation. Every typification, of course, entails incipient anonymity" (Berger and Luckmann 1966, p. 31). The process of typification involves a socially communicated body of values, roles, rules and norms—one need only think of the maxims cited in everyday actions—that ascribe normality and abnormality to situations, actions and appearances, as well as to spatial objects and clusters. It follows that, far from being isolated interpretive schemata, typifications rely for their genesis and meaning on mutual interconnectedness (Schütz and Luckmann 1973). Here, objects play a central role, for "every object and every utensil points to those anonymous people who produced it so that other anonymous people could use it to attain typical ends with typical means" (Schütz 1971 [1962], p. 20). Typifications of this sort construct a familiar, routine world without our having to reflect on the conditions and processes through which it has come about (Berger and Luckmann 1966; Garfinkel 1967; Zahavi 2007). For the point of familiarity and routine is not that we should know some truth but that we should be able to act in a social context.

Knowledge comes about in a threefold network of familiarity, recognition and belief (Schütz 1971 [1962], 1971):

- Something is familiar if we "understand not only what it is and how it is constituted but also why it is so" (Schütz 1971, p. 157)—for instance, those houses are built of wood so that people can live in them.
- We recognize something if we "know what it is but do not ask why" (ibid.)—for example, we can recognize a blast furnace if we see one, without (unless we are specialists in the field) knowing how it works.
- If we do not know, or only vaguely know, what something is, we might say 'I believe it is a such and such'. Belief in this sense may be more or less well founded, or based on trust in authority, or be the fruit of ignorance (ibid.). Thus,

a shield volcano may be recognized as a volcano without any knowledge of how it came to have its particular form, or it may be thought of simply as an elevation in the landscape.

Knowledge grows and is communicated through social interactions and continuously changing processes of interpersonal relation and exchange. It is these, rather than immediate individual perceptions, that are responsible for most of what a normal adult knows. Social processes of negotiation and mediation are also constitutive in the development of the individual. For, as Mead and Morris (1967) observed, one does not primarily perceive oneself directly but indirectly, from the viewpoint of other members of one's social group, either individually communicated or generalized. By taking over the attitudes of others, one objectifies oneself (ibid.). Knowledge, in other words, is socially communicated (Schütz and Luckmann 1973). We do not 'naturally' know what is and is not a volcano: we learn this in various ways, from family, friends, teachers etc.

Two sorts of social interaction can in this context be distinguished: 'non-symbolic'—instinctual interactions that require no reflection; and 'symbolic'—interactions whose meaning is generated in a process of social negotiation or definition using signs (Mead and Morris 1967; Blumer 1973). Signs become symbols when the same meaning is made for the recipient as for the giver of the sign: for example, when concertgoers clap their hands the pianist regards this as a sign of approbation for the performance; or where a flowerbed is marked off from the path with a low hedge, this is taken to mean that one should not walk on the bed—i.e. it communicates the will of the person who planted the hedge. Symbolic communication involves what Berger and Luckmann (1966) describe as processes of externalization and internalization. Externalization is the ascription of meaning to objects, making them into signs—e.g. a knife as a commonly accepted sign of aggression (Burr 2005; see also Costonis 1982). Conversely, internalization refers to the socialization of objects and actions in the symbolic world of a particular society with its everyday meanings, institutions, rules and typifications (Burr 2005; for greater detail see below: Sect. 4.1—Socialization of Landscape Constructs).

Symbolic interactions are generally connected with 'things' in the broadest sense of whatever a person perceives in their world—physical objects, like a tree or a chair; other people, like friends or enemies; institutions, like school or the government; ideals and principles, like independence or honesty; the actions of others, like commands or wishes; and whatever situations an individual encounters in daily life (Blumer 1973). For in relation to things, people act on the basis of the meaning these things have for them, which originates in their mutual social interactions (Blumer 1973), one meaning adding to another. Such meanings are not stable or irreversible: they "are negotiated and changed in the interpretive process engaged in by the individual with the things they encounter" (Blumer 1973, p. 81). Their inherent reversibility opens these meanings to the power processes of discourse, which may be deeply interested in their preservation or revision; and this introduces the question as to who has this power, who can promote or defend one meaning against another (see e.g. Weber 2013; Hannigan 2014; Kühne and Weber 2015).

## 2.1 Social Constructivism

Social constructivist landscape research regards physical objects like trees, houses and fields "as symbols [...]—concrete material embodiments of social ideas, relations, habits, lifestyles etc.—in which the social is abstracted from its physical embodiments by a process of interpretation" (Hard 1995, p. 52). This perspective distinguishes social constructivism from other constructivist viewpoints like radical constructivism or the discourse theory of Laclau and Mouffe (1985—see also Glasze and Mattissek 2009; Weber 2013, 2015), which focuses on social communication to the (virtual) exclusion of the material dimension of communicative power processes.

To return to the question of knowledge: the acquisition of knowledge, and hence knowledge itself—especially in the case of socially communicated meanings—is not equally available to all. It is both socially and culturally differentiated. The reality of a Tibetan monk—to cite the example given by Berger and Luckmann (1966)—is quite different from that of an American businessman. Moreover, this variation in the construction of 'reality' and the distribution of knowledge indicates the historical nature of thought (Berger and Luckmann 1966; Stearns 1995). The sociocultural differentiation of knowledge is in many cases also geographical—as the example of the monk and the businessman already suggests. Berger and Luckmann (1966) argue, however, that among the many realities of science, business, administration, politics etc. one stands out as the cornerstone and point of relation of all the others: the reality of everyday life. This is characterized by pre-arranged patterns; it seems already objectivized, for its meanings have already been inscribed before the individual subject encounters and acts within it: We are born into a world whose conceptual framework and categories are already given in the inherited culture" (ibid.; see also Burr 2005). The reality of the everyday world seems to be ordered around the 'here' of the body, as the condition of all spatial experience of the world in which we live (Merleau-Ponty 1945), and the 'now' of the bodily present. This 'here and now' is the starting point for the construction of the individual's world: it is accepted as real and requires no further verification (Berger and Luckmann 1966). In this sense, the reality of the world is socially mediated: "The reality of everyday life further presents itself to me as an inter-subjective world, a world that I share with others" (Berger and Luckmann 1966, p. 23).

The everyday world is, then, perceived as real and objective (Berger and Luckmann 1966). This involves the objectivization of essentially subjective experiences—i.e. their embodiment in the processes and objects of daily life: "The reality of everyday life is not only filled with objectivations [sic]; it is only possible because of them" (Berger and Luckmann 1966, p. 35). Three stages of objectivization can be distinguished (Schütz and Luckmann 1973; Rammert 2007). In the first stage, it takes the form of actions in shared situations—e.g. the discursive agreement to (constructively) judge a landscape beautiful. In the second stage, the term is applied to the physical products of human action—technical devices and constructs such as gardens, woods, cultivated land, settlements etc. The third stage is characterized by the situative decoupling effected by the abstract sign systems of language. These introduce a new level of anonymization and idealization (Schütz and Luckmann 1973)—for example, the word 'landscape' releases specific associations irrespective of whether or not it is used in a physical space that might normally be referred to as landscape. Abstraction and anonymity increase with increasing distance (social as well as spatial) from the point of reference of an individual's 'here and now' (Berger and Luckmann 1966). Individualized personal perceptions will then often yield to generalizations like 'they are typical nouveau riches/on benefits/Americans/East Germans etc.' (see Schlottmann 2005).

The structures of everyday reality remain unproblematic so long as the routine typifications that generate them are not disturbed by the need for adjustment; or as Schütz and Luckmann (1973) put it, so long as the unbroken chain of everyday structures can be taken for granted. This also applies to persons: the stranger who conforms to the expected roles of my world is unproblematic, he 'fits in', he is 'integrated'; he only becomes problematic when I have to adjust my own perspective to fit him (Berger and Luckmann 1966). Or in other words: "The world is accepted as familiar and in that sense 'real' up to the point, at least, when it becomes problematic and needs to be questioned" (Werlen 2000, p. 39).

The social division of knowledge which has accompanied its rapid expansion in the course of the modernization of society has created more or less closed provinces of meaning that are clearly marked off from everyday knowledge—one need only think of academic sociologists and philosophers, or of mechatronic engineers, designers and planners. All such groups tend to develop their own symbols of authority, ranging from working apparel and dress codes to specialist language, which distinguish them from laypeople (Berger and Luckmann 1966; for greater detail see below: Sect. 4.2). These symbolic worlds are in constant competition with each other—as evidenced, for example in the question, whether landscape is a physical object or a social construct. However, the development of alternative symbolic worlds shows that the interpretations inherent in earlier worlds of meaning were neither conclusive nor mandatory; for absolute knowledge of the world (including oneself) is unattainable. As Berger and Luckmann (1966, p. 116) observe: "Because they are historical products of human activity, all socially constructed universes change, and the change is brought about by the concrete actions of human beings". The application of this insight to landscape, with its implications of continuing historical change, is a central concern of this book.

## 2.2 Landscape

From a social constructivist perspective landscape is not an objective, univocally definable entity existing within a physical, material world: it is the sociocultural product of a process of mediation (Wojtkiewicz and Heiland 2012; see also Cosgrove 1985; Greider and Garkovich 1994; Graham 1998; Ipsen 2002; Mitchell 2002; Soyez 2003; Kaufmann 2005; Ahrens 2006; Kühne 2006a, b, c, d, 2008a, 2009; Chilla 2007; Backhaus et al. 2007; Lingg et al. 2010; Micheel 2012). Greider and Garkovich (1994, p. 2) explain this process as follows: "Through sociocultural phenomena, the physical environment is transformed into landscapes that are the reflections of how we define ourselves". Cosgrove (1984, p. 13) succinctly characterized the transition from an objectivist to a constructivist understanding of landscape in visual terms: "Landscape is not merely the world we see; it is a construction, a composition of that world. Landscape is a way of seeing the world." This reflected the approach of W.G. Hoskins in *The Making of the English Landscape* (2006 [1956]), which examined not only the development of physical structures but also the historical process of affective appropriation in which the concept of England and its landscape came into being. Commenting on the academic dimension of this insight, Cosgrove (1985, p. 47) warned that "the landscape idea is a visual ideology; an ideology all too easily adopted unknowingly into geography when the landscape idea is transferred as an unexamined concept into our discipline". Ideological blindness of this sort has not been restricted to geography; it was (and to some extent still is) to be found in neighbouring disciplines as well; but, as already observed—and thanks not least to Dennis Cosgrove's work—the situation in geography has improved.

Landscapes are the product on the one hand of abstraction, on the other of emotive projection (Goodman 1951; see also Daniels and Cosgrove 1988; Greider and Garkovich 1994; Agnew 1997; Cosgrove 1998; Graham 1998; Howard 2011); the interpretation of sense impressions that generates them is—as indicated in the previous section on social constructivism—based on what the individual learns in and from the long social evolutionary process of normative acculturation. Landscape, then, is experienced not as a visual image of the world but as an integral aspect of its meaning (see also Burr 1995; Assmann 1999; Wöhler 2001); or as Helmut Rheder already observed in the early 1930s, "that nature is grasped as landscape is the work of thought" (1932, p. 1). It is this that produces forms and distinctions between the similar and dissimilar, between "what human vision separates and moulds into separate entities, [...] into the individuality of 'landscape'" (Simmel 1996 [1913], p. 95).

The process is unconscious—which is why "it does not appear to us as a social construction, but as reality" (Ipsen 2006, p. 31). Gailing (2012, p. 3) describes it as follows: landscapes become "more or less distinct spatial units, reified into an ontological synthesis"—reification (or hypostasization) being what Werlen (2000) has called the mental transformation of a concept of landscape or space into a thing. Following Schlottmann (2005), Gailing (2012, p. 149) understands the reification of 'spaces' or 'landscapes' as the "unequivocal conception of spatial units as independent of human action and the human observer, and hence as in principle non-negotiable" (see also Miggelbrink 2002); this represents, he continues, "the treatment of abstractions as substantial entities" (ibid.). The construct 'landscape' is thought of as an immediately perceptible object. This entails, however, a linguistic process commonly observable in phrases such as 'the Camargue/Cape Breton etc. landscape', which are normative as well as descriptive, for in everyday usage they associate with terms denoting beauty, wildness or other values connected with the desirability of conservation. In any investigation of landscape—as in all constructivist research—the "context of language usage" (Strüver and Wucherpfennig

2009, p. 117) plays an indispensable role. This immediately poses the question of power: Whose language? Who constructs landscape, in what social context, and how? Who establishes the construct as normative?

The conscious perception of 'landscape' as the inherent form of specific objects entails more than merely recognizing those objects, "just as understanding a text entails more than recognizing the meaning of its words" (Berendt 2005, p. 29). In the terminology of systems theory it can be seen as an act of complexity reduction, or what Miggelbrink (2009, p. 191) calls "a reaction to a perceived need for structuring" (see also in a landscape context Papadimitriou 2010; Kühne 2004, 2014)—an act that enables (and is indispensable for) orientation. Without such reduction, "every tree would have to be encountered anew, for none is quite the same as any other" (Eibl-Eibesfeldt 1997, p. 901), and no number of adjacent trees would be recognizable as a wood (see also Tuan 1974; Kaplan et al. 1998; Nohl 1997; Kühne 2008b). Burckhardt (2006 [1991], p. 82) draws these considerations together when he calls landscape a "trick of our perceptions that allows us to collate heterogeneous factors into a picture, excluding other factors".

It would, however, be wrong to think that in a social constructivist perspective objects like trees, houses or grass do not exist—that they are mere fictions or products of illusion. What that perspective maintains is something quite different: that the reality of those things is the fruit of a process of social production. Edley (2001) states clearly that language is not the only reality social constructivists accept (see also Newman and Paasi 1998; Chilla 2007; Marshall 2008): "They do not suppose that, say, Nottingham appears in the middle of the M1 motorway because it says so on the page and neither do they imagine it somehow springs into existence at the moment it is mentioned. The way that constructivism upsets our common-sense understandings is much more subtle than this. Instead, a constructivist might point out that Nottingham is a city by virtue of a text (i.e. by royal decree) and that boundaries—where it begins and ends—are also matter for negotiation and agreement. The argument is not, therefore, that Nottingham doesn't exist, but that it does so as a socially constructed reality" (Edley 2001, p. 439).

Our stock of knowledge of landscape also has a history. Following the general principles outlined by Schütz and Luckmann (1973), this is part of our 'sedimentary experience' (this will be treated in greater detail in Chap. 3). The experience of landscape does not derive only—or even principally—from direct sensory confrontation with physical objects designated as 'landscape'. Other elements of the socialization process—interaction with parents, peer group, teachers, films and books (see Kühne 2008a)—play a more crucial role in cultivating the ability to construct physical space as landscape, and hence in the genesis of individual concepts of landscape. This ability depends on one's socially mediated individual stock of knowledge—a cultural mediation that is "as a rule a guideline to [a process of] selection, a filtering out of impressions" (Burckhardt 2006 [1995], p. 257; see also Jacks 2004).

As shown, the socialization process conveys specific social interpretations, assessments and symbolic occupations of objects. Accordingly, "spatial images play so important a role in the collective memory" (Halbwachs 1992 [1939], p. 2).

## 2.2 Landscape

The starting point for the symbolic attachment to and occupation of objects is the inescapable spatiality of social activities: "Every group and every kind of collective activity is linked to a specific place, or segment of space" (Assmann and Czaplicka 1995, p. 7). Through physical objects, social action is structured—a house becomes a private retreat, emotionally charged and thus becomes a home (cf. Häußermann and Siebel 1996). Deep social bonds, ritualized in their spatial anchors, lead to the fact that people describe physical objects that they see as landscapes as their homeland (Kühne and Spellerberg 2010; cf. also Petermann 2007).

Certain places are attributed special significance for the formation of collective identities (see for example Graham et al. 2000; Ashworth et al. 2007; Legg 2007). In Switzerland, for example, the Rütli meadow, where representatives of Uri, Schwyz and Unterwalden are said to have conjured up a pact of support, has become the central site of the national founding myth (Kreis 2004). A place can be anchored very different in the (here national) myths. Stalingrad (in relation to the battle there between 1942 and 1943) was—not least because of the name of the city after the Soviet dictator—particularly mythically charged and symbolically linked to the turning point of the Second World War. In the Soviet Union, the city was associated with the beginning of the victory in the 'Great Patriotic War'. From a German perspective, the battle is now a symbol of a dubious military action in which hundreds of thousands of soldiers were sacrificed senselessly (in conservative to nationalist interpretation: 'heroically fought to the last'; cf. Ebert 1989; Troebst 2005; in general concercing national iconography Paasi 1996).

From a social constructivist perspective, things and social constructs stand to each other in a mutually conditioning relation—and this applies to both the social and individual construction of landscape (see e.g. Muir 2003; Marshall 2008; Kost 2013). In other words, a constructed landscape goes hand in hand with a 'not-landscape'—which exemplifies the process of complexity reduction noted above. For 'landscape' says far less than might be said about spatial complexes in general, making it a special case of the more general concept of space.

Like all other sign systems, social patterns of interpretation and ascription of landscape have to be learned: "There is no naïve relationship to landscape prior to society. The naïve individual cannot perceive landscape, for he has not learned its language" (Burckhardt 2006 [1977], p. 20). The process of construction of landscape has recourse to culturally rooted, socially mediated and temporally conditioned typifications; for "[we see] in general only what we have learned to see, and we see it as the style of our age requires" (Lehmann 1973 [1950], p. 48). Landscape arises, then, as a cultural image in which our surroundings are structured and symbolized (Daniels and Cosgrove 1988; see also Berendt 2005; Matless 2005; Kühne 2008b); and the acquisition of knowledge of landscape represents the sedimentation of current experience in structures of meaning along already acquired lines of relevance and typification (see in general: Berger and Luckmann 1966; Schütz and Luckmann 1973; specifically on this point: Kühne 2008a). These structures, in turn, consist for the most part in adopted patterns of socially current interpretation and evaluation. In addition, this process is rarely continuous: the immediate encounter with physical objects in terms of 'landscape' is in modern and/or postmodern society more often sporadic. For, in our urban society, human settlements are not commonly thought of as 'landscape': the experience of landscape is generally confined to weekend trips and vacations (see e.g. Kühne 2006a).

As with other such constructs, the concept of landscape is initially taken for granted. Only when deviations from the unreflected sediment of meaning and typification are noticed will it be questioned—although an exception must be made for those engaged professionally in this area. As observed in the preceding section, knowledge of landscape characteristically takes the form of familiarity, recognition or belief:

- 'familiarity' is knowledge of e.g. the climate, geology, vegetation, patterns of housing, agricultural usage and aesthetics of a specific area
- 'recognition' is knowledge of the discursive existence of an area commonly considered a 'landscape' (for example, the San Bernadino Mountains, the Lake District)
- 'belief' is knowledge e.g. of the existence of landscapes other than those I already know.

Individual knowledge is generally restricted, but can combine all these forms. A person familiar with the geology of their own region may merely 'recognize' other aspects of that region in the same way as they do the geology of other regions. Moreover, of course, knowledge may grow step by step from one level to another (Schütz and Luckmann 1973). The growth of knowledge about the physical structure of the British Isles through the classification of fossils, and with it the launch of geology as a discipline, exemplifies this process, for the first geological mapping took place in the context of discourse on stones and fossil finds in the houses of their genteel collectors—a discourse, incidentally, in which the Bible was increasingly thrown on the defensive (see Kühne 2013a).

Social constructivist landscape research—like social constructivist research in general (Gergen and Gergen 2009)—takes place essentially on a meta-level. It "investigates and casts light on what people mean when they speak of 'landscape'" (Haber 2001, p. 20; see also Leibenath and Gailing 2012); it is concerned, in other words, with what brings us to share our generalized concepts of reality and value with others in respect of certain spatial complexes (Gergen and Gergen 2009, p. 100). On the research level, two further aspects can be distinguished: the microsocial aspect of an individual's anchoring in social and communal contexts (see Berger and Luckmann 1966; Gergen 1999); and—in the wake of deconstructivism—the macrosocial aspect of the structuring power and function of language (Myers 1996; Burr 2005).

The prime interest of social constructivist landscape research is not ontology (how things exist, what landscape *is*) but language usage and how this expresses its psychological and ideological roots (Mitchell 2002, p. 1; and see Lacoste 1990; Potter 1996; Leibenath and Gailing 2012; Kühne 2013a). Questions arising from this focus include:

## 2.2 Landscape

- How, why and in what circumstances is the term 'landscape' ascribed to a specific space?
- What is excluded from such ascriptions and why?
- How are such ascriptions communicated? How and in what circumstances does landscape become thematized?
- What patterns of power and knowledge distribution are in play here? Who defines landscape, and decides where and when it should be declared worth protecting?

The factors underlying the designation of an agglomeration of material objects as landscape may be as varied as an active person or persons, a body of social knowledge, or the material objects themselves. To enable analytic consideration of such disparate dimensions, I have—following Pierre Bourdieu, Karl Popper and Martina Löw—developed the concept of four distinct but related levels of landscape (see Kühne 2006c, d, 2013a, b). The structuring in levels is indebted to Bourdieu (1991), who developed the threefold categorization of 'social', 'physical' and 'appropriated physical' space. For Bourdieu social space is a metaphor for society as an arena of the fight for status and position; physical space arises "when one intentionally ignores the fact that space is appropriated and inhabited, as e.g. in physical geography" (Funken and Löw 2002, p. 85); and physical space is appropriated when it is inscribed with social relations (as, for instance, a fence symbolizes restricted rights of access). Popper (1973; see also Hard 2002 [1987]) contributed his 'three worlds' hypothesis, distinguishing 'World 1': the world of physical objects and events; 'World 2': the world of individual perceptions and awareness; and 'World 3': the world of cultural and scientific knowledge. The third theoretical angle derives from Löw's (2001) reflections on the relational ordering of objects, which, she argues, "*depends on the relational system of the observer*" (Löw 2001, p. 34; original emphasis). Against this theoretical background the four levels (or perspectives) of landscape are as follows:

(a) *Social landscape* is the socially constructed, aesthetic dimension of landscape, a "socially defined object and ensemble of signs" (Hard 2002 [1987a], p. 227) that belongs to Popper's 'World 3' (1973). It comprises what in a society or social grouping is understood as 'landscape'—see the various cultural perspectives described above.

(b) *Individually appropriated social landscape* has symbolic, aesthetic, emotional and cognitive dimensions, inasmuch as it is characterized by personal knowledge, preferences, emotions, and patterns of interpretation and evaluation, these latter also entailing the distinction between actual and target values (see Agnew 1997; Ipsen 2002; Kühne 2006a; Kühne and Spellerberg 2010). As a personally appropriated perspective it goes beyond the social landscape on which it is based (see de Certeau 1990). It belongs to Popper's 'World 2'.

(c) *Physical space* in general can be defined as the ordering of material objects in a spatial relationship irrespective of their social or individual designation as landscape. As the material substratum of the other perspectives it belongs to Popper's 'World 1'.

(d) *Appropriated physical space* comprises physical entities infused with individual, socially molded meanings that are assumed into the construction of social space and its individual actualizations. Without using this terminology, Ipsen (1997, p. 7) explains this perspective as "the intensification of a spatial agglomeration of objects into an image that we can interpret and evaluate". Appropriated physical space is what specialists in the field mean when they speak of 'landscape'. It belongs to Popper's 'World 1'. As the "largely unstable" (Mrass 1981, p. 29) consequence of social acts, appropriated physical space can be further described as "a juxtaposition of long and short term, latent and manifest developments" (Békési 2007, p. 23). Its instability has two aspects: in their physical manifestation material objects are subject to the impact of socio-historical developments (e.g. 19th century industrialization, 21st century expansion of renewably sourced energies), and as social landscapes they are subject to social as well as individual constructions of meaning. That these are inherently changeable (Kühne 2006a; Olwig 2009) can be seen, for example, in the fact that up to the 1970s heavy industry plants in Germany were generally considered ugly, whereas today they rank as a valuable cultural heritage attracting frequent events and continuous tourism (see e.g. Hauser 2001; Kühne 2006c).

The constitutive factor in appropriated physical space is not a relation immanent to the object, as positivist approaches to landscape would have it; nor is it the 'essence' posited by essentialists as immanent within a specific topography (on these two trends see Kühne 2013a; Chilla et al. 2015). The concept is best understood as deriving from that of the individually appropriated social landscape, whose sign system is assimilated in the course of socialization by a process of selection and synthesis (at least logically) before individual actualization can occur.

## 2.3 Power

Power has already been mentioned as a prime factor in the definition of landscape, but the following section will focus on this multifaceted issue in order to approach the conceptual development of landscape in a more differentiated fashion. The concept of power itself is of almost universal application, but perhaps for this very reason its definition is wrapped in what Han (2005, p. 7) has called a "theoretical chaos", and its evaluation in insoluble contradictions. On the one hand, it is associated with freedom, on the other with oppression; here with order, there with social pressure; in one instance with right and rectitude, in another with willful arbitrariness. Power, then, is a many-sided, chameleon-like, contradictory concept, but—as Sofsky and Paris (1994, p. 11) observe—it remains a "central form of socialization", omnipresent within any context concerning social beings (see also

## 2.3 Power

Imbusch 1998). From a systems theory perspective, more or less stable power relations can be described as an essential element of the stability of society. They would satisfy the need for reliability and security (Parsons 1951; cf. also Anter 2012). A view that Dahrendorf (1963) vehemently contradicts by referring to the social productivity of conflicts arising from different power distributions.

From the point of view of power, human interaction knows "no sterile circumstances" (Popitz 1992, p. 272). Power cannot be grasped in a simple cause-effect relation: it is a recursive process "immersed in the rules of social intercourse and at the same time following those rules" (Sofsky and Paris 1994, p. 11). Power in this sense is, then, a normal and normative function of human relationships (Paris 2005; Imbusch 1998). It produces order, but not without conflict; in fact power struggles are "an integral aspect of the ongoing establishment of normality" (Paris 2005, p. 7). Popitz (1992, p. 12) underlines the anthropogenesis of power when he writes that "power structures are not God-given, not bound up in myth, not ordained by natural necessity nor hallowed by immutable tradition; they are the work of man". As such, they are in principle as reversible as the interpretations of the world in which they operate (Popitz 1992; and see previous sections of this chapter).

Weber (1976 [1922], p. 29) described power as "sociologically amorphous"; for "every conceivable human quality and circumstance" can lead to "the assertion of the [individual] will in a given situation". For Weber (1976 [1922], p. 28) power implies "the capability, in whatever circumstances, to assert one's will within a human relationship even against the will of others". This definition formulates four criteria (Anter 2012):

1. The category of 'chance' refers to the potentiality of power.
2. 'Social opportunity' refers to the personal character of power.
3. The 'own will' refers to the voluntaristic element of power.
4. The word 'reluctance' refers to a potential resistance that opposes 'one's own will'.

The definition of Max Weber includes, in other words, the capacity to directly or indirectly compel others to give up their own aims (Foucault 2006 [1976]). Power is not, however, the property of a specific individual or group; nor is it stable. On the contrary, it is, at least at times, eminently reversible and—depending on place, time and situation—discontinuous. It may also, as Latour (2002, p. 218) dramatically illustrates, depend on actual (or at least potential) use of an instrument:

> With a weapon in your hand you are someone else, and the weapon in your hand is not the same either. You are a different subject because you are holding a weapon; the weapon is a different object because of its relation to you. It is no longer a weapon in the safe, or in the drawer, or in your pocket. No, it's a weapon in your hand now, and it's pointing at someone who is yelling for their life.

The social context in which the acquisition, establishment and loss of power evolve generally includes other factors than just the more and the less powerful (Sofsky and Paris 1994). This means, according to Butler (2001), that the principal actors are neither fully determined by power nor fully determine it. The game, in fact, has various roles: that of observer or ally as well as that of protagonist. Sofsky and Paris (1994, p. 14) describe such configurations as "a complex weft of mutual, asymmetrical relations involving several persons, groups or parties, in which change in one relation effects change in the others".

The sociologist Popitz (1992) distinguishes four basic types of power:

- *Active* (coercive) power is the power to hurt others. Based on the one hand on the vulnerability of the human body, on the other on the ability to withdraw means of subsistence and social participation, it operates typically through physical superiority and the threat of violence.
- *Instrumental* (persuasive) power is the capacity to ordain reward and punishment, to grant or withdraw privileges in order to produce conformity; it operates typically by instilling hope and fear.
- *Authoritative* (directive) power directs the behavior and attitude of others; it operates typically by engendering unconscious, willing obedience based on unquestioning submission.
- *Technical* (structural) power—also called by Popitz the 'power of data constitution'—resides in the creation and possession of technical artifacts that can structure the actions of others; it operates typically through technological dominance.

The actual incidence of these four types is variable in composition and dosage, as well as in place and time. In the course of social and technological development, the emphasis has shifted from active and instrumental to authoritative and technical power. Central to my argument here is the role of artifacts in the mediation of power. As developed by Popitz, this contradicts both the treatment of artifacts as banalities and their over-dramatization (Fohler 2004, p. 42). The former position was taken by Koelle (1822) and Kapp (1978 [1877]), for whom artifacts were simply a means to self-knowledge and the perfection of the human being—or what Fohler (2004, p. 40) describes as "the projection of human organs and functions as an external condition". Gehlen (1986) pursues the same thought when he characterizes the human condition as essentially lacking. The opposite position of over-dramatization of artifacts sees "the unparalleled hegemony of science and technology as [threatening] to destroy the foundations of society" (Fohler 2004, p. 40), for humankind has unleashed upon itself objects capable of autonomous development (see e.g. Jünger 1946).

Power involves knowledge (in what forms and ways will be detailed later); indeed for Foucault (2006 [1976]) it *is* the 'will to knowledge', and as such reciprocally related to the production of truth (Foucault 1983; see also Leibenath 2015). The symbiosis of these two forces can be seen in their mutual conditioning: power does not exist without knowledge, nor does knowledge without power. Alternatively, put negatively, power resides in the person "who can afford not to have to learn anything" (Deutsch 1969, p. 171). As Foucault stated in 1977, power systems generate knowledge systems and vice versa. Not that knowledge per se

## 2.3 Power

implies power; it only does so in the case of "exclusive knowledge—knowledge that one person has and another needs but does not have" (Paris 2005, p. 42). In this sense, however, social differentiation entails a knowledge and information gap that also conditions power. Far from being centrally focused either spatially or functionally, power operates decentrally in an omnipresent network of tension between governor and governed (Bourdieu 2001). In opposition to Weber, Foucault (2006 [1976]) describes it not as an objectively measurable resource (see Leibenath 2015), but as the multiple interrelation of forces operating within and organizing a specific area; and he characterizes these relations as a game which in never-ending struggle and confrontation transforms, reinforces and inverts those relations and the systems of support they create with each other—or the displacements and opposing tendencies with which they isolate each other (ibid.).

These relations can often be felt at the edge of spatially and temporally restricted knowledge milieus and networks (see Castells 2001; Matthiesen 2006). Called 'KnowledgeScapes' by Matthiesen (2006, p. 167), these hybrid structures are characterized by a specific topography of knowledge which, despite rapid fluctuation, facilitates their identification and location in social (and even physical) space (see Honneth 1989). The differentiated monopolization of knowledge brings with it the growth of intricate self-multiplying hierarchies in many areas and subsystems of society (see Willke 2005; Ortner 2006). These, however, are inimical to the purpose of a democracy, with its reliance on the formation of collective knowledge for collective decisions (see Willke 2002). For a democracy not only "fosters the growth of collective knowledge, but cultivates dissent, diversity and heterogeneity; it aims not to suppress social complexity, but to intensify it" (Willke 2005, p. 48).

Closely connected with the general concept of power is that of domination, which Luhmann (1984, p. 37)—from a systems-theory perspective—calls simply a "mode of systems description, [...] a manifestation of the immanent control exercised by a system over itself". For Weber (1976 [1922], p. 541) domination is "a specific instance of power", and he defines it in contradistinction to power as the probability that a specific command will be obeyed—or as Luhmann (1971, p. 92) puts it, "that one's goals will become the goals of others". Domination, for Weber, is legitimated when it is accepted by those subjected to it, irrespective of its impact on individual lives and opportunities; and he names three grounds for such acceptance: *reason*—the thoughtful weighing up of interests; *custom*—doing what has always been done; and *emotion*—acting from personal inclination (Weber 1976 [1922]). In contrast to power, domination is more specific: it does not include absolute control over others, but "is always limited to certain contents and identifiable persons" (Dahrendorf 1972, p. 33; following Max Weber). Dahrendorf (1983) sees this stronger organization of (undifferentiated) power to rule as a central aspect of peaceful social development (as can be found, for example, in the representative democracy he favored). According to Weber (1976 [1922]), the development of power into rule begins with the emergence of the modern state. This came to the multitude of local rulers, power relations were centralized and increasingly regulated in a generally binding manner (i.e. law applies equally to all and compliance with it is guaranteed by the state). The development of the state is accompanied by the monopolization of physical violence (which, of course, remains incomplete, because otherwise there would be no 'violent crimes') and the specific organisation of knowledge in the form of bureaucracy (cf. Anter 2012). Both developments are likely to generate security, which in turn is an essential basis for the acceptance of a central power (see Anter 2012). As, however, these grounds are relatively unstable, domination must itself ensure stabilization of the subject's belief in its legitimacy. On the basis of the above-named grounds, Weber (1982) distinguishes three types of domination:

- *legal* domination is legitimated by the rational belief in the right of governance invested in those who exercise it
- *traditional* domination is legitimated by the common belief in the hallowed nature of established traditions and the right of governance invested in its officials
- *charismatic* domination is legitimated by the exceptional qualities (holiness, heroism, nobility etc.) invested in a particular person and the order of governance they create or reveal.

Domination, as opposed to power, is relatively stable. Following Weber (1976 [1922]), Imbusch (2002, p. 172) defines it as "a permanent, institutionalized power invested in a person or group in relation to another person or group that possesses at least a minimal will to acknowledge and obey". In the same tradition, Neuenhaus sees this will, backed by reason and discipline, as "elevating social awareness to the level of an automatism, and the corresponding domination to that of an end in itself, devoid of meaningful purpose and demanding universal obedience. Its 'case-hardened cage of conformity' (Weber 1976 [1922]) quells the struggle for power and leads to a uniformity of social action" (Neuenhaus 1998, p. 78).

Treiber (2007, pp. 54–55) describes the development of the concept of domination in the following terms: "the more precise definition of the concept of domination sought by Weber consists in positing the power relation as permanently institutionalized through processes of depersonalization, formalization and integration. That this leads to the typical asymmetry of ruler and ruled is a function of the conditional programming inherent in the horizon of expectations of a probabilistic concept of causality."

Along these lines, Popitz (1992) further defines the three processes of institutionalization mentioned by Treiber as the subjection of power to increasing:

1. *depersonalization*: "Power no longer stands and falls with the individual who has the say at the moment, but is connected sequentially with specific functions of a suprapersonal character" (Popitz 1992, p. 233);
2. *formalization*: the exercise of power is governed by ever more rules, procedures, norms and rituals;
3. *integration* into a superior system of order: power "binds itself and is bound up in a social structure that supports and is supported by it" (Popitz 1992, p. 234).

## 2.3 Power

The domination uses bureaucracy. Bureaucracy does not only serve to enforce rule, it is itself the wearer of power (Weber 1988 [1918]). On the one hand, this brings together the specialist knowledge acquired by the staff in a wide variety of training courses, and on the other hand, the staff members have exclusive knowledge of the official organisational processes (where they are able to exert an influence on the society). The official organisation is characterised by hierarchy, division of labour, effectiveness and predictability (more on Fukuyama 2013). Once introduced, "a virtually unbreakable form of power relations has emerged" (1976 [1922], p. 570). In the sense of Popitz, with the perfecting of the 'data-setting power' "in the late 20th century, a supervisory state has thus emerged, which—theoretically speaking—can find out everything about its citizens" (Anter 2012, p. 73).

A critique of power and domination has been mounted from two otherwise mutually antagonistic positions: liberalism and critical theory. Liberalism is committed to maximizing individual freedom, which it sees as jeopardized by increasing state power. Hence classical liberalism in particular confines the tasks of the state to "protection of property, freedom and peace" (Mises 1927, p. 33). State intervention in the marketplace not only causes economic imbalance, and with it an inefficient distribution of production resources; it also concentrates power in a few hands, enhancing the danger of its abuse (Mises 1927)—in liberalism a deep-seated, skeptical thesis. The other side of this coin is the loss of power and increasing conformism of the governed (Paris 2005). Inevitably "power crystallizes into domination", for "it is not the law itself but the subject's compliance with it that causes institutionalization" (Sofsky 2007, p. 17), robbing the individual of freedom. Unwittingly and without resistance "the conformist lives and moves in a rigid mental prison, without noticing the damage long since inflicted on the power of thought" (Sofsky 2007, p. 129). That power entails the will on the part of the governed to constitute themselves as subjects has been forcefully expressed by Hillebrandt (2000, p. 120): "Only the immanent self-subjection of the individual creates awareness of the disciplining eye of domination". Moreover, this can be seen in the example of prison, where physical power over the body works "to approximate the prisoner to a behavioral ideal, a model of obedience" (Butler 2001, p. 82). The prisoner, in Foucault's words (1977, p. 260), becomes "the principle of [his or her] own subjection"—a process that in Popitz's terminology exemplifies authoritative power.

It is for such reasons that liberalism rejects the hope—often associated with a deeper belief in the state—of attaining an optimal future societal condition, for this hope is undermined by the "intolerance and abuse of power" (Dahrendorf 2008, p. 76) that accompany it. The adherents of critical theory, on the other hand, reject domination, not so much from experience of its violent excesses, which in some cases have even instigated a reign of terror, but in principle; for domination, "to maintain itself in domination, tends always to totality" (Adorno 1969, p. 105). This is especially evident in its social effects, like maximizing disparities in the possession of symbolic capital and hence perpetuating inequalities of opportunity and, as a result, also of living standards (see Dubiel 1992; Imbusch 1998).

For Bourdieu (1979), symbolic capital is a central means of power and the exercise of power. He explains the concept as the opportunity, "perceived and accepted as legitimate", to gain and preserve social recognition and prestige. Concretely it can take the form of either economic, social, or cultural capital (Bourdieu 1979, 1982a, b, c):

- *economic* capital consists in material possessions that can be converted into money;
- *social* capital resides in social networks that engender recognition;
- *cultural* capital has three manifestations: *objectified* (books, artworks, technical devices); *incorporated* (education, cultural and creative abilities); and *institutionalized* (academic titles, certificates and memberships).

The social standing of an individual depends on "(a) the volume of their capital; (b) its structure—i.e. the relation of the different types of symbolic capital to each other; and (c) the relation of starting capital to actual capital" (Wayand 1998, p. 223). This also determines whether one is on a rising or falling social trajectory (Bourdieu 1979). Moreover, the three sorts of capital are in principle interchangeable, and economic capital in particular can be readily converted into the other two types. Differences in symbolic capital create a vertical differentiation in society that manifests itself above all in differences of taste between the social classes. According to Bourdieu (1972), these derive, not merely from individual and collective perceptions and intentions, but from the "appropriation of the same objective structures" within a single class.

Based on symbolic capital Bourdieu (1979) distinguishes three social classes:

(a) The *ruling class* consists of entrepreneurs (with high economic but low cultural capital) and intellectuals (with high cultural but low economic capital); this class determines 'legitimate' taste with its values of cultivated discrimination.
(b) The *middle class* (aka *petite bourgeoisie*) comprises the downwardly mobile (above all practitioners of traditional trades and crafts), the new middle class (new professions without higher qualification—e.g. salespersons, animateurs etc.), and middle management (employees who invest considerable time in 'improving themselves' educationally); this class determines conventional taste, which is marked by the values of education and industry.
(c) The *lower class* (the ruled) comprises the rest of society; it determines popular taste, which is governed by life's necessities and the values they impose.

In today's milieu-based society, the concept of class as a social determiner seems rather antiquated. Nevertheless, Bourdieu's aesthetic distinctions also hold for social milieus and are in this sense still valuable for sociological landscape research (see e.g. Hradil 1992; Zerger 2000; Kühne 2008a; Irrgang 2014). Taste remains a factor in the distribution of social opportunity: the ruling class is interested in setting aesthetic standards that will preserve its position in the asymmetric distribution of opportunity, the middle class is interested in evening out that asymmetry, and the ruled have generally come to terms with existing imbalances.

## 2.3 Power

A central factor in the perpetuation of social inequality is what Bourdieu calls 'habitus': the ingrained physical embodiment of socialization (Bourdieu 1996) specific to each class. It is "a system of boundaries" (1982a, p. 33); for "to know a person's habitus is to know intuitively what behavior is for that person forbidden" (1982a, p. 33). Habitus can be described as a transmission mechanism between mental and social structures and the daily life of society. The inscription of those structures within the individual creates a correspondence "which according to Bourdieu induces an immediate spontaneity in social actors to do what society requires of them" (Wayand 1998, p. 226; see also DeMarrais et al. 1996). In the respect habitus—especially that of the lower classes—is the key element in the perpetuation of political domination, because it is this that grounds people's readiness to be ruled (Bourdieu 2001). As a social mechanism for maintaining domination, its efficacy is a function of its covert nature; for habitus is rarely recognized and evaluated as what it is (Bourdieu 1977). Or as Han (2005, p. 56; see also Lenski 2013) starkly puts it: "The ruled even relish their deprivation. Poverty becomes a chosen lifestyle. Compulsion and repression are experienced as freedom."

The attitudes and values of habitus are transmitted through the educational system. It is striking that precisely economically disadvantaged families, whose children are barred by lack of social and cultural capital from higher educational establishments, believe most strongly in "talent and competence as decisive for scholarly success" (Bourdieu 1977, p. 16). The social function of schooling, according to Wayand (1998, p. 226), is to bring "the dialectic between subjective expectations and objective structures" to a standstill; its function as an instrument for maintaining established domination becomes all the clearer when one reflects that school as an institution binds "the organized learning processes of the coming generation [...] into an official apparatus governed by the state, so that learning can in this way be administratively controlled and politically influenced" (Tillmann 2007, p. 113). For Althusser (2011 [1970]), school can develop abilities, but it does so in ways that ensure compliance with the ruling ideology, and/or mastery of its practice; and within any system of hierarchically skewed power relations (Tillmann 1976) this submission will prevail even in situations of competition between formal equals (Bernfeld 1925). For 'no ruling class', Althusser continues, can hold power for long without at the same time imposing its hegemony over and within the ideological channels of government (2011 [1970]).

School classes, then, "of set purpose, serve to mediate influence [and] are directed towards the acquisition of socially desired knowledge, abilities and values" (Tillmann 2007, p. 114). To stabilize the system and legitimate its concept of training (Masuch 1972) socially useful qualifications are granted, while at the same time the forms and modalities of capitalism are practised (albeit without serious reflection on their inevitable contingencies). Thus, concealed "under a cloak of neutrality" (Bourdieu 1973), the educational system reproduces established social structures, "instilling respect for the ruling culture in the children of the ruled, without granting them admittance to that culture" (Fuchs-Heinritz and König 2005, p. 42; see also Sidanius and Pratto 2001).

Education devalues the culture of the lower classes; its prime means of so doing is language. In the "system of signs and rules" (Werlen and Weingarten 2005, p. 192) that constitutes language, members of the lower class will "earn low marks as soon as they speak their own language; their pronunciation and grammar etc. are simply deemed wrong" (Bourdieu 1982a, p. 49). Without any need of physical intervention—Popitz's 'active power'—habitus produces, here too, unquestioned acceptance, "directing [personal] actions, so that established structures of domination are reproduced in a quasi-magical fashion that remains below the horizon of reasoning" (Han 2005, pp. 56–57). For Bourdieu (1984), the university incorporates similar power structures and establishes the same inequality of opportunity, reproducing in its own structures those of the arena of power and passing them on through its own communicative and selective functions. Althusser (2011 [1970]) sees state and private media, as well as churches, clubs and associations, as playing a role complementary to that of educational institutions, and Gramsci (2001) includes them all in what he calls the 'apparatus of hegemony'.

In sum, the entire educational system—reinforced by patterns of behavior and socialization mediated by parents, age-group, peer-group etc. (see Fromm 1936)—contributes massively to the conversion of external to internal constraints which Elias (1992) sees as characteristic of the development of civilization. For, as societies grow more complex (the process known as sociogenesis), its members are compelled "to regulate their behavior in a more balanced, stable and differentiated fashion" (Elias 1992, p. 117). This entails an internalization of norms, which in turn implies the internalization of what was initially an externally imposed domination. As Foucault observed (1974, p. 95): "Humanity does not move slowly from conflict to conflict until a universal state of understanding is achieved; it anchors its violence in rules and regulations, moving from one [manifestation of] domination to another."

Liberal thinkers tend also to see school (along with its sister institutions) as an instrument for the stabilization of domination, although with a stronger focus here on the domination of the state (Prollius 2014). Despite a fundamental belief in the opportunities afforded by education, school as an institution is viewed by liberals with critical differentiation, for the multiplying function of education multiplies the power of the state, whose existence and action—"including state television and radio and state schools" (Prollius 2014, p. 187; see also Kersting 2009; Sofsky 2007)—are, as a direct upshot of this, taken to lack any alternative. In such a perspective inequality of opportunity will, it is objected, pass without reflection or comment; it will simply be normalized or, where consciously perceived, earn only a shrug of resignation. The critique of state schooling goes back as far as the classical liberal thinker Wilhelm von Humboldt (1767–1835), for whom education was the prerequisite for a self-determined life in possession of its rights (Humboldt 1960). However, this ideal of the free and responsible individual could, Humboldt asserted, only be cultivated in a privately established environment. This, then, must be preferable to a state-run system intent only on producing obedient citizens.

The argument of this book will have frequent recourse to the concepts of power developed by Popitz and Bourdieu. Popitz sheds light on various dimensions of power that apply in particular to activities related to terrestrial spaces (Kühne 2015). Bourdieu laid down guidelines that facilitate investigation of the means and mechanisms through which social interests are defined and generalized. Among these—and hence, too, in the perpetuation of power structures and the unequal distribution of life's opportunities—aesthetic judgments play an essential role.

## 2.4 Aesthetics

The relation in which landscape stands to the world, however culturally and professionally differentiated, is commonly held to be aesthetic (for such relations in general see e.g. Sheppard 1987; Townsend 1997; Graham 2005; Schweppenhäuser 2007). The word 'aesthetic' derives from the ancient Greek, where it was used to denote the science (or philosophy) of sense perception (*aisthētikḗ episteme*), which was complementary to that of thinking (*logikḗ episteme*) and moral action (*ēthikḗ episteme*). In the early European thought of figures like Augustine (c. CE 390) and Pseudo-Dionysius the Areopagite (c. CE 500) the three areas of logic, ethics and aesthetics reflected and underpinned the unity of the true, the good and the beautiful (see Augustinus 1962; Pseudo-Dionysius 1988); it was, in fact, only in the Enlightenment that aesthetics became an independent branch of philosophy (see Gilbert and Kuhn 1953; Majetschak 2007). Alexander Gottlieb Baumgarten's *Aesthetica* (2007 [1750–1758]) introduced a new approach that "increasingly supplanted the ancient and medieval paradigm of an ontologically founded theory of beauty" (Schneider-Sliwa 2005, p. 7). The key relation for Baumgarten was between the aesthetics of art and the logic of science. Ritter (1996 [1962], p. 43) explains: "When the wholeness of nature—of heaven and earth as aspects of our existence—no longer falls within the ambit of science, the sensitive mind creates the aesthetic image, the poetic word, in which [that wholeness] is realized in its affinity with our existence and its truth can be felt." While analytic science unravels the world in order to examine its many contexts, the aesthetic perspective draws the threads together again (see Peres 2013). There was (and is) an undoubted clash of interests here; nevertheless—or perhaps for that very reason—aesthetics gained increasing attention during the Romantic period, when modern science, too, was burgeoning.

Five central, continually intertwining strands (or questions) can be distinguished in aesthetic discourse (Kühne 2012, 2013a):

(a) What is defining quality of the aesthetic: the beautiful, the sublime, the picturesque or the ugly?
(b) Has aesthetics to do with art or nature or both, and how and why is this so?
(c) Does the aesthetic reside in the object or its beholder: objectivist versus subjectivist aesthetics?
(d) How can/should we approach the aesthetic: rationally, emotionally, or via the senses?

(e) How is the aesthetic socially evaluated: high versus popular (trivial/kitsch) culture?

The following paragraphs of this section will consider these questions in turn, especially in their bearing on landscape research.

(a) Beauty remains a central issue in aesthetics; indeed Borgeest (1977, p. 100) sees the development of aesthetics as a "continuous reinterpretation of the idea of beauty" (see Fig. 2.1). Traditionally understood as "unity in multiplicity" (Schweppenhäuser 2007, p. 63), beauty was characterized by Kant (1956 [1790]) as what commonly pleases without any mediating concept or immediate (e.g. economic) interest on the part of the beholder. As far as landscape is concerned, the panoptic vision of objects to form an Arcadian landscape, accompanied by an aesthetic judgment of its beauty, presupposes that the viewer has no other interest (e.g. as a farmer or landowner) in the scene. Dewey (1929, 1988 [1934]) and others rejected Kant's separation of the aesthetic and practical worlds on the grounds that the construction of beauty also involved individual consumer interests, and that it was rooted in an interaction between object and subject.

The concept of the sublime is traditionally distinguished from that of beauty as having to do with "nature in relation to man or, more precisely, nature in its capacity to arouse moral ideas in the observer" (Gethmann-Siefert 1995, p. 90; see also Cronon 1996; Graham 2005; Loesberg 2005; Wicks 2011). Nature in the aesthetic context can be understood as the "world of human realities perceptible to the senses that has arisen (and continues to arise) without human agency" Seel (1996, p. 20). For Burke (1989 [1757]), the beautiful differed from the sublime in its emotional impact: where beauty stimulates love, sublimity arouses awe; beauty is accordingly associated with small and pleasant objects, sublimity with great and even threatening ones (e.g. volcanoes). Kant saw beauty as rooted "in the harmonious interplay of mind and sensible imagination" (Peres 2013, p. 38; and see Graham 2005)—to which Gethmann-Siefert (1995, p. 90) adds that this interplay is "free"—whereas "sublimity derived from a disharmony between mind and imagination" (Peres, p. 38). Kant (1974) further observed that sublimity might be found in a formless object if this indicates boundlessness or totality. Therefore, in contrast

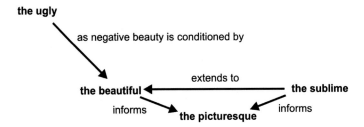

**Fig. 2.1** Relations between the beautiful, the ugly, the sublime and the picturesque

to Burke, the sublime was, for Kant, not just a matter of physical quantity: it could also imply a superordinate quality. From this, he derived the distinction between the mathematically sublime, inherent in the sheer size of an object, and the dynamically sublime, inherent in its power (see Kant 1974). The power of nature, being of this latter kind, can be found (and felt) in landscape. Yet the perceived sublimity of nature derives not from the 'things of nature' but from human creativity. Nature and reason are two different powers: the true source of sublimity is the knowing subject in its capacity for reflection—a function Butler (2001, p. 27) regards as disposed to "assume ardent desire into the circle of self-awareness" (and see Casey 2006). Sublimity took on a new lease of life in the post-1970s discussion of the intangible/ineffable centered on Lyotard (1991) and postmodernism (see Peres 2013). At the same time "the aesthetically beautiful [...] was reduced uncritically to the field of design and the status of a mere consumer product" (Friesen 2013, p. 90). Nevertheless, contemporary philosophy is also concerned with the 'aesthetics of the living environment', and with a critique leveled—e.g. by Rüdiger Bubner, Odo Marquard and Wolfgang Welsch—at "the threat of excess and satiety, numbing sameness, and social desensitization" (Recki 2013, p. 229).

A further aesthetic dimension—first extensively treated by Rosenkranz (1996 [1853])—is that of ugliness. Far from possessing significance in its own right, ugliness is regarded as "negative beauty", and as such "essentially secondary" (Rosenkranz 1996 [1853], pp. 14–15). Rosenkranz grounds this thesis in an analogy with good and evil, pointing out that evil, too, has a place in ethics: "Hell is not just a religious and ethical concept, it is also aesthetic. Ugliness surrounds us". According to Rosenkranz it does so in three basic forms:

- *amorphousness* (formlessness, indefiniteness of shape)—lack of appropriate boundaries, of unity in required diversity (Pöltner 2008);
- *asymmetry*—lack of balance between opposites, lopsidedness;
- *disharmony*—disproportion between parts and whole; lack of unity, with false contrasts where there should be agreement.

According to Rosenkranz (1996 [1853]) the ugly can be elevated aesthetically by transformation into the comical; for the comical "takes the ugly (which always contains an element of compulsion) back into the freedom of beauty, combining beauty and ugliness, freeing each from its (pseudo-ideal) one-sidedness" (Hauskeller 2005, p. 61). This is true above all of caricature, where exaggeration and imbalance become a virtue.

It should be noted in any consideration of aesthetics that the sublime is not a "median value between the beautiful and the ugly" (Seel 1996, p. 132) but a category in its own right. Complementing that 'aesthetic triad' (Seel 1996), is the picturesque (Kühne 2013a). If the beautiful is characteristically small, fine and subtly various, and the sublime big, strong, intense and awe-inspiring, the picturesque falls somewhere between those poles. Encountered frequently in landscape painting, where the fore- and middle-ground are marked by 'beautiful' tree, bush or flower motifs and the background by 'sublime' mountain ranges or wild seas, it is typically complex, irregular, and differentiated (Carlson 2009; see Fig. 2.1).

Whether a landscape can also be ugly depends on the breadth of the concept. In German usage, a narrow view of landscape judges quality solely in terms of 'historical and cultural maturity' (see e.g. Hokema 2013), allowing room for the attributes 'beautiful', 'picturesque' and 'sublime'. But to predicate ugliness of a landscape as an element of its social construction, the concept would have to be extended to any space that can be grasped as an aesthetic unity—if this is not to contravene norms of usage long since recognized in international discussion (see e.g. Hartz and Kühne 2009; Hokema 2013, 2015; and Chap. 3 below).

(b) Closely bound up with the development of philosophical aesthetics is the question of the relation between nature and art as aesthetic dimensions. Kant (1956 [1790]) ranks nature above art in this respect, for the beauty of nature lacks intentionality. Hegel (1970 [1835–1838]) for that very reason takes the opposite line: "artistic beauty, *a beauty born and reborn of mind*, is superior to nature in the same measure as mind and its products stands higher than nature and its manifestations" (Hegel 1970 [1835–1838], p. 14; original emphasis). Beauty, for Hegel, arises in the concord of concept and external reality; it is, then, an ideal value and can as such be termed the 'truth of appearances'. Unconscious nature cannot effect such truth: "Mind lies beyond [nature]; there it is only hinted at. Nature lacks the unconditioned unity and autonomy of the concept" (Hauskeller 2005, p. 52). Hegel's aesthetics, therefore, "belongs wholly to the philosophy of art" (Peres 2013, p. 32); it no longer, as Baumgarten had proposed, has to do with the "emancipation of the senses" (Friesen 2013, p. 80). Croce (1930) took Hegel's argument to a new level of intensity when he asserted that beauty lay in expression, and expression was a function of the mind; nature, therefore, because it is passive and mindless, must be excluded altogether from the ascription of beauty.

In recent decades, with the growing crisis of environmental pollution, issues of nature and natural beauty have gained new topicality—not least in the question of an aesthetic revaluation of nature in the wake of its colonization by a technologically driven society (see Adorno 1970; Lundmark 1997; Tiezzi 2005). The contemplation of natural beauty remains a human concern, but nature is not simply thought of as a physical resource for aesthetic appreciation (see van Noy 2003; Haber 2006), and the dichotomy between natural and artistic beauty is accordingly fading. "Aesthetically speaking, both are 'unitary phenomena'" (Seel 1996, p. 269), evoking correspondences between object and observer which are, however, (generally) ascribed to the object (Hartmann 1953). The issue of nature versus art (or artifact) is also relevant to the distinction (or construct) between natural and cultural landscapes—for example in the question how natural and/or cultural objects should, for aesthetic (and other) reasons, be preserved; or whether the aesthetic designation of a specific space as landscape depends on a judgment of its quality as nature and/or artifact.

## 2.4 Aesthetics

(c) Another aesthetic issue is whether such a judgment is rooted in objective qualities or is purely subjective. Shusterman (2001) calls the first position 'naturalism', the second 'historicism'. Plato (4th century BCE) was in this sense a 'naturalist', teaching that the 'idea' inherent in the object was the ground of its being (Platon 2005), and the more intense the imprint of this idea in the material object, the greater its beauty. Francis Hutcheson (1694–1747), on the other hand, propagated the subjectivist position that beauty, while grounded in the consonance of unity and diversity within the object, depended particularly on the ability of the observer to respond to it (Hutcheson 1988 [1725])—or as Friedrich Theodor Vischer (1922, p. 438) later succinctly put it: "Beauty is an act, not a thing." As the "product of the mental disposition and abilities of the [human] subject" (von Hartmann 1924, p. 3), beauty has social foundations—an aspect to which Kant (1956 [1790], 1959 [1781]) had already drawn attention, aesthetic judgment being based on "sociocultural values, acquired norms, personal experience, and qualities of character, imagination and desire" (Frohmann 1997, p. 175).

Kant (1956 [1790]) also taught that aesthetic judgments were not concerned with knowledge; they were judgments of taste, not logic, and as such "based exclusively in subjective reflection" (Peres 2013, p. 35). Nevertheless, taste, as "a synonym for aesthetic judgment" (Illing 2006, p. 8), was a matter of social discourse and negotiation. What should be deemed aesthetic, when, and by whom, without loss of social kudos, was, therefore, variable? Borgeest (1977, p. 100) makes the same point with regard to the concept of beauty: "There is no single point of view, accepted at all times and by all, from which beauty can be ascribed, that does not equally justify the opposite assertion." While naturalism takes an essentialist position in this matter, historicism takes a constructivist view. Empirically, a preference for certain object-centered structures can be established and at the same time the question asked by whom, how and in what circumstances such preferences are expressed (see further Chap. 4 and Peres 2013).

(d) Individual aesthetic taste operates largely at an unconscious emotional level, but express judgments of taste often also involve cognitive processes. Satter (2000) illustrates the two aspects with the example of music, where basic sensitivity to sad and happy sounds is a prerequisite for any emotional response, but a more highly cultured sensitivity will require a qualitatively higher music to elicit such a response. Satter (ibid.) concludes that, on a scale of values, 'neutral' taste is purely emotional, but any judgment of musical quality presupposes cognitive input. Both depend on sense perception, but an aesthetic judgment, as an act of cognition requires the combination of sense with intellect. This view elevates aesthetics above the merely sensory plane and replaces it precisely where Baumgarten had set it, as the 'science of sense perception/knowledge' (Satter ibid.). Nelson Goodman (1992) likewise rejects the reduction of aesthetic experience to sense perception and emotion: Any picture of aesthetic experience as a sort of emotional bath or orgy is plainly preposterous. Compared with the fear, grief, depression or enthusiasm induced by a real battle or loss, defeat or victory, the emotions at work there are usually quiet and oblique, and generally no more intense than the excitement, despair or pleasure that accompany scientific research and discovery.

The interpenetration of emotional and cognitive inputs in the genesis of 'world' is reflected in the relation of art to science. Goodman (1978) sees these as two ultimately inseparable cultures, as mutually conditioned ways of creating the world in which we live and think—with the result that aesthetics can be seen as a particular form of epistemology. As a synthesis of art and science, the aesthetic attitude has been characterized as 'restless, searching, testing' (Goodman 1968), and its knowledge component as less concerned with producing truth than with generating "various models of the world (in science as in art)" (Gethmann-Siefert 1995, p. 110). Goodman (1978) emphasizes the contextual dependency of art when he suggests that asking 'What is art?' invariably entails asking 'When is art?'. In this sense, too, it can be asked whether a narrow view of landscape, bound to the norm of historically developed rural cultural space (see Sect. 4.2.1), is to be preferred to a broader view that would include, for example, suburban settlements. The issue can, in fact, be interpreted as a matter of taste (see Hokema 2009, 2013; Wojtkiewicz and Heiland 2012; Kühne 2013a).

(e) The creation of different models of the world and their acceptance as equally valid is a central feature of what has been called the 'radical tolerance' of postmodern aesthetics. One aspect of this is the transcending of the dichotomy between high and popular culture or 'kitsch' (see Welsch 1988a, b). The stigma of bad taste attached to the latter by adherents of the former amounted to the accusation of an evolutionary (or educational) shortfall in relation to standards regarded as universally valid (Illing 2006; for the architectural context see Stevens 2002). In contrast to modernity, however, postmodern culture, far from being conceptually "bound to universal authorities […] and norm-giving hierarchies" (Kastner 2002, p. 232), is marked by "interpretive polyvalence" (ibid.). The attempt to establish and discursively preserve a single cultural standard has been increasingly undermined by the social shift from class-based to milieu-based differentiation, which entails and encourages the multiplication of cultural discourses, and at the same time dismantles the values on which the distinction between high and popular culture was based. Kitsch is no longer seen as the "false expression of false needs, nor as the false expression of true needs: for the tolerant aesthetics of our day, it is a true expression of true needs" (Liessmann 2002, pp. 26–27).

Postmodern aesthetics reflects the perspectivity and relativity of postmodern thought. It can be seen as the fruit of a process that "has been increasingly moving since Kant towards the insight that the basis of what we call reality is fictional" (Welsch 2006, p. 8). What in modernity was a matter of enlightened reason is in postmodernity a matter of aesthetic judgment. Reality reveals itself increasingly as "not realistic, but aesthetically constructed" (Welsch 2006, p. 7; see also Trigg 2009), and postmodern culture is accordingly marked, not by innovation, but by

## 2.4 Aesthetics

reshaping and recombining, quoting and ironically plagiarizing (see e.g. Federman 1991). Here simulation and so-called 'playgiarism' have replaced the real and authentic (Seidman 2012). As far as landscape is concerned, this provides fertile ground for constructivist approaches and the broadening of the concept of landscape mentioned above (see Hokema 2013).

That the foregoing five strands of aesthetic discourse are closely connected with the construction of landscape can be illustrated as follows:

(a) The question of the aesthetic values of beauty, picturesqueness, sublimity and ugliness as applied to appropriated physical landscapes is central to landscape research. What is the physical basis of such differentiation, for example, between the picturesque and the ugly? Why, if all are appropriated, are some landscapes preferred to others? What combination of beauty and sublimity informs the judgment that a set of physical objects is picturesque?

(b) The question of the aesthetics of art versus that of nature informs the definition of cultural versus natural landscape, and even more so the conscious shaping of landscape. When should a landscape be 'read' as art and when as nature—when as a hybrid of nature and culture?

(c) The question of the objectivity or subjectivity of aesthetic judgments is crucial to landscape research in determining whether the values predicated of a landscape are grounded in its physical structures or constructed subjectively on the basis of social interpretations and evaluations.

(d) The question of the role of reason, emotion and the senses in aesthetic judgment has a multiple bearing on landscape research: What sense impressions are synthesized into landscape? Can landscape aesthetics be explained rationally? How does emotion inform the construction of landscape? Given the dual role of emotion and cognition, are these elements in harmony or in competition?

(e) The question of the social aspect of aesthetic values (high vs. popular culture) bears on the discursive hegemony of landscape definitions and values, as well as on the social mechanisms of their construction, and their temporal and cultural dependency.

In the social constructivist approach of this book, the answer to Question (c) must be that the appropriation of landscape is subjective: its aesthetic qualities lie not (as the essentialist view would maintain) in the physical object but in the judgment of the beholder, made in the context of social norms and values. This position conditions the answers to the other four questions, transferring these from object- to meta-level—which means, in general, examining socio-historical contexts rather than physical objects.

- This is the case in determining, for example, in what circumstances physical landscapes are described as beautiful, picturesque, sublime or ugly (Question (a))—see Sect. 4.2: Distinction and landscape, and 5.4: Conflicts of power—landscape and the extension of renewably sourced energy).

- Question (b), too, is addressed through a research approach that asks in what contexts natural and cultural landscapes are respectively valued more highly—a question that runs right through this book.
- The issue of rationality, sense and emotion (Question (d)) is examined by asking in what contexts people respond cognitively and/or emotionally to what they conceive as landscape (see especially Sect. 4.1: Socialization). (This rules out the reduction of 'landscape' to an emotive entity on the object level, for if the aesthetic appeal of landscape is purely emotional, an (objectively!) cognitive research approach is inevitably redundant.)
- Finally, the dichotomy between high and popular culture (Question (e)) resolves in landscape research into asking who defines and decides the terms of the argument—i.e. the power question in Bourdieu's sense (see Sect. 4.2: Distinction and landscape).

In this perspective, social constructivist landscape research applies the general principles of philosophical aesthetics—or what Peres (2013, p. 36) calls the "conception and usage of aesthetic predicates, value judgments and standards"—on two levels, focusing on the one hand on landscape (rather than e.g. sculpture), on the other on social processes. For the aesthetic construction of landscape this means that it is no longer enough simply to recreate "the context of place, ethics, and spirituality (loosely, 'meaning')", for this "is inevitably embedded in a social, economic and political matrix" (Porteous 2013, p. 10). Rooted in social and individual processes of construction, as well as in physical objects, this 'theoretically impure' (Leibenath 2014) approach is in line largely not only with the philosophical research tradition in aesthetics but also with that of positivist landscape research (Kühne 2013a)—which makes its results also suitable for use in planning processes (see Bruns and Kühne 2013b).

## References

Adorno, T. (1969). Diskussionsbeitrag. In T. Adorno (Ed.), *Spätkapitalismus oder Industriegesellschaft? Verhandlungen des 16. Deutschen Soziologentages* (pp. 100–106). Stuttgart: Enke.

Adorno, T. (1970). *Ästhetische Theorie*. Suhrkamp: Frankfurt a. M.

Agnew, J. (1997). Representing space: space, scale and culture in social science. In J. Duncan & D. Ley (Eds.), *Place, culture, representation* (pp. 241–271). London: Routledge.

Ahrens, D. (2006). Zwischen Konstruiertheit und Gegenständlichkeit – Anmerkungen zum Landschaftsbegriff aus soziologischer Sicht. In Institut für Landschaftsarchitektur und Umweltplanung – Technische Universität Berlin (Ed.), *Perspektive Landschaft* (pp. 229–240). Berlin: Wissenschaftlicher Verlag Berlin.

Althusser, L. (2011 [1970]). Idéologie et appareils idéologiques d'État: (Notes pour une recherche). In L. Althusser (Ed.), *Sur la reproduction* (pp. 263–306). Paris: Presses Universitaires de France.

Anter, A. (2012). *Theorien der Macht zur Einführung*. Hamburg: Junius.

Ashworth, G., Graham, B., & Tunbridge, J. (2007). *Pluralising pasts: heritage, identity and place in multicultural societies*. London: Pluto Press.

# References

Assmann, J. (1999). *Das kulturelle Gedächtnis. Schrift, Erinnerung und politische Identität in frühen Hochkulturen*. München: C. H. Beck.
Assmann, J., & Czaplicka, J. (1995). Collective memory and cultural identity. *New German Critique, 65,* 125–133.
Augustinus, (1962). *Theologische Frühschriften (De libero arbitrio, De vera religione)*. Zürich: Artemis.
Backhaus, N., Reichler, C., & Stremlow, M. (2007). *Alpenlandschaften – von der Vorstellung zur Handlung*. Zürich: vdf.
Bailer-Jones, D. (2005). The difference between models and theories. In Ch. Nimtz & A. Beckermann (Eds.), *Philosophie und/als Wissenschaft* (pp. 339–353). Paderborn: Mentis.
Baumgarten, A. G. (2007 [1750–1758]). *Ästhetik. 2 Volumes*. Hamburg: Meiner.
Békési, S. (2007). *Verklärt und verachtet. Wahrnehmungsgeschichte einer Landschaft: Der Neusiedler See*. Frankfurt a. M et al.: Peter Lang Verlag.
Berendt, B. (2005). Kognitionswissenschaft. In K. Sachs-Hombach (Ed.), *Bildwissenschaft. Disziplinen, Themen, Methoden* (pp. 21–36). Frankfurt a. M.: Suhrkamp.
Berger, P., & Luckmann, T. (1966). *The social construction of reality*. New York: Penguin Books.
Bernfeld, S. (1925). *Sisyphos oder die Grenzen der Erziehung*. Leipzig: Internationaler Psychoanalytischer Verlag.
Blumer, H. (1973). Der methodologische Standort des symbolischen Interaktionismus. In A. B. Soziologen (Ed.), *Alltagswissen, Interaktion und gesellschaftliche Wirklichkeit* (Vol. 1, pp. 80–146). Reinbek bei Hamburg: Rowohlt.
Borgeest, C. (1977). *Das sogenannte Schöne. Ästhetische Sozialschranken*. Frankfurt a. M.: S. Fischer.
Bourdieu, P. (1972). *Esquisse d'une théorie de la pratique: Précédé de 'Trois études d'ethnologie kabyle'*. Genève, Suisse: Librairie Droz.
Bourdieu, P. (1973). Kulturelle Reproduktion und soziale Reproduktion. In P. Bourdieu & J.-C. Passeron (Eds.), *Grundlagen einer Theorie der symbolischen Gewalt* (pp. 89–127). Frankfurt a. M.: Suhrkamp.
Bourdieu, P. (1977). Politik, Bildung und Sprache. In M. Steinrücke (Ed.), *Die verborgenen Mechanismen der Macht* (pp. 13–30). Hamburg: VSA.
Bourdieu, P. (1979). *La distinction: critique sociale du jugement*. Paris: Les Editions de Minuit.
Bourdieu, P. (1982a). Die feinen Unterschiede. In M. Steinrücke (Ed.), *Die verborgenen Mechanismen der Macht* (pp. 31–48). Hamburg: VSA.
Bourdieu, P. (1982b). Die verborgenen Mechanismen der Macht enthüllen. In M. Steinrücke (Ed.), *Die verborgenen Mechanismen der Macht* (pp. 81–87). Hamburg: VSA.
Bourdieu, P. (1982c). *Leçon sur la leçon*. Paris: Les Editions de Minuit.
Bourdieu, P. (1984). *Distinction: A social critique of the judgement of taste*. Cambridge: Harvard university press.
Bourdieu, P. (1991). Physischer, sozialer und angeeigneter physischer Raum. In M. Wentz (Ed.), *Stadt-Räume. Die Zukunft des Städtischen* (pp. 25–34). Frankfurt a. M.: Suhrkamp.
Bourdieu, P. (1996). Die Praxis der reflexiven Anthropologie. Einleitung zum Seminar an der École des hates études en sciences sociales. Paris, Oktober 1987. In P. Bourdieu & L. Wacquant (Eds.), *Reflexive Anthropologie* (pp. 251–294). Frankfurt a. M.: Suhrkamp.
Bourdieu, P. (2001). *Meditationen. Zur Kritik der scholastischen Vernunft*. Suhrkamp: Frankfurt a. M.
Burckhardt, L. (2006 [1977]). Landschaftsentwicklung und Gesellschaftsstruktur. In M. Ritter & M. Schmitz (Eds.), *Warum ist Landschaft schön? Die Spaziergangswissenschaft* (pp. 19–33). Kassel: Schmitz.
Burckhardt, L. (2006 [1991]). Ästhetik der Landschaft. In M. Ritter & M. Schmitz (Eds.), *Warum ist Landschaft schön? Die Spaziergangswissenschaft* (pp. 82–90). Kassel: Schmitz.
Burckhardt, L. (2006 [1995]). Spaziergangswissenschaft. In M. Ritter & M. Schmitz (Eds.), *Warum ist Landschaft schön? Die Spaziergangswissenschaft* (pp. 257–301). Kassel: Schmitz.
Burke, E. (1989 [1757]). *Philosophische Untersuchung über den Ursprung unserer Ideen vom Erhabenen und Schönen*. Hamburg: Meiner.

Burr, V. (1995). *An introduction to social constructionism.* London: Routlegde.
Burr, V. (1998). Realism, relativism, social constructivism and discourse. In I. Parker (Ed.), *Social Constructivism, Discourse and Relativism* (pp. 13–26). London: Routlegde.
Burr, V. (2005). *Social Constructivism.* London, New York: Busch-Lüty.
Butler, J. (2001 [1997]). *Psyche der Macht. Das Subjekt der Unterwerfung.* Frankfurt a. M.: Suhrkamp.
Carlson, A. (2009). *Nature and landscape. An introduction to environmental aesthetics.* New York: Columbia University Press.
Casey, E. (2006). *Ortsbeschreibungen – Landschaftsmalerei und Kartographie.* München: Fink.
Castells, M. (2001). *Der Aufstieg der Netzwerkgesellschaft. Teil 1 der Trilogie: Das Informationszeitalter.* Opladen: Leske + Budrich.
Certeau, M. D. (1990). *L'invention du quotidien. Arts de faire 1.* Paris: Gallimard.
Chilla, T. (2007). Zur politischen Relevanz raumbezogener Diskurse. Das Beispiel der Naturschutzpolitik der Europäischen Union. *Erdkunde, 61*(1), 13–25.
Chilla, T., Kühne, O., Weber, F., & Weber, F. (2015). 'Neopragmatische' Argumente zur Vereinbarkeit von konzeptioneller Diskussion und Praxis der Regionalentwicklung. In O. Kühne & F. Weber (Eds.), *Bausteine der Regionalentwicklung* (pp. 13–24). Wiesbaden: Springer VS.
Cosgrove, D. E. (1984). *Social formation and symbolic landscape.* London, Sydney: University of Wisconsin Press.
Cosgrove, D. (1985). Prospect, perspective and the evolution of the landscape idea. *Transactions of the Institute of British Geographers, 10*(1), 45–62.
Cosgrove, D. (1998). Cultural Landscapes. In T. Unwin (Ed.), *A European Geography* (pp. 65-72). London: Routledge.
Costonis, J. (1982). Law and aesthetics. A critique and a reformulation of the dilemmas. *Michigan Law Review, 80*(3), 355–461.
Croce, B. (1930). *Aesthetik als Wissenschaft vom Ausdruck und allgemeine Sprachwissenschaft.* Tübingen: J. C. B. Mohr.
Cronon, W. (1996). Introduction. In search of nature. In W. Cronon (Ed.), *Uncommon ground. Rethinking the human place in nature* (pp. 23–68). New York, London: W. W. Norton.
Dahrendorf, R. (1963). *Die angewandte Aufklärung. Gesellschaft und Soziologie in Amerika.* München: Piper.
Dahrendorf, R. (1972). *Konflikt und Freiheit. Auf dem Weg zur Dienstklassengesellschaft.* München: Piper.
Dahrendorf, R. (1983). Gespräch mit Ralf Dahrendorf. In R. Dahrendorf, F. V. Hayek, & F. Kreuzer (Eds.), *Franz Kreuzer im Gespräch mit Friedrich von Hayek und Ralf Dahrendorf.* Wien: Franz Deuticke.
Dahrendorf, R. (2008). *Die Versuchungen der Unfreiheit. Die Intellektuellen in Zeiten der Prüfung.* München: C. H. Beck.
Daniels, S., & Cosgrove, D. (1988). Introduction: Iconography and landscape. In D. Cosgrove & S. Daniels (Eds.), *The iconography of landscape. Essays on the symbolic representation, design and use of environments* (pp. 1–10). Cambridge et al.: Cambridge University Press.
DeMarrais, E., Castillo, L., & Earle, T. (1996). Ideology, materialization, and power strategies. *Current Anthropology, 37*(1), 15–31.
Deutsch, K. (1969). *Politische Kybernetik. Modelle und Perspektiven.* Freiburg: Rombach.
Dewey, J. (1929). *Experience and nature.* Victoria: W. E. Norton and Company.
Dewey, J. (1988 [1934]). *Kunst als Erfahrung.* Frankfurt a. M.: Suhrkamp.
Dubiel, H. (1992). *Kritische Theorie der Gesellschaft. Eine einführende Rekonstruktion von den Anfängen im Horkheimer-Kreis bis Habermas.* Weinheim: Juventa.
Ebert, J. (1989). *Zwischen Mythos und Wirklichkeit: die Schlacht um Stalingrad in deutschsprachigen authentischen und literarischen Texten.* Berlin: Humboldt-Universität.
Edley, N. (2001). Unravelling social constructivism. *Theory and Psychology, 11*(3), 433–441.
Egner, H. (2010). *Theoretische Geographie.* Darmstadt: WBG.
Eibl-Eibesfeldt, I. (1997). *Die Biologie des menschlichen Verhaltens. Grundriß der Humanethologie.* München: Piper.

# References

Elias, N. (1992 [1939]). *Über den Prozess der Zivilisationen*. Frankfurt a. M.: Suhrkamp.
Federman, R. (1991). *Surfiction: Der Weg der Literatur. Hamburger Poetik-Lexikon*. Frankfurt a. M.: Suhrkamp.
Fohler, S. (2004). Zur Alterität der Artefakte in sozialen Prozessen. In W. Eßbach, S. Kaufmann, D. Verdicchio, W. Lutterer, S. Bellanger, & G. Uerz (Eds.), *Landschaft, Geschlecht, Artefakte. Zur Soziologie naturaler und artifizieller Alteritäten* (pp. 39–48). Würzburg: Ergon.
Foucault, M. (1974). *Von der Subversion des Wissens*. München: Hanser.
Foucault, M. (1977 [1975]). *Surveiller et punir. Naissance de la prison*. Paris: Éditions Gallimard.
Foucault, M. (2006 [1976]). *La Volonté de savoir: Droit de mort et pouvoir sur la vie*. Paris: Folio.
Foucault, M. (1983 [1976]). *La volonté de savoir: Histoire de la sexualité 1*. Paris: Gallimard.
Friesen, H. (2013). Philosophische Ästhetik und die Entwicklung der Kunst. In H. Friesen & M. Wolf (Eds.), *Kunst, Ästhetik, Philosophie. Im Spannungsfeld der Disziplinen* (pp. 71–106). Münster: mentis.
Frohmann, E. (1997). *Gestaltqualitäten in Landschaft und Freiraum: abgeleitet von den körperlich-seelisch-geistigen Wechselwirkungen zwischen Mensch und Lebensraum*. Wien: Österreichischer Kunst- und Kulturverlag.
Fromm, E. (1936). Sozialpsychologischer Teil. In M. Horkheimer & E. Fromm (Eds.), *Studien über Autorität und Familie* (pp. 474–947). Paris: Klampen.
Fuchs-Heinritz, W., & König, A. (2005). *Pierre Bourdieu. Eine Einführung*. Konstanz: utb.
Fukuyama, F. (2013). What Is Governance? CGD Working Paper 314. http://www.cgdev.org/content/publications/detail/1426906. Accessed May 17, 2017.
Funken, C., & Löw, M. (2002). Ego-shooters container. Raumkonstruktionen im elektronischen Netz. In R. Maresch & N. Werber (Eds.), *Raum – Wissen – Macht* (pp. 69–91). Frankfurt a. M.: Suhrkamp.
Gailing, L. (2012). Sektorale Institutionensysteme und die Governance kulturlandschaftlicher Handlungsräume. Eine institutionen- und steuerungstheoretische Perspektive auf die Konstruktion von Kulturlandschaft. *Raumforschung und Raumordnung, 70*(2), 147–160.
Garfinkel, H. (1967). *Studies in ethnomethodology*. Englewood Cliffs: Prentice-Hall.
Gehlen, A. (1986). *Anthropologische und Sozialpsychologische Untersuchungen*. Reinbeck bei Hamburg: Rowohlt Taschenbuch Verlag.
Gergen, K. (1985). The social constructionist movement in modern psychology. *American Psychologist, 40*(3), 266.
Gergen, K. (1999). *An invitation to social construction*. London: Routledge.
Gergen, K., & Gergen, M. (2009). *Einführung in den sozialen Konstruktivismus*. Heidelberg: Carl Auer.
Gethmann-Siefert, A. (1995). *Einführung in die Ästhetik*. München: Fink.
Gilbert, K., & Kuhn, H. (1953). *A history of aesthetics*. Bloomington: Indiana University Press.
Glasersfeld, Ev. (1995). *Radical constructivism. A way of knowing and learning*. London: Routledge.
Glasze, G., & Mattissek, A. (2009). Die Hegemonie- und Diskurstheorie von Laclau und Mouffe. In G. Glasze & A. Mattissek (Eds.), *Handbuch Diskurs und Raum. Theorien und Methoden für die Humangeographie sowie die sozial- und kulturwissenschaftliche Raumforschung* (pp. 153–179). Bielefeld: transcript.
Goodman, N. (1951). *The structure of appearance*. Cambridge: Harvard University Press.
Goodman, N. (1968). *Languages of art: An approach to a theory of symbols*. Indianapolis: Hackett publishing.
Goodman, N. (1978). *Ways of worldmaking*. Indianapolis: Hackett Publishing.
Goodman, N. (1992). Kunst und Erkenntnis. In D. Henrich & W. Iser (Eds.), *Theorien der Kunst* (pp. 569–591). Frankfurt a. M.: Suhrkamp.
Graham, B. (1998). The past in Europe's present: Diversity, identity and the construction of place. In B. Graham (Ed.), *Modern Europe. Place. Culture. Identity* (pp. 19–49). London: Routledge.
Graham, G. (2005). *Philosophy of the arts: An introduction to aesthetics*. London, New York: Routledge.

Graham, B., Ashworth, G. J., & Tunbridge, J. E. (2000). *A geography of heritage: Power, culture, and economy*. London et al.: Arnold.
Gramsci, A. (2001). *Further selections from the prison notebooks*. Cape Town: Electric Book Company.
Greider, T., & Garkovich, L. (1994). Landscapes: The social construction of nature and the environment. *Rural Sociology, 59*(1), 1–24.
Haber, W. (2001). Kulturlandschaft zwischen Bild und Wirklichkeit. In A. für Raumforschung & Landesplanung (Eds.), *Die Zukunft der Kulturlandschaft zwischen Verlust, Bewahrung und Gestaltung = Forschungs- und Sitzungsberichte der ARL, Nr. 215* (pp. 6–29). Hannover: Verlag der ARL.
Haber, W. (2006). Kulturlandschaften und die Paradigmen des Naturschutzes. *Stadt + Grün, 55*, 20–25.
Hacking, I. (1999). *The social construction of what?* Cambridge: Harvard University Press.
Halbwachs, M. (1992 [1939]). *On collective memory*. Chicago: University of Chicago Press.
Han, B.-C. (2005). *Was ist Macht?* Stuttgart: Reclam.
Hannigan, J. (2014). *Environmental sociology*. London, New York: Routledge, Taylor & Francis Group.
Hard, G. (1995). *Spuren und Spurenleser – zur Theorie und Ästhetik des Spurenlesens in der Vegetation und anderswo*. Osnabrück: University Press Rasch.
Hard, G. (2002 [1987a]). Auf der Suche nach dem verlorenen Raum. In G. Hard (Ed.), *Landschaft und Raum. Aufsätze zur Theorie der Geographie* (pp. 211–234). Osnabrück: University Press Rasch.
Hartmann, Ev. (1924). *Philosophie des Schönen*. Berlin: Wegweiser-Verlag.
Hartmann, N. (1953). *Ästhetik*. Berlin: Alfred Kröner.
Hartz, A., & Kühne, O. (2009). Aesthetic approaches to active urban landscape planning. In A. J. J. van der Valk & T. van Dijk (Eds.), *Regional Planning for Open Space* (pp. 249–278). London: Routledge.
Hauser, S. (2001). *Metamorphosen des Abfalls. Konzepte für alte Industrieareale*. Campus: Frankfurt a. M.
Hauskeller, M. (2005). *Was ist Kunst? Positionen der Ästhetik von Platon bis Danto*. München: C. H. Beck.
Häußermann, H., & Siebel, W. (1996). *Soziologie des Wohnens: eine Einführung in Wandel und Ausdifferenzierung des Wohnens*. Weinheim: Beltz Juventa.
Hegel, G. (1970). Vorlesungen über die Ästhetik I-III. In E. Moldenhauer & K. Michel (Eds.), *Werke in 20 Bänden* (pp. 13–15). Frankfurt a. M.: Suhrkamp.
Hillebrandt, F. (2000). Disziplinargesellschaft. In G. Kneer, A. Nassehi & M. Schroer (Eds.), *Soziologische Gesellschaftsbegriffe* (pp. 101–126). München: utb.
Hokema, D. (2009). Die Landschaft der Regionalentwicklung: Wie flexibel ist der Landschaftsbegriff? *Raumforschung und Raumordnung, 67*(3), 239–249.
Hokema, D. (2013). *Landschaft im Wandel? Zeitgenössische Landschaftsbegriffe in Wissenschaft, Planungspraxis und Alltag*. Wiesbaden: Springer VS.
Hokema, D. (2015). Landscape is everywhere. The construction of landscape by US-American Laypersons. *Geographische Zeitschrift, 103*(3), 151–170.
Honneth, A. (1989 [1986]). *Kritik der Macht. Reflexionsstufen einer kritischen Gesellschaftstheorie*. Frankfurt a. M.: Suhrkamp.
Hoskins, W. (2006 [1956]). *The making of the English Landscape*. London: Little Toller Books.
Howard, P. (2011). *An introduction to landscape*. Farnham, Burlington: Routledge.
Hradil, S. (1992). Alte Begriffe und neue Strukturen. Die Milieu-, Subkultur-und Lebensstilforschung der 80er Jahre. In S. Hradil (Ed.), *Zwischen Bewußtsein und Sein* (pp. 15–55). Opladen: Leske + Budrich.
Humboldt, Wv. (1960). *Ideen zu einem Versuch, die Grenzen der Wirksamkeit des Staates zu bestimmen*. Stuttgart: Reclam.
Husserl, E. (1973 [1929]). *Cartesianische Meditationen und Pariser Vorträge. Husserliana I*. Den Haag: Martinus Nijhoff.

# References

Hutcheson, F. (1988 [1725]). *Über den Ursprung unserer Ideen von Schönheit und Tugend.* Hamburg: Meiner.

Illing, F. (2006). *Kitsch, Kommerz und Kult. Soziologie des schlechten Geschmacks.* Konstanz: UVK.

Imbusch, P. (1998a). Macht und Herrschaft in der Diskussion. In P. Imbusch (Ed.), *Macht und Herrschaft – sozialwissenschaftliche Konzeptionen und Theorien* (pp. 9–26). Opladen: Leske + Budrich.

Imbusch, P. (1998b). Macht und Herrschaft. In H. Korte & B. Schäfers (Eds.), *Einführung in die Hauptbegriffe der Soziologie* (pp. 161–182). Opladen: Leske + Budrich.

Ipsen, D. (1997). *Raumbilder: Kultur und Ökonomie räumlicher Entwicklung* (Vol. 8). Pfaffenweiler: Centaurus.

Ipsen, D. (2002). Raum als Landschaft. In D. Ipsen & D. Läpple (Eds.), *Soziologie des Raumes – Soziologische Perspektiven* (pp. 86–111). Hagen: Centaurus.

Ipsen, D. (2006). *Ort und Landschaft.* Wiesbaden: Springer VS.

Irrgang, B. (2014). Architekturethik oder Gestaltung von Wohnen. Über das Bauen und die Einbettung von Architektur in Natur und Kultur. *Ausdruck und Gebrauch, 12,* 10–29.

Jacks, B. (2004). Reimagining walking: Four practices. *Journal of Architectural Education, 51*(3), 5–9.

Jünger, F. (1946). *Die Perfektion der Technik.* Frankfurt a. M.: Vittorio Klostermann.

Kant, I. (1956 [1790]). *Kritik der Urteilskraft.* Hamburg: Anaconda.

Kant, I. (1959 [1781]). *Kritik der reinen Vernunft.* Hamburg: Anaconda.

Kant, I. (1974 [1790]). *Kritik der Urteilskraft.* Frankfurt a. M.: Suhrkamp.

Kaplan, R., Kaplan, S., & Ryan, R. (1998). *With people in mind. Design and management of everyday nature.* Washington: Covelo.

Kapp, E. (1978 [1877]). *Grundlinien einer Philosophie der Technik. Zur Entstehungsgeschichte der Cultur aus neuen Gesichtspunkten.* Braunschweig: Mainer.

Kastner, J. (2002). Existenzgeld statt Unsicherheit? Zygmunt Bauman und die Krise globaler Politik angesichts der neoliberalen Globalisierung. In M. Junge & T. Kron (Eds.), *Zygmunt Bauman. Soziologie zwischen Postmoderne und Ethik* (pp. 225–254). Opladen: Leske + Budrich.

Kaufmann, S. (2005). *Soziologie der Landschaft.* Wiesbaden: Springer VS.

Kersting, W. (2009). *Verteidigung des Liberalismus.* Hamburg: Murmann.

Kloock, D., & Spahr, A. (2007). *Medientheorien: Eine Einführung.* München: utb.

Kneer, G. (2009). Jenseits von Realismus und Antirealismus. Eine Verteidigung des Sozialkonstruktivismus gegenüber seinen postkonstruktivistischen Kritikern. *Zeitschrift für Soziologie, 38*(1), 5–25.

Koelle, A. (1822). *System der Technik.* Berlin: Carl Friedrich Amelang. https://books.google.de/books?id=67ZRAAAAMAAJ&pg=PA430&dq=System+der+Technik.+Berlin.&hl=de&sa=X&ved=0ahUKEwjKzKLb39nUAhWHRhQKHeSyBHMQ6AEIJzAA#v=onepage&q=System%20der%20Technik.%20Berlin.&f=false. Accessed June 26, 2017.

Kost, S. (2013). Landschaftsgenese und Mentalität als kulturelles Muster. Das Landschaftsverständnis in den Niederlanden. In D. Bruns & O. Kühne (Eds.), Landschaften: Theorie, Praxis und internationale Bezüge (pp. 55–70). Schwerin: Oceano.

Kreis, G. (2004). *Mythos Rütli: Geschichte eines Erinnerungsortes.* Zürich: Orell Füssli.

Kühne, O. (2004). *Monetarisierung der Umwelt - Chancen und Probleme aus raumwissenschaftlich-systemtheorietischer Perspektive. Beiträge zur kritischen Geographie 3.* Wien: Verein kritische Geographie.

Kühne, O. (2006a). *Landschaft in der Postmoderne. Das Beispiel des Saarlandes.* Wiesbaden: Deutscher Universitätsverlag.

Kühne, O. (2006b). Landschaft und ihre Konstruktion – theoretische Überlegungen und empirische Befunde. *Naturschutz und Landschaftsplanung, 38,* 146–152.

Kühne, O. (2006c). *Landschaft, Geschmack, soziale Distinktion und Macht – von der romantischen Landschaft zur Industriekultur. Eine Betrachtung auf Grundlage der*

*Soziologie Pierre Bourdieus. Beiträge zur Kritischen Geographie 6*. Wien: Verein kritische Geographie.
Kühne, O. (2006d). Soziale Distinktion und Landschaft. Eine landschaftssoziologische Betrachtung. *Stadt + Grün, 56*(12), 40–43.
Kühne, O. (2008a). *Distinktion, Macht, Landschaft. Zur sozialen Definition von Landschaft.* Wiesbaden: Springer VS.
Kühne, O. (2008b). Die Sozialisation von Landschaft – sozialkonstruktivistische Überlegungen, empirische Befunde und Konsequenzen für den Umgang mit dem Thema Landschaft in Geographie und räumlicher Planung. *Geographische Zeitschrift, 96*(4), 189–206.
Kühne, O. (2009). Heimat und Landschaft – Zusammenhänge und Zuschreibungen zwischen Macht und Mindermacht. Überlegungen auf sozialkonstruktivistischer Grundlage. *Stadt + Grün, 58*(9), 17–22.
Kühne, O. (2012). *Stadt – Landschaft – Hybridität. Ästhetische Bezüge im postmodernen Los Angeles mit seinen modernen Persistenzen.* Wiesbaden: Springer VS.
Kühne, O. (2013a). *Landschaftstheorie und Landschaftspraxis. Eine Einführung aus sozialkonstruktivistischer Perspektive.* Wiesbaden: Springer VS.
Kühne, O. (2013b). Landschaftsästhetik und regenerative Energien – Grundüberlegungen zu De- und Re- Sensualisierungen und inversen Landschaften. In L. Gailing & M. Leibenath (Eds.), *Neue Energielandschaften – Neue Perspektiven der Landschaftsforschung* (pp. 101–120). Wiesbaden: Springer VS.
Kühne, O. (2014). Landschaft und Macht: von Eigenlogiken und Ästhetiken in der Raumentwicklung. *Ausdruck und Gebrauch, 12*, 151–172.
Kühne, O. (2015). The streets of Los Angeles—About the integration of infrastructure and power. *Landscape Research, 40*(2), 139–153.
Kühne, O., & Spellerberg, A. (2010). *Heimat und Heimatbewusstsein in Zeiten erhöhter Flexibilitätsanforderungen. Empirische Untersuchungen im Saarland.* Wiesbaden: Springer VS.
Kühne, O., & Weber, F. (2015). Der Energienetzausbau in Internetvideos – eine quantitativ ausgerichtete diskurstheoretisch orientierte Analyse. In S. Kost & A. Schönwald (Eds.), *Landschaftswandel – Wandel von Machtstrukturen* (pp. 113–126). Wiesbaden: Springer VS.
Laclau, E., & Mouffe, C. (1985). *Hegemony and socialist strategy. Towards a radical democratic politics.* London: Routledge.
Lacoste, Y. (1990). *Geographie und politisches Handeln.* Berlin: Wagenbach.
Latour, B. (2002 [1999]). *Die Hoffnung der Pandora.* Frankfurt a. M.: Suhrkamp.
Legg, S. (2007). Reviewing geographies of memory/forgetting. *Environment and Planning A, 39*(2), 456–466.
Lehmann, H. (1973 [1950]). Die Physiognomie der Landschaft. In K. Paffen (Ed.), *Das Wesen der Landschaft* (pp. 39–70). Darmstadt: Wege der Forschung.
Leibenath, M. (2014). Landschaft im Diskurs: Welche Landschaft? Welcher Diskurs? Praktische Implikationen eines alternativen Entwurfs konstruktivistischer Landschaftsforschung. *Naturschutz und Landschaftsplanung, 46*(4), 124–129.
Leibenath, M. (2015). Landschaften und Macht. In S. Kost & A. Schönwald (Eds.), *Landschaftswandel – Wandel von Machtstrukturen* (pp. 17–26). Wiesbaden: Springer VS.
Leibenath, M., & Gailing, L. (2012). Semantische Annäherung an 'Landschaft' und 'Kulturlandschaft'. In M. Schenk, M. Kühn, M. Leibenath, & S. Tzschaschel (Eds.), *Suburbane Räume als Kulturlandschaften* (pp. 58–79). Hannover: Verlag der ARL.
Lenski, G. (2013). *Power and privilege: A theory of social stratification.* Chapel Hill: UNC Press Books.
Liessmann, K. (2002). *Kitsch! Oder warum der schlechte Geschmack der gute ist.* Wien: Christian Brandstätter Verlag.
Lingg, E., Reutlinger, C., & Fritsche, C. (2010). Landschaft. In N. Günnewig (Ed.), *Raumwissenschaftliche Basics* (pp. 119–127). Wiesbaden: Springer VS.
Loesberg, J. (2005). *A Return to Aesthetics. Autonomy, Indifference, and Postmodernism.* Stanford: Stanford Universitiy Press.

# References

Löw, M. (2001). *Raumsoziologie*. Frankfurt a. M.: Suhrkamp.
Luhmann, N. (1971). Zweck – Herrschaft – System. Grundbegriffe und Prämissen Max Webers. In N. Luhmann (Ed.), *Politische Planung. Aufsätze zur Soziologie von Politik und Verwaltung* (pp. 90–112). Opladen: Westdeutscher Verlag.
Luhmann, N. (1984). *Soziale Systeme. Grundriß einer allgemeinen Theorie*. Frankfurt a. M.: Suhrkamp.
Lundmark, T. (1997). *Landscape, recreation, and takings in German and American Law*. Stuttgart: Verlag H.-D. Heinz.
Lynch, M. (2016). Social constructivism in science and technology studies. *Human Studies, 39*, 101–112.
Lyotard, F. (1991). *Leçons sur l'Analytique du sublime*. Paris: Galilée.
Majetschak, St. (2007). *Ästhetik zur Einführung*. Hamburg: Junius.
Marshall, J. (2008). Toward phenomenology. A material culture studies approach to landscape theory. In R. DeLue & J. Elkins (Eds.), *Landscape Theory* (pp. 195–203). New York, London: Routledge.
Masuch, M. (1972). *Politische Ökonomie der Ausbildung*. Reinbeck bei Hamburg: Mauke.
Matless, D. (2005). *Landscape and englishness*. London: Reaktion Books.
Matthiesen, U. (2006). Raum und Wissen. Wissensmilieus und KnowledgeScapes als Inkubatoren für zukunftsfähige stadtregionale Entwicklungstechniken? In D. Tänzler, H. Knoblauch, & H.-G. Soeffner (Eds.), *Zur Kritik der Wissensgesellschaft* (pp. 101–138). Konstanz: UVK Verlagsgesellschaft mbH.
Maturana, H., & Varela, F. J. (1987). *The tree of knowledge. The biological roots of human understanding*. Boston: New Science Library/Shambhala Publications.
Mead, G., & Morris, C. (1967). *Mind, self & society from the standpoint of a social behaviorist*. Chicago: University of Chicago Press.
Merleau-Ponty, M. (1945). *Phénoménologie de la perception*. Paris: Gallimard.
Micheel, M. (2012). Alltagsweltliche Konstruktionen von Kulturlandschaft. *Raumforschung und Raumordnung, 70*(2), 107–117.
Miggelbrink, J. (2002). Konstruktivismus? 'Use with caution'. Zum Raum als Medium der Konstruktion gesellschaftlicher Wirklichkeit. *Erdkunde, 56*(4), 337–350.
Miggelbrink, J. (2009). Verortung im Bild. Überlegungen zu 'visuellen Geographien'. In J. Döring & T. Thielmann (Eds.), *Mediengeographie. Theorie – Analyse – Diskussion* (pp. 179–202). Bielefeld: transcript.
Mises, L. V. (1927). *Liberalismus*. Jena: Gustav Fischer.
Mitchell, W. (2002). Introduction. In W. Mitchell (Ed.), *Landscape and power* (pp. 1–4). Chicago, London: University of Chicago Press.
Mrass, W. (1981). Ökologische Entwicklungstendenzen im ländlichen Raum und ihre Auswirkungen auf die Flurbereinigung. In Bayrisches Staatsministerium für Ernährung, Landwirtschaft und Forsten (Ed.), *Berichte aus der Flurbereinigung 37* (pp. 29–40). München: Bayerisches Staatsministerium für Ernährung, Landwirtschaft und Forsten.
Muir, R. (2003). On change in the landscape. *Landscape Research, 31*(4), 383–403.
Myers, G. (1996). Naming and placing the other: Power and the urban landscape in Zanzibar. *Tijdschrift voor economische en sociale geografie, 87*(3), 237–246.
Neuenhaus, P. (1998). Max Weber: Amorphe Macht und Herrschaftsgehäuse. In P. Imbusch (Ed.), *Macht und Herrschaft – sozialwissenschaftliche Konzeptionen und Theorien* (pp. 77–93). Opladen: Westdeutscher Verlag.
Newman, D., & Paasi, A. (1998). Fences and neighbours in the postmodern world: boundary narratives in political geography. *Progress in Human Geography, 22*(2), 186–207.
Nohl, W. (1997). Bestimmungsgründe landschaftlicher Eigenart. *Stadt + Grün, 46*, 805–813.
Olwig, K. (2009). Introduction to part one law, polity and the changing meaning of landscape. In K. Olwig & D. Mitchell (Eds.), *Justice, power and the political landscape* (pp. 5–10). London, New York: Routledge.
Ortner, S. B. (2006). *Anthropology and social theory: Culture, power, and the acting subject*. Durham: Duke University Press.

Paasi, A. (1996). *Territories, boundaries and consciousness: The changing geographies of the Finnish-Russian Border*. Chichester: Wiley.

Papadimitriou, F. (2010). Conceptual modelling of landscape complexity. *Landscape Research, 35* (5), 563–570.

Paris, R. (2005). *Normale Macht. Soziologische Essays*. Konstanz: UVK Verlagsgesellschaft mbH.

Parsons, T. (1951). *The social system*. Glencoe: Free Press.

Peres, C. (2013). Philosophische Ästhetik. Eine Standortbestimmung. In H. Friesen & M. Wolf (Eds.), *Kunst, Ästhetik, Philosophie. Im Spannungsfeld der Disziplinen* (pp. 13–69). Münster: mentis.

Petermann, S. (2007). *Rituale machen Räume: Zum kollektiven Gedenken der Schlacht von Verdun und der Landung in der Normandie*. Bielefeld: Transcript.

Platon. (2005). *Werke in acht Bänden, griechisch und deutsch*. Darmstadt: ZVAB.

Pöltner, G. (2008). *Philosophische Ästhetik*. Stuttgart: W. Kohlhammer Verlag.

Popitz, H. (1992). *Phänomene der Macht*. Tübingen: Mohr Siebeck.

Popper, K. (1973). *Objektive Erkenntnis. Ein evolutionärer Entwurf*. Hamburg: Hoffmann und Campe.

Porteous, J. (2013). *Environmental aesthetics: Ideas, politics and planning*. London, New York: Routledge.

Potter, J. (1996). *Representing Reality. Discourse, rhetoric ans social construction*. London: Sage.

Prollius, M. V. (2014). Legitime Staatsausgaben. In H. Krebs (Ed.), *Klassischer Liberalismus. Die Staatsfrage – gestern, heute, morgen* (pp. 183–213). Norderstedt: Forum freie Gesellscahft.

Psyeudo-Dionysius Areopagita. (1988). *Über die göttlichen Namen*. Stuttgart: Suchla.

Rammert, W. (2007). *Technik - Handeln - Wissen: zu einer pragmatistischen Technik- und Sozialtheorie*. Wiesbaden: Springer VS.

Recki, B. (2013). Stil im Handeln oder die Aufgaben der Urteilskraft. In H. Friesen & M. Wolf (Eds.), *Kunst, Ästhetik, Philosophie. Im Spannungsfeld der Disziplinen* (pp. 221–244). Münster: mentis.

Reusswig, F. (2010). Klimawandel und Gesellschaft. Vom Katastrophen-zum Gestaltungsdiskurs im Horizont der postkarbonen Gesellschaft. In M. Voss (Ed.), *Der Klimawandel* (pp. 75–97). Wiesbaden: Springer VS.

Rheder, H. (1932). *Die Philosophie der unendlichen Landschaft*. Halle: M. Niemeyer.

Ritter, J. (1996 [1962]). Landschaft. Zur Funktion des Ästhetischen in der modernen Gesellschaft. In G. Gröning & U. Herlyn (Eds.), *Landschaftswahrnehmung und Landschaftserfahrung* (pp. 28–68). Münster: Lit-Verlag.

Rosenkranz, K. (1996 [1853]). *Ästhetik des Häßlichen*. Leipzig: Reclam.

Satter, E. (2000). Ästhetik. In J. Bretschneider (Ed.), *Lexikon freien Denkens*. Neustadt am Rübenberge: Angelika Lenz Verlag.

Schlottmann, A. (2005). *RaumSprache. Ost-West-Differenzen in der Berichterstattung zur deutschen Einheit. Eine sozialgeographische Theorie*. Stuttgart: Franz Steiner.

Schneider-Sliwa, R. (2005). *USA*. Darmstadt: Wissenschaftliche Buchgesellschaft.

Schütz, A. (1971). *Gesammelte Aufsätze 3. Studien zur phänomenologischen Philosophie*. Den Haag: Martinus Nijhoff.

Schütz, A. (1971 [1962]). *Gesammelte Aufsätze 1. Das Problem der sozialen Wirklichkeit*. Den Haag: Martinus Nijhoff.

Schütz, A. (2004 [1932]). *Der sinnhafte Aufbau der sozialen Welt. Eine Einleitung der sozialen Welt*. Konstanz: UVK Verlagsgesellschaft mbH.

Schütz, A., & Luckmann, T. (1973). *The structures of the life-world* (Vol. 1). Evanston, Illinois: Northwestern University Press.

Schweppenhäuser, G. (2007). *Ästhetik. Philosophische Grundlagen und Schlüsselbegriffe*. Frankfurt a. M., New York: Campus.

Seel, M. (1996). *Eine Ästhetik der Natur*. Frankfurt a. M.: Suhrkamp.

Seidman, S. (2012). *Contested knowledge: Social theory today*. Malden, Oxford: Wiley-Blackwell.

# References

Sheppard, A. (1987). *Aesthetics: An introduction to the philosophy of art*. Oxford, New York: Oxford University Press.

Shusterman, R. (2001). Tatort: Kunst als Dramatisieren. In J. Früchtl & J. Zimmermann (Eds.), *Ästhetik der Inszenierung. Dimensionen eines künstlerischen, kulturellen und gesellschaftlichen Phänomens* (pp. 126–143). Frankfurt a. M.: Suhrkamp.

Sidanius, J., & Pratto, F. (2001). *Social dominance: An intergroup theory of social hierarchy and oppression*. Cambridge: Cambridge University Press.

Simmel, G. (1996 [1913]). Philosophie der Landschaft. In G. Gröning & U. Herlyn (Eds.), *Landschaftswahrnehmung und Landschaftserfahrung. Texte zur Konstitution und Rezeption von Natur als Landschaft* (pp. 91–105). München: Minerva Publication.

Sofsky, W. (2007). *Verteidigung des Privaten. Eine Streitschrift*. München: C. H. Beck.

Sofsky, W., & Paris, R. (1994). *Figurationen sozialer Macht. Autorität, Stellvertretung, Koalition*. Frankfurt a. M.: Suhrkamp.

Soyez, D. (2003). Kulturlandschaftspflege: Wessen Kultur? Welche Landschaft? Was für eine Pflege? *Petermanns Geographische Mitteilungen, 147*, 30–39.

Stearns, P. (1995). Emotion. In R. Harré & P. Stearns (Eds.), *Discursive psychology* (pp. 37–54). London: Sage.

Stevens, G. (2002). *The favored circle: The social foundations of architectural distinction*. Cambridge, Massachusetts: MIT Press.

Strüver, A., & Wucherpfennig, C. (2009). Performativität. In G. Glasze & A. Mattissek (Eds.), *Handbuch Diskurs und Raum. Theorien und Methoden für die Humangeographie sowie die sozial- und kulturwissenschaftliche Raumforschung* (pp. 107–128). Bielefeld: transcript.

Tiezzi, E. (2005). *Beauty and science*. Southampton: WIT Press.

Tillmann, K.-J. (1976). *Unterricht als soziales Erfahrungsfeld*. Frankfurt a. M.: Suhrkamp.

Tillmann, K.-J. (2007). *Sozialisationstheorien*. Reinbeck bei Hamburg: Rowohlt.

Townsend, D. (1997). *An introduction to aesthetics*. Malden (Massachusetts): Blackwell.

Treiber, H. (2007). Macht – ein soziologischer Grundbegriff. In P. Gostmann & P.-U. Merz-Benz (Eds.), *Macht und Herrschaft. Zur Revision zweier soziologischer Grundbegriffe* (pp. 49–62). Wiesbaden: Springer VS.

Trepl, L. (2012). *Die Idee der Landschaft. Eine Kulturgeschichte von der Aufklärung bis zur Ökologiebewegung*. Bielefeld: transcript.

Trigg, D. (2009). *The Aesthetics of Decay. Nothingness, Nostalgia, and the Absence of Reason*. New York et al.: Peter Lang Publishing Group.

Troebst, S. (2005). Jalta versus Stalingrad, GULag versus Holocaust. *Berliner Journal für Soziologie, 15*(3), 381–400.

Tuan, Y.-F. (1974). *Topophilia: A study of environmental perception, attitudes and values*. Englewood Cliffs: Prentice Hall Inc.

van Noy, R. (2003). *Surveying the interior*. Reno, Las Vegas: University of Nebraska Press.

Vischer, F. (1922). *Kritische Gänge*. München: Meyer & Jessen.

Wayand, G. (1998). Pierre Bourdieu: Das Schweigen der Doxa aufbrechen. In P. Imbusch (Ed.), *Macht und Herrschaft – sozialwissenschaftliche Konzeptionen und Theorien* (pp. 221–237). Opladen: Leske + Budrich.

Weber, M. (1976 [1922]). *Wirtschaft und Gesellschaft. Grundriß der verstehenden Soziologie*. Tübingen: Mohr Siebeck.

Weber, M. (1982). *Gesammelte Aufsätze zur Wissenschaftslehre*. Tübingen: utb.

Weber, M. (1988 [1918]). Parlament und Regierung im neugeordneten Deutschland. In J. Winckelmann (Ed.), *Gesammelte Politische Schriften von Max Weber* (pp. 306–443). Tübingen: J.B.C. Mohr.

Weber, F. (2013). *Soziale Stadt – Politique de la Ville – Politische Logiken. (Re-)Produktion kultureller Differenzierungen in quartiersbezogenen Stadtpolitiken in Deutschland und Frankreich*. Wiesbaden: Springer VS.

Weber, F. (2015). Diskurs – Macht – Landschaft. Potenziale der Diskurs- und Hegemonietheorie von Ernesto Laclau und Chantal Mouffe für die Landschaftsforschung. In S. Kost &

A. Schönwald (Eds.), *Landschaftswandel – Wandel von Machtstrukturen* (pp. 97–112). Wiesbaden: Springer VS.

Welsch, W. (1988a). Einleitung. In W. Welsch (Ed.), *Wege aus der Moderne. Schlüsseltexte der Postmoderne-Diskussion* (pp. 1–46). Weinheim: VCH, Acta Humaniora.

Welsch, W. (1988b). *Postmoderne – Pluralität als ethischer und politischer Wert*. Köln: Bachem.

Welsch, W. (2006). *Ästhetisches Denken*. Stuttgart: Reclam.

Werlen, B. (2000). *Sozialgeographie. Eine Einführung*. Bern, Stuttgart, Wien: utb.

Werlen, B., & Weingarten, M. (2005). Tun, Handeln, Strukturieren—Gesellschaft, Struktur, Raum. In M. Weingarten (Ed.), *Strukturierung von Raum und Landschaft. Konzepte in Ökologie und der Theorie gesellschaftlicher Naturverhältnisse* (pp. 177–221). Münster: transcript.

Wetherell, M., & Still, A. (1998). Realism and relativism. In R. Sapsford, M. Still, D. Wetherell, D. Miell, & R. Stevens (Eds.), *Theory and social psychology* (pp. 99–114). London: Sage in association with the Open University.

Wicks, R. (2011). *European aesthetics. A critical introduction from Kant to Derrida*. Richmond: The Massachusetts Institute of Technology.

Willke, H. (2002). *Atopia. Studien zur atopischen Gesellschaft*. Frankfurt a. M.: Suhrkamp.

Willke, H. (2005). Welche Expertise braucht die Politik? In A. Ogner & H. Torgersen (Eds.), *Wozu Experten? Ambivalenzen der Beziehung von Wissenschaft und Politik* (pp. 45–63). Wiesbaden: Springer VS.

Wöhler, K. (2001). Pflege der Negation. Zur Produktion negativer Räume als Reiseauslöser. In A. Keul, R. Bachleitner, & H. Kagelmann (Eds.), *Gesund durch Erleben? Beiträge zur Erforschung der Touristengesellschaft* (pp. 29–36). München, Wien: Profilverlag.

Wojtkiewicz, W., & Heiland, S. (2012). Landschaftsverständnisse in der Landschaftsplanung. Eine semantische Analyse der Verwendung des Wortes 'Landschaft' in kommunalen Landschaftsplänen. *Raumforschung und Raumordnung, 70*(2), 133–145.

Zahavi, D. (2007). *Phänomenologie für Einsteiger*. Paderborn: UTB.

Zerger, F. (2000). *Klassen, Milieus und Individualisierung*. Frankfurt a. M., New York: Campus.

# Chapter 3
# The Genesis of Social Landscapes and Their Physical Manifestations

Following Berger and Luckmann (1966), symbolic 'worlds of meaning', of which landscape is one can be viewed as social—and accordingly as historical—products. A consideration of the social meaning of landscape must, therefore, take account of the developing historical understanding of that term. Given the co-evolutionary bond between a social construct and its objective correlate, reference will at the same time be made to the development of the physical spaces we know as 'landscape'. This dual approach will first consider the German language area before turning to the development of the concept of landscape in other languages and culture (further insights into this topic are provided by Müller 1977; Piepmeier 1980; Eisel 1982; Kühne 2013, 2015b; and Kirchhoff and Trepel 2009).

## 3.1 The Genesis of Social Landscapes in Co-evolution with the Development of Physical Spaces in the German Language Area

### 3.1.1 Etymological Origins of the German Concept of Landschaft ('Landscape')

The suffix '-*schaft*' in Germanic languages, common in words like *Landschaft* (landscape), is derived from the Gothic *skapjan* (cognate with the Old English verb *scieppan* and other related forms such as *\*skapi-, \*skapja- \*skafti-*), in the general sense of 'to make/shape/create'. Corresponding nouns were relatively constant, with the meaning of 'form/shape/quality/nature/condition/manner'. According to Müller (1977) the German variants can be divided into three groups:

(a) abstract designations: *Meisterschaft* (championship—literally 'mastership'), *Herrschaft* (lordship, dominance) etc.

(b) collective designations for groups of persons: *Mannschaft* (team), *Genossenschaft* (cooperative society—literally 'comradeship') etc.
(c) spatial designations: *Grafschaft* (county), *Landschaft* (landscape) etc.

Common to all these terms is the sense of something that belongs together as a result of human activity. This is the sense of the modern German verbs *schaffen* and *schöpfen*, as well as of the English verb 'to shape', all of which mean 'to make/create/form' and can also include an aesthetic dimension (Haber 2007).

The Old High German word *Lantscaf* is first recorded in the early 9th century (Gruenter 1975 [1953]), when it designated "something that in almost every case possessed the quality of a largish area of settlement" (Müller 1977, p. 6). This was a time of continuous population growth (Fig. 3.1), in which "economic, governmental, and religious powers […] together aimed to make Central Europe a region of stable local settlement, for that alone presented a calculable basis for economic growth and effective rule" (Küster 1999, p. 172). Rather than possessing an immediate reference to physical space or its delimitation, the term retained its derivation from the collective word for persons and groups, and was used at that time in the sense of the social and behavioral norms of those who lived in a particular area. Only in the following centuries did the meaning shift from "the social norms in a stretch of land" to "the

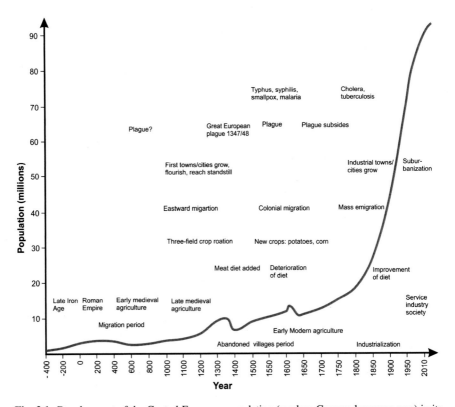

**Fig. 3.1** Development of the Central European population (modern German language area) in its historical context (Schenk 2011, modified)

stretch of land harboring those norms" (Müller 1977, p. 7). In the course of the 12th century '*Landschaft*' gained a dual political connotation as a legally defined space constitutive of a larger political entity (Müller 1977) in which politically active elements (as opposed to peasants) were, as a group, considered the "representatives of the 'whole *Landschaft*'" (Hard 1977, p. 14).

In the High Middle Ages, the concept of *Landschaft* began to denote the area cultivated and governed by a town or city, as opposed to the virgin forest—in other words a space distinct from the untamed wilderness (Müller 1977; see also Haber 2007). This developed in the late medieval period into a precise term for a specific spatial entity governed by the institutions of law (Müller 1977; see also Olwig 1996). Thus, in the course of the Middle Ages, in addition to its earlier initially descriptive and then gradually normative reference to a social grouping within its settled locality, the German word *Landschaft* gained political connotations which went hand in hand with the development and delimitation of regional and local lordships.

## 3.1.2 Landscape Painting from the Renaissance to the German Romantics

The aesthetic vision of physical space found in antiquity (see esp. Appleton 1986; Büttner 2006) was abruptly discontinued in the Middle Ages, when "eyes turned to heaven rather than to earthly appearances" (Lehmann 1968, p. 9). In medieval painting the motifs of what would later be known as landscape withdrew into the background in favor of biblical scenes, legends of the saints and visualizations of doctrine; for in a Christian world "the function of art [was] to express the divine work" of salvation history (Büttner 2006, p. 36). Medieval painting "did not arise from the desire to present a colorful world in the wealth and diversity of its relationships, but to keep before people's eyes the history and symbols of their salvation" (Böheim 1930, p. 82).

Not until the Renaissance—and this also applied to Germany—was landscape painting seen as a discipline in its own right (Andrews 1989). A crucial step in this direction was the development of a centralized perspective (Piepmeier 1980; Eisel 1982; Cosgrove 1984, 1985). Another factor, whose roots also lay in the Renaissance, was the socially idealized vision of nature that "emerged in seventeenth century European painting and [...] found its definitive expression in the work of Claude Lorrain" (Riedel 1989, p. 45; see also esp. Cosgrove 1993). Lorrain (1600–1682) and Nicolas Poussin (1594–1665) were the major influences on German landscape painting of that period (Roters 1995).

The desire to reconnect with classical antiquity found expression in the 'Italian journey', which "became a stable component in the training of artists from north of the Alps" (Büttner 2006, p. 125), resulting in a repertoire of idealized, Arcadian images: "utopian landscape in which the human and natural worlds are imagined to co-exist harmoniously" (Howard 2003, p. 53). Far from seeking to represent specific stretches of countryside, these compositions were, as Burckhardt (2006a [1998], p. 116) pointedly remarks, "pieced together in the studio in accordance with idealized patterns". Conversely, as Olwig (2008a) observes, the creative process was all the more evident: painters created a 'painted land', and it was this that passed into the expectations that conditioned social ways of seeing. In Herbert Lehmann's words (1968, p. 7) the fine arts served as "the pacemakers of our vision and experience of landscape"; and this vision—in Germany as in the neighboring countries of Central and Northern Europe—extended to the material spaces of 'real' countryside (Cosgrove 1993; Haber 2007; Czepczyński 2008; Schenk 2013; Kühne 2013). It was in this sense that Oppel (1884, p. 36) could define landscape as "terrestrial space presented to a specific viewpoint as a single whole". Attributed to Alexander von Humboldt, the designation of landscape as 'the total character of a terrestrial area' (Hard 1970) goes even further, postulating perception of an inherently cultural 'character' in a stretch of physical land: an act of cognition that clearly transcends the visual-aesthetic process.

Nevertheless, the habits of vision cultivated by the landscape artists of the day found no counterpart in the real physical spaces of Central and Northern Europe, where the pressure of growing populations (see Fig. 3.1) was changing the aspect not only of agrarian land but also of the native forests. The intrusion of pastures into sylvan areas, together with the clearance of the forest floor to garner bedding material for cattle, diminished the regenerative capacity of woodlands (see Radkau and Schäfer 1987; Urmersbach 2009), and the indirect use of solar energy in the form of charcoal, wind, and water power, as well as agricultural products, reached its pre-fossil limits (Fig. 3.2):

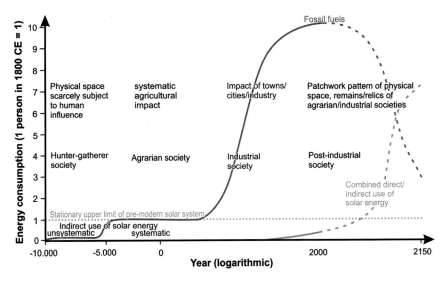

**Fig. 3.2** Energy consumption and growth of Central and West European society (Schenk 2005, modified)

3.1 The Genesis of Social Landscapes in Co-evolution … 53

The landscape construct of German Romanticism and the Biedermeier period was both intense and persistent (Safranski 2007). In the classical era of 18th century Germany, landscape had either served as a background to the presentation of historical events or, infused with reason, was itself elevated to the status of an *objet d'art* (Hohl 1977). It was the Romantic movement of the early 19th century that finally "accorded it its highest valuation, widening the concept of landscape to embrace both historical and mythological aspects" (Hohl 1977, p. 45; see also Piepmeier 1980). For German Romantic artists—among whom Caspar David Friedrich (1774–1840) holds a leading position—painting was no longer simply a matter of artistic practice: it extended to "the innermost moral and religious sentiments of its practitioners" (Büttner 2006, p. 262). Drawing consciously on medieval antecedents for its profound allegorical and symbolic force, Romantic art added a post-Reformation awareness of the modern individual as an isolated figure cut off from a paradisiacal nature (Zink 2006). In this sense Caspar David Friedrich's landscapes (Fig. 3.3; see also Fig. 3.6) are archetypical "landscapes of

**Fig. 3.3** Caspar David Friedrich's The Lonely Tree projects a Romantic world view, with its central symbols of the German oak reaching from past to present, the shepherd leaning upon it in harmony with nature, and the village in the middle ground framed by the light from the heavens, which falls not only on its fields but is reflected in the still waters of its ponds as a metaphor of a fundamentally religious cosmic-historical order (Hannessen 1985). An ecologist would see the gnarled—"actually crippled" (Urmersbach 2009, p. 34)—oak as the victim of constant feeding by domestic animals. (bpk/Nationalgalerie, SMB /Jörg P. Anders)

the soul, [...] saturated with spiritual and religious pathos" (Spanier 2006, p. 33). The aesthetic and emotional landscapes of the Romantics re-endowed physical space and its natural phenomena with the magic of which the Enlightenment had robbed them. In that sense Romanticism was the dark "reverse [...] of the Enlightenment" (Illing 2006, p. 48), diametrically opposed to—but by the same token intimately dependent upon—its ideal of enlightened rationalism.

Throughout the 18th century rational measures had made crucial inroads on the agrarian sector where, for instance, erstwhile local differences had been erased by the systematic selection of seed types and methods of dunging, irrigation and drainage, the introduction and wide dissemination of new field crops (beets and turnips, clover, oilseed rape, potatoes), and the abolition of common land (see e.g. Beck 1996; Konold 1996; Häcker 1998; Job 1999; Gudermann 2005). Improvement measures of this sort had been taken not only in the interests of feeding a growing population and hence filling state coffers—for which reason they were ardently supported by the fiscal authorities (Abel 1967)—but also to facilitate the immediate exercise of power. Thus "the draining of marshland removed the last hideaways of deserters and provided unhindered passage for the well-drilled armies of the king [i.e. Frederick the Great (O.K.)] on their forward progress". The cartographic surveys undertaken at this time not only established "a basis for property tax" (Blackbourn 2007, p. 57), they were also an indispensable instrument of military strategy. Above all, perhaps, they were the expression of a rational, scientifically enlightened vision of the world that considered it "necessary to conscientiously survey both dwellings and fields and to prepare special maps with detailed descriptions of soil types" (Däumel 1963, p. 346). Executed with precision, the cartographic campaigns of the age provided an objective image of the state's territorial possessions, and as such were "a benchmark of political power" (Blackbourn 2007, p. 14).

The cultural divide between the Enlightenment and the Romantics is evident in the Romantic vision of a holistic science as a combined cognitive, moral, aesthetic, and intuitive understanding of the world (Eisel 2009; see Sect. 2.4 above)—an approach that has persisted to the present day in some essentialist perceptions of science. In the Biedermeier period, landscape took on a humanitarian symbolism as a counter-force to the omnipresent utilitarian belief in the 'civilizing' progress of science and technology and the correlative expansion of cities and industries into an increasingly threatened countryside (Kortländer 1977). The age-old bond between physical space and its economic use was being loosened: "With every technical advance, the connection with the land became more remote, the autonomous, self-glorifying status of technology grew, until the point came when landscape was thought of merely as a location to be selected for reasons of industrial expediency" (Freyer 1996 [1966], p. 81). The impact of the Romantic Movement on the social landscape in the German language area will be treated in greater detail in Sect. 4.2.1.1: 'Landscape as a class signifier in the wake of social modernization'.

## 3.1.3 The Sociocritical Dimension of 'Landscape'—Life Beyond the City Streets

As elsewhere in Europe, the aesthetic view of landscape was cultivated in Germany above all by the educated urban classes who enjoyed the necessary economic and social distance to the countryside in its physical reality (Ritter 1996 [1962]; Bourassa 1991; etc.), governed as this was by the labor of the agricultural year and the ever-present risk of failed harvests (Kühne 2013). The urban bourgeois were in a position to "see reality with the eye of the painter and nature with the eye of the landscape painter", and it was "under the influence of that vision that the term 'landscape' began, in educated circles, to be applied to external reality, viewed as an 'artistic segment of nature'" (Hard 1977, p. 14). Burckhardt (2006b [1994], p. 94) points up the reciprocal perspective operating in the construction of the city-country dichotomy: "The 'landscape' image is a product of the interface between city and country. The country is 'landscape' because of the Easter walks taken there by the city dwellers; city is city because of the astonished faces of the market women and hawkers of firewood." Against this background, landscape assumed a compensatory note: the everyday pressure of urban life enhanced the value of the immediate confrontation with what was thought of as unspoiled nature (Ritter 1996 [1962]; see also Trepl 2012).

The transition from an agrarian to an industrial society tore asunder rural society's age-old bond with nature. Instead of living by the rhythm of the seasons (marked by sowing and harvesting) and the days (marked by the hours of sunlight), the burgeoning city populations were ruled by the economy and logic of technology (shifts, manufacturing processes etc.; see Kühne 2013). For Bätzing (2000, p. 197) the concept of landscape was also conditioned by this changing context; for a world that was becoming more complex by the day, with its different work processes, functions, and professions, "was still available in the hours of Sunday leisure in the pristine integrity of a 'beautiful landscape'".

Increasing complexity also characterized the physical aspects of urban development (Krabbe 1989; Bertels 1997; Kühne 2011; Ueköttter 2007; Winiwarter and Knoll 2007; Schott 2008), as the cities spread out along roads and railways into the local countryside. Animal power yielded to steam, electricity, and the internal combustion engine; the electric streetcar replaced the horse-drawn tram, the delivery van the horse and cart. Fresh water was collected in remote reservoirs, making the cities independent of their own ground-water supplies, and wastewater was disposed of in flushable sewers rather than allowed to flow down city streets.

With the growth of science and technology came a growing sense of mastery over nature (Körner 2005, 2006a, b), of which the city was the most visible manifestation. There rationality had triumphed, erasing the bond of nurture—and with it of dependence—on nature. Conversely, the non-urban environment—nature, countryside, landscape—began increasingly to symbolize freedom from the workaday city: "What drives the urban population out of doors and into the country is precisely that: to escape from the social and spatial constrictions of city streets"

(Kaufmann 2005, p. 59). This was a perfect reversal of the medieval view, in which breathing city air was a liberating experience. Against this background, the concept of 'landscape' took on aesthetic connotations as an "expression of the 'good life' in harmony with nature and a 'natural' social order: an anti-democratic perspective that led via Counter-Enlightenment and Romanticism to a conservative political program" (Körner and Eisel 2006, p. 46).

### 3.1.4 Landscape as Cultural Heritage—the Concept of Heimat *(Home Environment)*

A central component in the "semantic train" (Hard 1969) of what we today know as landscape is the concept of *Kulturlandschaft* (cultural landscape), a term derived from the conservative cultural anthropologist and social theorist Riehl (1854) that soon spread from the German language area into worldwide use. Based on the postulate of an indissoluble bond between a particular *Volk* (people) and 'their' landscape (Eisel 1982; Lekan and Zeller 2005; Körner and Eisel 2006), it suggested that "a locally specific; organic harmony of culture and nature was perceptible" (Körner 2006a, p. 6). Dependent "not on aesthetic impact […] but rather on the clarity with which contexts and patterns are perceived" (Gradmann 1924, p. 134), landscapes were classified as 'harmonious' to the extent that their historical development had been 'organic' (Eisel 2009; see also Hard 2002 [1987b]).

Ernst Rudorff (1994 [1897]) took up Riehl's concept of the union of nature and people in a cultural landscape and developed it into a modern critical principle of preservation of the *Heimat* (home environment) which, leaning heavily on the Romantic tradition, rejected the Enlightenment faith in abstract reason, the formally individualist beliefs of Liberalism ('everyone is equal before the law'), and the economic pragmatism underlying the industrial doctrine of ever greater efficiency. Rudorff sketched out a historical-political philosophy of "concretely functioning reason and qualitatively enhanced individuality" (Körner 2006a, p. 6) that incorporated the idea of 'monadic spaces'. What he meant by this was inherently integral landscapes embodying individual stretches of land and their 'native' cultures (see e.g. Eisel 2009; Körner 2006a; Zutz 2015). Specific to German-speaking landscape research of the late nineteenth and early 20th century, the understanding of landscape as a concrete physical space informed by a unique 'essence' deriving from its particular symbiosis of culture and nature had a sustained impact on scientific approaches to the subject (see Sect. 3.2—The genesis of social landscapes outside the German language area).

In the second half of the 19th century, the city—and especially the metropolis— had become for its inhabitants a symbol of their uprootedness from anything that could be called a 'cultural landscape'. A cauldron of social leveling, the city created "a jumble of humanity in which all trace of nature is lost" (Körner 2006a, p. 7; Rudorff 1994 [1897]). As an antidote to this process, Rudorff called on the one hand for the preservation of the historical testimonies and artifacts of the *Heimat*, and on the other for an end to the use of machinery in agriculture. The anti-modernist, anti-urban tradition he founded—a typical product of bourgeois agrarian Romanticism (Knaut 1993)—has remained an indelible feature of much of the German nature conservation movement.

The *Heimatschutz* (home environment preservation) movement was not the only current critical of modernization that influenced late 19th and early 20th century German constructs of landscape. The art reform movement, with its aims of preserving 'authentic' culture by aesthetic education and the cultivation of good taste —especially among consumers and the common people—was another spearhead against the inroads of mass culture (Maase 2001; see Vicenzotti and Trepl 2009). Committed to 'regionally typical', original artworks, as opposed to mass produced prints and copies, art reform found in the home environment movement a kindred spirit which it could embrace and utilize as "a rich source of stimulus" (Pazaurek 2007 [1912], p. 119). In this context, it should be noted that "'regions' are based at times on collective social classifications/identifications, but more often on multiple practices in which the hegemonic narratives of a specific regional entity and identity are produced, become institutionalized and are then reproduced (and challenged) by social actors within a broader spatial division of labour" (Paasi 2002, p. 185), whereby the social construction of 'region' differs from the one of 'landscape' in which aestheticization is generally unavoidable (see also Chilla et al. 2016).

## 3.1.5 A Specifically German Tale—The 'Wild Woods'

Compared with other nations, Germans attach great cultural value to their forests. Woodland—more precisely German woodland—is regarded as contributing to German cultural identity (Lehmann 2001)—a cult that started with the Romantic mystique of the *Battle* of the Teutoburg Forest (9 CE), when the Cheruscan leader, Hermann (Arminius), inflicted a resounding defeat on the Roman legions, the tenor being that the Germans, or their tribal forebears, in spiritual union with the forests, were invincible. Above all the German oak was elevated into "a symbol of the eternity of the so-called autochthonous Germanic people" (Urmersbach 2009, p. 76). The beech also took on considerable symbolic force through its association with writing, the German words *Buch* (book) and *Buchstabe* (letter) deriving from the beech-wood panels on which the Germanic tribes scored their runes (*Buche* = beech). Some 1500 place names in Germany also derive from the beech tree.

The woods are still today associated on the one hand with a state of untarnished nature, on the other with traditional fairytale figures of poachers, robbers, witches, fairies, and other social deviants (Urmersbach 2009). (Michael Ende's popular fantasy novel *The Neverending Story* is a modern example of the same tradition— here the wood is the backdrop for a plot in which every aspect and every living creature is a product of the human imagination, and is accordingly threatened by the Nothing that represents the abandonment of dreams and hopes). In the wake of 19th century German Romanticism, reinforced in the Biedermeier period, the woods took on the political and pedagogical undertone of a "force that could improve humanity and its world" (Urmersbach 2009, p. 85). Together with the *Heimatschutz* movement, the late 19th century back-to-nature youth organization of the *Wandervogel* (literally 'hiking birds', because of their habit of singing on their hikes) saw in the woods a counter-symbol to the industrialization, individualism and rationalism of the age. Ironically, however, these attempts to compensate for the loss of the old order inflicted by the Enlightenment and Industrial Revolution enjoyed some of the effects of those changes, for the massive shift from regenerative to fossil energy sources—i.e. from wood and charcoal to coal and coke—made room for extensive programs of reforestation (albeit often in drab monocultures) and woodland conservation (Schenk 2006; Uekötter 2007; Winiwarter and Knoll 2007; Zutz 2015).

The National Socialists used the mythical connotations of the German woods to feed their fantasies of racial superiority (see immediately below and Körner 2006a, b, c). Post-war, the sentimental homeland films of the 1950s were on the whole successful in their attempt to counter Nazi propaganda by presenting the woods as nature in its wild and noble state, "simple, romantic, beautiful, and thoroughly non-political" (Urmersbach 2009, p. 105). Some decades later, the high social sensibility to the 'dying forest' syndrome can also be explained by the special relationship between the Germans and their woods. A result of pollution, the widespread impairment of the forest stock "brought about a cultural crisis that profoundly affected contemporary political awareness" (Lehmann 1996, p. 145). Today the woods remain an aesthetic symbol of harmonious coexistence—albeit more objectively as the coexistence of trees of different ages and species (Lehmann 2001). A final point, however, is that for all their traditional, emotional-aesthetic commitment to their woods, modern Germans have little real knowledge about either trees or forests (Lehmann 2001; Kühne 2014; Schönwald 2015).

### *3.1.6 The Rejection of Romantic Concepts of Cultural Landscape and Home Environment Under the National Socialists, and the Post-war Rise of a Scientific Approach to Landscape*

The National Socialists subjected not only the native woods to their ideological program. Their absorption of social Darwinist principles led to the transformation of the concept of cultural landscape—itself derived from the *Heimatschutz* movement—into an offensive weapon. The conservative tradition of the 'unity of land and people' was infused with the ideals of the 'blood and soil' theory to create—along with racism and a euphoria for technology—a powerful engine of territorial expansion. The German cultural landscape became an apotheosis of the 'superiority of the Nordic race' (Trepl 2012; Eissing and Franke 2015), and dichotomies circulated between 'Germany's fruitful landscapes' and the 'desolate wilderness' of Slavic regions (Blackbourn 2007)—a state of affairs that manifestly resulted "from the neglect by the Polish regime of landscapes formed by German hands" (Fehn 2007, p. 44). The eastern territories became known as the Wild East—a play on the Wild West stereotypes of Karl May's popular adventure stories—and were as such the playground of fantasies for colonially minded landscape and political planners: German technological expertise would turn these expanses once again into 'blossoming landscapes' (Blackbourn 2007; Fehn 2007; Trepl 2012; Eissing and Franke 2015). Apart from such excesses, there were other "undeniable affinities" (Blackbourn 2007, p. 341) between National Socialism and the *Heimatschutz* movement. Both shared "an antipathy to the 'cold' materialism of big cities, and saw untrammeled liberal capitalism as responsible for the threat to the natural beauty of the landscape. Moreover they were at one in a whole series of spontaneous aversions: to concrete as an un-German material, to advertisement hoardings as a disfigurement of the landscape, and to the encroachment of alien flora in the form of 'non-native' trees and bushes" (Blackbourn 2007, p. 341; see also Zutz 2015).

Conservative views persisted into the post-war nature conservation and *Heimatschutz* movements (see e.g. Böhm 1955), which were now directed against the inroads of Communism and what was thought of as an American lifestyle. But in the 1950s—and increasingly in the 1960s—the understanding of landscape, in particular of nature conservation, became gradually informed by the ecological sciences (Engels 2003; Körner 2006c; Blackbourn 2007; Berr 2014), and the concept of *Heimatschutz* duly gave way to that of the preservation of natural species, ecosystems, and biocenoses.

The ecological approach to landscape is fundamentally positivist: as an ecosystem, landscape is a physical entity whose structures and functions are independent of the observer and can be investigated with empirical methods and defined in a neutral and value-free way (see Chilla 2007). Positivist landscape research is accordingly based on the observation of individual phenomena and subsequent "inductive abstraction of the collected evidence to form a general picture" (Eisel 2009, p. 18). The original rationale for this approach, however, was strategic rather than scientific, for "democracy has a systematic bias toward objectively convincing argumentation" (Körner 2006c, p. 137; Berr 2014); and—given that many people are neither scientifically nor politically minded—beneath the transition to more rational views the values associated with local cultural landscapes persisted (Körner 2005, p. 112). Thus the regional specificity of flora and fauna—and, frequently derived from this, a sense of its unique aesthetic value—"continued to play a central role in the evaluation of biotopes". The duality underlying the burgeoning ecological awareness of the later 20th century issued in open contradiction when, for example, the construction of landscape as a natural resource was "set up as a counter-model to a scientifically and hence rationally accessible nature" (Weber 2007, p. 22).

*Heimatschutz*-oriented and ecological nature conservation share an opposition to the physical and spatial manifestations of Fordist modernity, whose essence lay in a "radical transformation of efficiency standards and patterns of consumption" (Ipsen 2006, p. 81; and see Hirsch and Roth 1986; Moulaert and Swyngedouw 1989; Lipietz 1991; Ipsen 2000; Eissing and Franke 2015). The modernist architectonic principles of exploiting on the one hand advantages of scale and on the other the separation of functions (in line, for example, with the Athens Charter) were extended from the planning of individual buildings to that of larger spatial complexes. Residential areas were separated from working (i.e. commercial and industrial) areas, inner city shopping and service areas from leisure and recreation facilities. The resulting spatial patchwork of monostructures, interconnected by mass means of initially public and then private transportation, gave rise between the wars to a rapid expansion of settled areas.

Fordist modernization entailed an increasing dependence of country areas on the city, both as a producer of foodstuffs and other raw materials and for the disposal of waste, with processing facilities like dairies and slaughterhouses sited close to the city and refuse dumps further away. There was, however, also a rural market for factory products: not only ready-made consumer goods but also agricultural machinery and other means of production that could no longer be hand-made by local craftspeople but were manufactured on the assembly lines of the major industrial centers (see Ipsen 2006)—for agriculture, too, was subjected to the same laws of scale and function. Field sizes were being maximized, cattle barns enlarged, and mechanization was increasing all round.

The ensuing physical changes in the landscape had an aesthetic dimension, inasmuch as they were "embedded in a specific developmental model of spatial organization and bound up with a belief in the purpose and utility of this model" (Ipsen 1997, p. 70; see also Nassauer and Wascher 2008). However, the Fordist model often had unintended side effects that radically changed the life of its rural practitioners, assimilating their daily round to that of the city dweller. As Lucius Burckhardt (2006c [1977], p. 29) remarks, "what the farm itself cannot produce, the farmer's wife, like her urban counterpart, buys in the shop".

### 3.1.7 Post-industrial Landscapes and the Contemporary Understanding of Landscape in Germany

Since the mid 1960s, the limitations of the Fordist economic model have become ever clearer. The developing affluence of the societies of East and South-East Asia, Australasia, Europe, and North America have created a demand for individualized articles that Fordist industrial structures, with their standardized product range, could no longer satisfy. The gap has been filled by specialist enterprises with small, flexible production runs based on a combination of computer-aided manufacturing processes, widely networked supply chains, just-in-time delivery, and low manufacturing input of their own. Against this background, range diversification combined with shorter product cycles has boosted the role of market research, and the demand for individuality that of the designer. Here as elsewhere in the economy, knowledge has become the central resource of the active economic player—and knowledge must be constantly renewed if it is to remain functional (Rifkin 2007).

The side effects of the Fordist conglomerate included serious ecological impacts on biodiversity, pollution, and waste. This led in the later 20th century to intense social discussion, which in turn stimulated an economic turn away from major centralized production facilities to decentralized structures (Hirsch and Roth 1986; Moulaert and Swyngedouw 1989; Soja 1989; Lipietz 1991; Zukin et al. 1992; Zukin 1993; Ipsen 2000; Kühne 2012). At the same time the approach to history changed: "While Modernism in every respect sought liberation from history" (Klotz 1985, p. 423), postmodern architecture and landscape architecture is concerned to integrate regional, ethnic and historical aspects in its work (Graham et al. 2000). The transition from an industrial to a post-industrial society has brought with it a new romanticism in the form of industrial archaeology. Former industrial sites, factory buildings, and production plants have taken on a symbolic value as witnesses to the "hard, simple life of the workers" (Vicenzotti 2005, p. 231)—a mirror image of the Romantic ideal of the 'hard, simple, life of the rural laborer' that accompanied the earlier transition from an agrarian to an industrial society. This interest in the relics of what has become known as the Second Industrial Age can be seen both as a reaction to the de-standardization and fragmentation of contemporary post-industrial society (Kühne 2008), and as a way of recalling the aesthetic interpretive patterns of an earlier period. Derelict urban industrial landscapes "with their decaying blast furnaces and towering ruins convey associations of the Baroque and the picturesque gardens of the 18th century" (Hauser 2004, p. 154). The cult of ruins, both modern and Romantic, expresses a deep skepticism in the value of progress. Fordism has been replaced by the new pragmatism of leisure, and in this context the industrial facilities of a bygone age—as Chilla (2005, p. 184) observes of the North Duisburg Landscape Park on the western edge of the Ruhr Area—have become the classical parks of our own day, where "intensive planting and other aspects of the traditional park combine to hold at arm's length the industrial heritage and simultaneously to endow it with new visual and recreational values" (see also Kühne 2007).

The underlying aesthetic of industrial archaeology is closely related to an extension of the concept of landscape to include urban structures (see Apolinarski et al. 2006)—a change of focus that has brought landscape research in the German language area into line with the broader international interest in vernacular as well as cultural landscapes. Theoretical discussion, too, has taken a lead from research in English, increasingly adopting constructivist perspectives whose fundamental insight is that landscape is neither an object of empirical analysis external to human perceptions (the positivist position), nor an organism with an essence of its own (the essentialist position), but a socially generated and mediated construct (see e.g. Kühne 2008, 2012, 2013; Wojtkiewicz and Heiland 2012; Kost 2013; Schönwald

2013; Gailing 2015; Kühne and Schönwald 2015). This construct, as it appears within the German language area, is the result of the process of development described in this chapter, and it governs the question (to be discussed later in greater detail) as to who, in what setting, and in what terms is entitled to speak of landscape (see Kühne 2008).

Current developments in the understanding of landscape are largely concerned with expert interpretations of the construct (see Kühne 2008, 2013; Weber 2015). Among non-experts, traditional interpretations of a rural idyll still dominate (see Fig. 3.4), where wind farms, cities, and highways play no (or at least no significant) role. Such views reflect to a great extent the image of landscape in the media. Given the Internet's function as a central medium of expertise (Münker 2009), Internet 'hits' are an important source for determining the social understanding of specific concepts. Figure 3.5 shows in tabular form the relative incidence of stereotypical elements in the first 120 pictorial images found in a Google search triggered by the

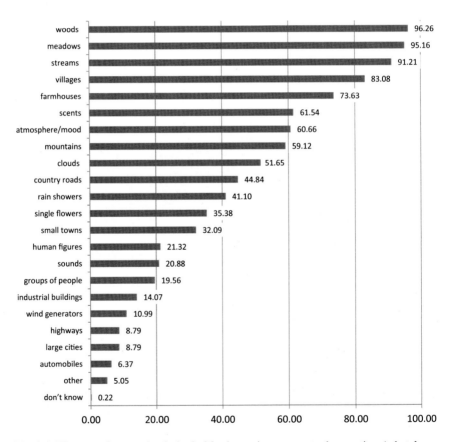

**Fig. 3.4** Elements of appropriated physical landscape in response to the question: 'what do you associate with landscape?' (n = 455; results in %; survey conducted in Saarland, Germany; adapted from Kühne 2006)

3.1 The Genesis of Social Landscapes in Co-evolution … 63

word *Landschaft* (Kühne 2015a). Especially notable is the near ubiquity of countryside and sky, giving these something like 'minimum feature' status in the current German understanding of landscape. Third place is taken by clouds, and meadows

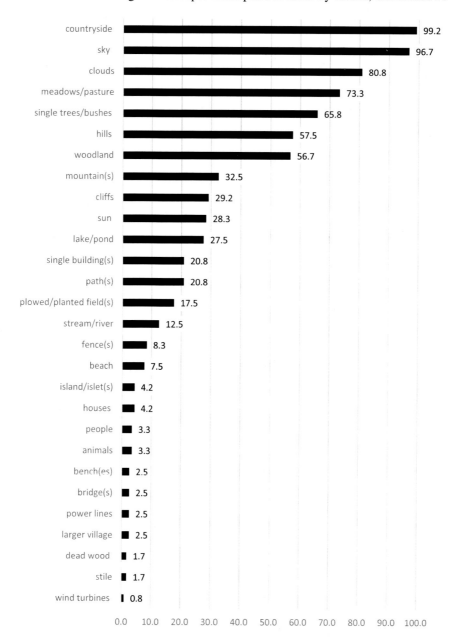

**Fig. 3.5** Google pictorial images for keyword 'landscape' (n = 120; results in %; survey date July 7, 2014)

and pastures—interestingly enough not the plowed fields indicative of more intense cultivation—as well as individual trees and bushes; on a larger scale, hills and woodland also occur in many of the Google images. Considerably less frequent are ponds and lakes, and even less so streams and rivers.

## 3.2 The Genesis of Social Landscapes Outside the German Language Area

The different perspectives and values observed in the social construction of landscape within the German language area become magnified once one takes other cultural contexts into consideration. A cultural studies approach—following Hall (2002)—sees a culture as the sum of its classification systems and discursive formations, which give rise to, and are in turn influenced by, specific social practices. Hence, whenever a segment of physical space is called 'landscape', this enacts and at the same time confirms a socially defined system of classification.

The following remarks on culture prescind altogether from the essentialist construction of what Hall (1994, p. 199) calls "cultural identities", with their stereotypical distinctions (*the* Americans, *the* Japanese, *the* Turks etc.). On the contrary, the approach taken here, with its comparative focus on the perspectives and values evident in the social construction of physical space, may simultaneously reveal distinctive paths of development in the descriptions of self and other that underlie the concept of culture. It is in this sense that cultural specificity enters the present analysis of the construct known in German as *Landschaft*. Although recent years have seen an increasing number of (frequently comparative) studies on culturally differentiated concepts of landscape (e.g. Wypijewski 1999; Makhzoumi 2002; Olwig 2002; Tolia-Kelly 2004; Gehring 2006; Drexler 2009, 2010, 2013; Küchler and Wang 2009; Taylor 2009; Ueda 2010, 2013; Özgüner 2011; Bruns 2013; Kost 2013; Kühne 2013, 2015b; Mels 2013; Zhang et al. 2013; Kühne 2015c), these allow as yet only general statements to be made about the development of that concept, and above all about the interface and influence of its cultural variants (see also Corner 1999).

In JoAnn Wypijewski's *Painting by Numbers* (1999)—a comparative questionnaire-based survey of pictorial landscapes illustrated by the pop artists Vitaly Komar and Alexander Melamid—the author-editor determines the landscape styles and motifs preferred in various countries and cultures worldwide. According to her, Americans prefer muted colors and realistic (frequently autumnal) images of lakes, rivers or oceans, as well as animals in their natural habitat, and groups of appropriately appareled people. The French, in contrast, prefer agrarian motifs with clear signs of human activity, which may well be "similar in many respects to the paintings of Lorrain" (Spanier 2008, p. 284). Russian preferences are for heavily wooded landscapes with hermits; in Finland the preferred motifs are forest workers and fauna; in Iceland landscapes in strong blue and green colors are popular, while

## 3.2 The Genesis of Social Landscapes Outside the German Language Area

Turks prefer pastel tones, and motifs of children playing. In Ukraine a landscape should show half-open countryside with traditional rural buildings, while Danes prefer a stormy twilight with human figures and a (naturally Danish) flag. Finally, in China popular landscapes feature mountains towering behind a plain with rice paddies and water buffaloes and, in the foreground, historical figures (Wypijewski 1999; Spanier 2008; Kühne 2013). Across all these culturally specific images, despite their contextual differences, there is a discernible preference for idyllic rural scenes near water.

Parallel to these visual-pictorial preferences, semantic analyses have demonstrated the existence of a range of cultural differences in the aesthetically unified conception of spatial objects we know as landscape. In numerous publications Drexler (e.g. 2009, 2010, 2013) has investigated the denotations and connotations of 'landscape' (English), *paysage* (French), *Landschaft* (German) and *táj* (Hungarian) in their "different conceptions of the world around and within us" (Drexler 2009, p. 120). *Landschaft* and *táj* connect objective and aesthetic references that are absent from 'landscape' and *paysage*: in English the meaning of 'land' and 'country' differs from that of 'landscape' just as in French *pays* and *campagne* differ from *paysage*; in both languages 'country'/*pays* has the additional meaning of 'state'. The objective references of the German and Hungarian terms go back to the Middle Ages (see Sect. 3.1 for the history of *Landschaft*), whereas 'landscape' and *paysage* only came into use at the end of the 16th century. During the 19th century, *Landschaft* and *táj* took on the emotional associations of *Heimat* in the Romantic sense of 'home environment' (Drexler 2009, 2010, 2013; see also Rodewald 2001). In 17th century France and England 'landscape' and *paysage* were used of physical spaces formed by human hand to meet the aesthetic criteria of their noble landowners, a usage found in Germany (and even more so in the Netherlands) only in court circles. Here "the old conception of a naturally 'mature' landscape" (Drexler 2009, p. 127; see also Olwig 2002, 2008a; Howard 2011) remained dominant until the 19th century, and has persisted in scientific discourse in the conceptual form of the 'appropriated physical landscape' until the present day.

The conflicting interests of court and county in 17th century England gave the concept of landscape a peculiar significance (Olwig 1996; Drexler 2010; Trepl 2012). The county party, consisting largely of landed gentry, was concerned to maintain the old Anglo-Saxon/Celtic systems of customary (or 'common') law which guaranteed its own standing and influence; the court party, on the other hand, led by the king and nobles at court, sought to establish a centralized absolutist state and consequently insisted on the hegemony of Roman law, with its basis in pragmatic reason. Paintings and stage scenery became an important instrument in this dispute, and it was in this context that the term 'landscape' first made its appearance—only later was it applied to the physical countryside. The county party favored Dutch representations of the daily life of men and women in their traditional environment—an environment shaped by customary law. To this the court party opposed Italian and Spanish landscapes "depicting formally beautiful, cultivated scenes with architectural motifs, whose symbolism referred not to the traditional life of the country but to its rational molding—something that demanded central planning and direction" (Trepl 2012, p. 164).

In mid-18th century Hungary the notion of *táj* was informed with a similarly powerful political component: here the normative symbolism of a traditional way of life in autochthonous class structures was evoked in opposition to the absolutist ambitions of the Habsburg state and its concept of a landscape that reflected the formal qualities of the French *paysage*. Toward the end of the 18th century the symbolic force of "enlightened aristocratic, bourgeois liberal, and democratic social ideas" (Drexler 2009, pp. 128–129) was added to this mixture.

The development of the American concept of landscape demonstrates the impact on the English language of new dimensions of experience, in this case the confrontation with physical spaces virtually untouched by human influence, which could, therefore, be freely subjected to the appropriative notions of rural space brought with them by the European settlers, who saw landscape in contrast not so much with the city as with the native wilderness—a contrast of the fruitfully cultivated and socialized with the wild, awesome, and untamed (see Tuan 1979a; Cronon 1996a; Keck 2006). Accordingly, landscape was initially conceived as a "picturesque spectacle" of land use and settlement (Clarke 1993, p. 9; see also Cosgrove 1985; Olwig 2008b; Hirsch 2003), an aesthetic perspective largely confined in the 16th and 17th centuries to the Virginia and New England colonies, where it also served to mark off the territories of the English crown from those of the French and Spanish colonists (Clarke 1993; Mills 1997a, b). From about the time of the Civil War (1861–1865), the "carefully manicured" (Hugill 1995, p. 157) English garden landscape established itself as the ideal form for the molding of rural space (Hayden 2004) Simultaneously with this classicizing tendency a romantic movement arose, especially in connection with the Hudson River School, whose work "combined detailed depictions of landscape with moral themes" (Campbell 2000, p. 63), presenting America as a Garden of Eden (Campbell 2000; see also Mills 1997a, b; Mitchell 2009). The ideas informing this movement differed, however, from those of European (especially German) Romanticism, the defining (conservative) construct of *Heimat* being replaced in America by that of a sublime wilderness that embraced not only the forests, swamps, mountains and coasts but also the deserts and semi-deserts, steppes and volcanoes (Clarke 1993; see also Osborne 1988; Schein 1993; Zapatka 1995; Kotkin 2006; Kühne 2008; Mitchell 2009; Megerle 2015) as positively charged symbols of "the idealized American values of independence, self-reliance and undaunted honesty" (Pregill and Volkman 1999, p. 436; Cronon 1996b; Kirchhoff and Trepel 2009). Belief in the special destiny of the American people was cultivated in images of vast and awesome "natural beauty, from the giant redwood to the canyon" (Spanier 2008, p. 278; see also Kühne 2012). From the 1920s onward, some of the positivist and even essentialist aspects of the contemporary German understanding of landscape were introduced into the American university tradition by the German–American geographer Carl O. Sauer, who taught at Berkeley. Sauer (1963 [1926]) understood landscape as an object that "was to be studied by the chorological method and its results transmitted descriptively in prose and above all by the map" (Cosgrove

## 3.2 The Genesis of Social Landscapes Outside the German Language Area

1985, p. 57; see also Mitchell 2008; Kühne 2013; Winchester et al. 2003; Antrop 2015) and "celebrated the idea of historical and geographical particularism and the unique qualities of diverse regions" (Heffernan 2003, p. 17). Figure 3.6 shows, in contrast, the endurance of a Romantic vision in a postmodern Western cityscape.

Chinese and Japanese constructs of landscape developed for many centuries without the polarizing influence of Western thought: indeed, according to Küchler and Wang (2009), the aesthetic perception of space began in China some 1000 years earlier than in Europe. Notable points of difference from the West are on the one hand the lack of the typically European distinction between the beautiful and the sublime (Ueda 2013), and on the other the non-dichotomous relation of human life with nature: "Man and nature are not opposite poles: man is integrated in the *ch'i* that is common to all beings, the life-force that penetrates all" (Lehmann 1968, p. 15). The range of Chinese vocabulary concerned with what we call landscape is correspondingly broad and differentiated, containing the following terms (Bianca 2009; Küchler and Wang 2009; Bruns 2013; Zhang et al. 2013):

**Fig. 3.6** A postmodern play (or 'playgiarism') on texts and places—here Caspar David Friedrich's Wanderer above a Sea of Fog (1818) and San Francisco city center. The mountain peaks of an idealized German landscape have been replaced with Californian skyscrapers, the traditional German walking coat with leisure apparel, the contemplative gaze into physical space with critical scrutiny of a smart-phone picture. Yet almost two centuries later (especially in Germany), the underlying Romantic perspective of Friedrich's painting still defines the socio-visual construction of space. (Photo Olaf Kühne; on Caspar David Friedrich and his impact see also Hofmann 2013)

- *shan shui* is used of the genre of landscape painting, generally in ink, sometimes complemented with colored woodcuts; this embraces natural elements, frequently accompanied with (male) figures; the perspective is neither centralized nor as clearly focused as in European landscape painting (see also Lehmann 1968);
- *tian yuan* describes landscape as a metaphor for freedom from compulsion represented in an "idyllic, harmonious rural scene created by human hand" (Küchler and Wang 2009, p. 205);
- *fen jing* is used of a place that "reflects the aesthetic ideals of *shan shui*" (Küchler and Wang 2009, p. 206)—i.e. a balanced arrangement of physical space that meets Chinese social criteria of beautiful scenery (Zhang et al. 2013);
- *jing se* is a combination of the words *jing* (scene) and *se* (color) and is used of "a (beautiful) overall impression of a stretch of land" (Küchler and Wang 2009, p. 206);
- *yuan lin* is used of the shaping of confined physical space (e.g. a garden) in accordance with the prescriptions of *shan shui* painting;
- feng shui (=wind and water) is an evaluative procedure for the "quality of the relationships informing a specific place or landscape with regard to the success of human activity" (Küchler and Wang 2009, p. 207); especially important is "the orientation of a planned object within its detailed topographical and cosmic setting (ibid.);
- shui tu (=water and earth) traditionally describes the physical geographical conditions of a situation and is used today in environmental regulations concerning water and soil quality;
- feng tu (=wind and earth) is used (as a complement to shui tu) of "the close interrelations between land and people in a specific location" (Küchler and Wang 2009, p. 207–208);
- jing guan is used of Western concepts of landscape imported into China—first, in the early 20th century, "as a Japanese translation of the German Landschaft, which Japanese writers had learned from their reading of German literature" (Küchler and Wang 2009, p. 215), and then (in the 1930s) via Japanese literature; today jing guan is used of beautiful natural landscape, an individual landscape, landscape as type, and instances of the art of the garden; in recent decades jing guan has become a key concept for social/scientific landscape construction in China, largely replacing the traditional, aesthetically charged yuan lin in planning contexts, and used in other contexts to denote social distinctions, where its "slightly elitist aura" is discernible (Küchler and Wang 2009, p. 216).

The shaping of physical space in China was unmistakably connected with power: "Ever since the founding of the imperial state, intrusions into the natural landscape were regarded as a metaphor for the exercise of political authority" (Bianca 2009, p. 55).

Other regions outside the Western cultural discourse which have developed their own differentiated social aesthetic of landscape include Turkey (Türer-Baskaya

2012) and the Arab world (Makhzoumi 2002), as well as Japan (Karatani 1993; Gehring and Kohsaka 2007; Ueda 2013), although from the turn of the 20th century onward Western constructs have also gained gradual dominance in these regions—at least in academic circles—sometimes even resulting in direct loan words like the Turkish peyzaj (Türer-Baskaya 2012). This has created a stumbling block for communication between experts and lay people that can severely impede cooperation in participatory spatial planning projects (see Ueda 2010 for Japan; see also Taylor 2009). Given the divergence of indigenous cultural constructs from modern Western notions of appropriated space, it is not to be expected that a layperson from a non-Western culture (or for that matter even from a Western one) will immediately understand what an expert means by 'landscape'—it may, indeed, be unclear among the experts themselves what colleagues from different disciplines actually mean when they use that term (Kühne 2008).

Especially in view of today's increasing international migration, an intensive examination of different cultural interfaces with (and hence conceptions of) what we call landscape seems more than ever necessary. Özgüner (2011), for example, cites the different cultural appropriation of urban parks in the West, where they are thought of primarily as places for activities like jogging, walking, or playing games, and in Turkey, where passive concepts like resting and picnicking predominate. Research of this kind can be of practical use to planners faced with the need to coordinate different cultural conceptions and aesthetic requirements in the use of physical space—a situation that may involve acute problems when international projects are concerned. As an example of a transnational clash of conservationist philosophies, Rothfuss and Winterer (2008, p. 151) cite the Bavarian Forest National Park in south-east Germany and the neighboring Šumava National Park in the Czech Republic: "The major challenge in achieving a management consensus is the divergent cultural construction of nature and its protection."

Looked at globally, the culture-specific concept of landscape has, then, developed in unique and different ways, but these have been subject to interference from the assimilated Western construct (see e.g. Bunn 2002), which has had an ongoing impact on physical space (see e.g. Frohn and Rosebrock 2008). The process has not, for the most part, been a question of peaceful cultural symbiosis, but of the establishment of a discursive hegemony that bypasses—and by the same token leaves largely intact—native cultural contexts. From a critical point of view, this may well be seen as an instance of cultural colonization, both conceptually and on the ground. Mitchell (2002, p. 9) castigates the concept of landscape in this sense as "imperialist dream-work". After all, the discursive process as a matter of course has conceptual winners and losers, and traditional hierarchies (e.g. the subordination of *yuan lin* to *jing guan* in China) are regularly supplanted by new subjections, mostly to Western notions of 'landscape' or *paysage*. English, French and German concepts, on the other hand, are at most only marginally touched by Turkish, Chinese, or Japanese aesthetic spatial ideas.

Resistance among cultural—and indeed subcultural—groups (e.g. especially among landscape experts) is often all the greater when a longstanding concept of landscape, seemingly invulnerable to alien aesthetic impact, is challenged by new and potentially broadening ideas. A case in point is the current discussion about the appropriated physical landscape in its everyday form (in Germany known as the 'Landscape 3' debate). And we have already seen how a different kind of resistance—namely philosophical and political—arose in 18th century Hungary and Germany when the aristocracy sought to establish French and English social constructs of paysage and 'landscape' against the opposition of their more traditionally minded compatriots.

## References

Abel, W. (1967). *Geschichte der deutschen Landwirtschaft*. 2 Volumes. Stuttgart: Ulmer.
Andrews, M. (1989). *The search for the picturesque: Landscape aesthetics and tourism in Britain 1760–1800*. Stanford: Stanford University Press.
Antrop, M. (2015). Interacting cultural, psychological and geographical factors of landscape preference. In D. Bruns, O. Kühne, A. Schönwald, & S. Theile (Eds.), *Landscape culture-culturing landscapes: The differentiated construction of landscapes* (pp. 53–62). Wiesbaden: Springer VS.
Apolinarski, I., Gailing, L., & Röhring, A. (2006). Kulturlandschaft als regionales Gemeinschaftsgut. Vom Kulturlandschaftsdilemma zum Kulturlandschaftsmanagement. In U. Matthiesen, R. Danielzyk, St. Heiland & S. Tzschaschel (Eds.), Kulturlandschaften als Herausforderung für die Raumplanung. Verständnisse—Erfahrungen—Perspektiven (pp. 81–98). Hannover: Verlag der ARL.
Appleton, J. (1986). *The experience of landscape*. Chichester, New York, Brisbane, Toronto, Singapore: Wiley.
Bätzing, W. (2000). Postmoderne Ästhetisierung von Natur versus, Schöne Landschaft' als Ganzheitserfahrung—von der Kompensation der, Einheit der Natur' zur Inszenierung von Natur als, Erlebnis'. In A. Arndt, K. Bal & H. Ottmann (Eds.), Hegels Ästhetik. Die Kunst der Politik—die Politik der Kunst. Zweiter Teil (pp. 196–201). Berlin: Akademie Verlag.
Beck, R. (1996). Die Abschaffung der ‚Wildnis'. Landschaftsästhetik, bäuerliche Wirtschaft und Ökologie zu Beginn der Moderne. In W. Konold (Ed.), Naturlandschaft—Kulturlandschaft. Die Veränderung der Landschaft nach Nutzbarmachung des Menschen (pp. 27–67). Landsberg: Ecomed-Verlags-Gesesllschaft.
Berger, P., & Luckmann, T. (1966). *The social construction of reality*. New York: Penguin Books.
Berr, K. (2014). Zum ethischen Gehalt des Gebauten und Gestalteten. *Ausdruck und Gebrauch, 12*, 30–56.
Bertels, L. (1997). Die dreiteilige Großstadt als Heimat: Ein Szenarium. Opladen: Leske + Budrich.
Bianca, M. (2009). Landschaft Status und Macht. Zur politischen Dimension der Gartenkultur in China, Korea und Japan. *Stadt + Grün, 58*(10), 54–58.
Blackbourn, D. (2007). *Die Eroberung der Natur. Eine Geschichte der deutschen Landschaft*. München: Deutsche Verlags-Anstalt.
Böheim, J. (1930). *Das Landschaftsgefühl des ausgehenden Mittelalters*. Leipzig: B. G. Teubner.
Böhm, A. (1955). *Epoche des Teufels*. Stuttgart: Kilpper.
Bourassa, S. (1991). *The aesthetics of landscape*. London: Belhaven Press.
Bruns, D. (2013). Landschaft, ein internationaler Begriff? In D. Bruns & O. Kühne (Eds.), *Landschaften: Theorie, Praxis und internationale Bezüge* (pp. 153–168). Schwerin: Oceano.
Bunn, D. (2002). ‚Our Wattles Cot'. Mercantile and Domestic Space in Thomas Pringle's Landscapes. In W. J. T. Mitchell (Ed.), Landscape and power (pp. 127–174). Chicago, London: University of Chicago Press.

# References

Burckhardt, L. (2006a [1998]). Landschaft. In M. Ritter & M. Schmitz (Eds.), Warum ist Landschaft schön? Die Spaziergangswissenschaft (pp. 114–123). Kassel: Schmitz.

Burckhardt, L. (2006b [1994]). Landschaft ist transitorisch. In M. Ritter & M. Schmitz (Eds.), Warum ist Landschaft schön? Die Spaziergangswissenschaft (pp. 90–94). Kassel: Schmitz.

Burckhardt, L. (2006c [1977]). Landschaftsentwicklung und Gesellschaftsstruktur. In M. Ritter & M. Schmitz (Eds.), Warum ist Landschaft schön? Die Spaziergangswissenschaft (pp. 19–33). Kassel: Schmitz.

Büttner, N. (2006). *Geschichte der Landschaftsmalerei*. München: Hirmer.

Campbell, N. (2000). *The cultures of the American New West*. Edinbourgh: Psychology Press.

Chilla, T. (2005). Stadt und Natur—Dichotomie, Kontinuum, soziale Konstruktion? *Raumforschung und Raumordnung, 3*, 179–188.

Chilla, T. (2007). Zur politischen Relevanz raumbezogener Diskurse. Das Beispiel der Naturschutzpolitik der Europäischen Union. *Erdkunde, 61*(1), 13–25.

Chilla, T., Kühne, O., & Neufeld, M. (2016). *Regionalentwicklung*. Stuttgart: Ulmer.

Clarke, G. (1993). Introduction. A Critical and Historical Overview. In G. Clarke (Eds.), The American Landscape. Literary Sources and Documents. 2 Volumes (pp. 3–51). The Banks, Mountfield: Helm Information.

Corner, J. (1999). Introduction. Recovering Landscape as a Critical Cultural Practice. In J. Corner (Ed.), Recovering landscape: Essays in contemporary landscape theory (pp. 1–29). New York: Princeton Architectural Press.

Cosgrove, D. E. (1984). *Social Formation and Symbolic Landscape*. London, Sydney: University of Wisconsin Press.

Cosgrove, D. (1985). Prospect, perspective and the evolution of the landscape idea. *Transactions of the Institute of British Geographers, 10*(1), 45–62.

Cosgrove, D. (1993). *The Palladian landscape: geographical change and its representation in sixteenth century Italy*. Leicester: Leicester University Press.

Cronon, W. (1996a). Introduction. In Search of Nature. In W. Cronon (Ed.), Uncommon ground. Rethinking the human place in nature (pp. 23–68). New York, London: W. W. Norton.

Cronon, W. (1996b). The trouble with wilderness; or, getting back to the wrong nature. In W. Cronon (Ed.), Uncommon Ground. Rethinking the Human Place in Nature (pp. 69–90). New York, London: W. W. Norton.

Czepczyński, M. (2008). *Cultural landscapes of post-socialist cities: representation of powers and needs*. London: Ashgate Publishing Ltd.

Däumel, G. (1963). Gustav Vorherr und die Landesverschönerung in Bayern. In: K. Buchwald, W. Lendholdt & K. Meyer (Ed.), *Festschrift für Friedrich Wiepking. Beiträge zur Landespflege 1*. Stuttgart: Eugen Ulmer.

Drexler, D. (2009). Kulturelle Differenzen der Landschaftswahrnehmung in England, Frankreich, Deutschland und Ungarn. In Th. Kirchhoff & L. Trepel (Eds.), Vieldeutige Natur. Landschaft, Wildnis, Ökosystem als kulturgeschichtliche Phänomene (pp. 119–136). Bielefeld: transcript.

Drexler, D. (2010). *Landschaft und Landschaftswahrnehmung: Untersuchung des kulturhistorischen Bedeutungswandels von Landschaft anhand eines Vergleichs von England, Frankreich, Deutschland und Ungarn*. Saarbrücken: Südwestdeutscher Verlag für Hochschulschriften.

Drexler, D. (2013). Die Wahrnehmung der Landschaft—ein Blick auf das englische, französische und ungarische Landschaftsverständnis. D. Bruns & O. Kühne (Eds.), Landschaften: Theorie, Praxis und internationale Bezüge (pp. 37–54). Schwerin: Oceano.

Eisel, U. (1982). Die schöne Landschaft als kritische Utopie oder als konservatives Relikt. *Soziale Welt, 33*, 157–168.

Eisel, U. (2009). *Landschaft und Gesellschaft. Räumliches Denken im Visier*. Münster: Westfälisches Dampfboot.

Eissing, H., & Franke, N. (2015). Orte in der Landschaft. Anmerkungen über die Macht von Institutionen. In S. Kost & A. Schönwald (Eds.), Landschaftswandel—Wandel von Machtstrukturen (pp. 55–63). Wiesbaden: Springer VS.

Engels, J. (2003). Hohe Zeit' und, dicker Strich': Vergangenheitsdeutung und -bewahrung im westdeutschen Naturschutz nach dem Zweiten Weltkrieg. In J. Radkau & F. Uekötter (Eds.), *Naturschutz und Nationalsozialismus* (pp. 363–403). Frankfurt a. M.: Suhrkamp.

Fehn, K. (2007). Naturschutz und Landespflege im ‚Dritten Reich'. Zur Terminologie der ‚artgemäßen Landschaftsgestaltung'. In B. Busch (Ed.), Jetzt ist die Landschaft ein Katalog voller Wörter. Beiträge zur Sprache der Ökologie 5 (pp. 42–50). Göttingen: Wallstein Verlag.

Freyer, H. (1996 [1966]). Landschaft und Geschichte. In G. Gröning & U. Herlyn (Eds.), Landschaftswahrnehmung und Landschaftserfahrung (pp. 69–90). Münster: lit.

Frohn, H.-W., & Rosebrock, J. (2008). 'Bruno der Bär' und die afrikanische Megafauna. Zum Habitus internationaler Naturschutzakteure—eine historische Herleitung. In K.-H. Erdmann, J. Löffler & S. Roscher (Eds.), Naturschutz im Kontext einer nachhaltigen Entwicklung. Ansätze, Konzepte, Strategien (pp. 31–50). Bonn-Bad Godesberg: Bundesamt für Naturschutz.

Gailing, L. (2015). Landschaft und productive Macht. Auf dem Weg zur Analyse landschaftlicher Gouvernementalität. In S. Kost & A. Schönwald (Eds.), Landschaftswandel—Wandel von Machtstrukturen (pp. 37–41). Wiesbaden: Springer VS.

Gehring, K. (2006). Landscape. Needs and Notions. Preferences, expectations, leisure motivation, and the concept of landscape from a cross-cultural perspective. Birmensdorf: Swiss Federal Research Institute.

Gehring, K., & Kohsaka, R. (2007). 'Landscape' in the Japanese language: Conceptual differences and implications for landscape research. *Landscape Research, 32*(2), 273–283.

Gradmann, R. (1924). Das harmonische Landschaftsbild. *Zeitschrift der Gesellschaft für Erdkunde zu Berlin, 55,* 129–147.

Graham, B., Ashworth, G. J., & Tunbridge, J. E. (2000). *A geography of heritage: Power, culture, and economy.* London et al.: Arnold.

Gruenter, R. (1975 [1953]). Landschaft. Bemerkungen zu Wort und Bedeutungsgeschichte. In A. Ritter (Ed.), Landschaft und Raum in der Erzählkunst (pp. 192–207). Darmstadt: Wissenschaftliche Buchgesellschaft.

Gudermann, R. (2005). Conviction and constraint. Hydraulic engineers and agricultural projects in 19-century Prussia. In Th. Lekan, Th. Zeller (Eds.), Germany's Nature. Cultural Landscapes and Environmental History (pp. 33–54). New Brunswick, New Jersey, London: Rutgers University Press.

Haber, W. (2007). Vorstellungen über Landschaft. In B. Busch (Ed.), Jetzt ist die Landschaft ein Katalog voller Wörter. Beiträge zur Sprache der Ökologie 5 (pp. 78–85). Göttingen: Wallstein Verlag.

Häcker, S. (1998). Von der Kulturlandschaft zum Landschaftsdenkmal? Eine Diskussion über die Erhaltung der Heidereliktе. In W. Blümel (Ed.), *Beiträge zur physischen Geographie Südwestdeutschlands* (pp. 44–65). Stuttgart: Institut für Geographie der Universität Stuttgart.

Hall, S. (1994). Rassismus und kulturelle Identität. Ausgewählte Schriften 2. Hamburg: Argument.

Hall, S. (2002). Die Zentralität von Kultur. Anmerkungen über die kulturelle Revolution unserer Zeit. In A. Hepp & M. Löffelholz (Eds.), Grundlagentexte zur transkulturellen Kommunikation (pp. 95–107). Konstanz: UVK Verlagsgesellschaft mbH.

Hannessen, H. (1985). *Gemälde der deutschen Romantik in der Nationalgalerie Berlin*. Berlin: Frölich & Kaufmann.

Hard, G. (1969). Das Wort Landschaft und sein semantischer Hof. Zur Methode und Ergebnis eines linguistischen Tests. *Wirkendes Wort, 9,* 3–14.

Hard, G. (1970). Der 'Totalcharakter der Landschaft'. Re-Interpretation einiger Textstellen bei Alexander von Humboldt. In H. Wilhelmy & G. Engelmann (Eds.), Alexander von Humboldt. Eigene und neue Wertungen der Reisen, Arbeit und Gedankenwelt (pp. 49–71). Stuttgart: Steiner.

Hard, G. (1977). Zu den Landschaftsbegriffen der Geographie. In A. v. Wallthor & H. Quirin (Eds.), Landschaft' als interdisziplinäres Forschungsproblem (pp. 13–24). Münster: Aschendorff.

# References

Hard, G. (2002 [1987b]). ‚Bewußtseinsräume' Interpretationen zu geographischen Versuchen, regionales Bewußtsein zu erforschen. In G. Hard (Ed.), Landschaft und Raum. Aufsätze zur Theorie der Geographie (pp. 303–328). Osnabrück: University Press Rasch.

Hauser, S. (2004). Industrieareale als urbane Räume. In W. Siebel (Ed.), *Die europäische Stadt* (pp. 146–157). Frankfurt a. M.: Suhrkamp.

Hayden, D. (2004). *Green fields and urban growth 1820–2000*. New York: Pantheon Books.

Heffernan, M. (2003). Histories of geography. In S. Holloway, S. P. Rice, & G. Valentine (Eds.), *Key concepts in geography* (pp. 4–22). New York et al.: Sage.

Hirsch, E. (2003). Introduction. Landscape: Between Place and Space. In E. Hirsch & M. O'Hanlon (Eds.), The Anthopology of Landscape. Perspectives on Place and Space (pp. 1–30). Oxford: Clarendon Press.

Hirsch, J., & Roth, R. (1986). *Das neue Gesicht des Kapitalismus*. Hamburg: VSA.

Hofmann, W. (2013). *Caspar David Friedrich Naturwirklichkeit und Kunstwahrheit*. München: C. H. Beck.

Hohl, H. (1977). Das Thema Landschaft in der deutschen Malerei des ausgehenden 18. und beginnenden 19. Jahrhunderts. In A. v. Wallthor & H. Quirin (Eds.), Landschaft als interdisziplinäres Forschungsproblem (pp. 45–53). Münster: Front Cover.

Howard, W. (2003). Landscapes of memorialisation. In I. Robertson & P. Richards (Eds.), *Studying Cultural Landscapes* (pp. 47–70). London, New York: Arnold.

Howard, P. (2011). *An Introduction to Landscape*. Farnham, Burlington: Routledge.

Hugill, P. (1995). *Upstate Arcadi:. Landscape, aesthetics, and the triumph of social differentiation in America*. Lanham, London: Rowman & Littlefield Publishers.

Illing, F. (2006). *Kitsch, Kommerz und Kult: Soziologie des schlechten Geschmacks*. Konstanz: UVK.

Ipsen, D. (1997). Raumbilder: Kultur und Ökonomie räumlicher Entwicklung (Vol. 8). Pfaffenweiler: Centaurus.

Ipsen, D. (2000). Stadt und Land—Metamorphosen einer Beziehung. In H. Häußermann, D. Ipsen, T. Krämer-Badoni, D. Läpple, M. Rodenstein & W. Siebel (Eds.), Stadt und Raum. Soziologische Analysen (pp. 117–156). Hagen: Centaurus.

Ipsen, D. (2006). *Ort und Landschaft*. Wiesbaden: Springer VS.

Job, H. (1999). Der Wandel der historischen Kulturlandschaft und sein Stellenwert in der Raumordnung. Eine historisch-, aktual- und prognostisch-geographische Betrachtung traditioneller Weinbau-Steillagen und ihres bestimmenden Strukturmerkmals Rebterasse, diskutiert am Beispiel rheinland-pfälzischer Weinbaulandschaften. Flensburg: Deutsche Akademie für Landeskunde.

Karatani, K. (1993). *Origins of modern Japanese literature*. Durham et al.: Duke University Press.

Kaufmann, S. (2005). *Soziologie der Landschaft*. Wiesbaden: Springer VS.

Keck, M. (2006). *Walking in the wilderness: The peripatetic tradition in 19-century American literature and painting*. Heildelberg: Universitätsverlag Winter Heidelberg.

Kirchhoff, T., & Trepel, L. (2009). Landschaft, Wildnis, Ökosystem: zur kulturbedingten Vieldeutigkeit ästhetischer, moralischer und theoretischer Naturauffassungen. Einleitender Überblick. In T. Kirchhoff & L. Trepel (Eds.), Vieldeutige Natur. Landschaft, Wildnis, Ökosystem als kulturgeschichtliche Phänomene (pp. 13–68). Bielefeld: transcript.

Klotz, H. (1985). *Moderne und Postmoderne: Architektur der Gegenwart 1960–1980*. Braunschweig, Wiesbaden: Vieweg.

Knaut, A. (1993). *Zurück zur Natur! Die Wurzeln der Ökologiebewegung*. Greven: Gilda Verlag.

Konold, W. (1996). Von der Dynamik einer Kulturlandschaft. Das Allgäu als Beispiel. In W. Konold (Ed.), Naturlandschaft—Kulturlandschaft. Die Veränderung der Landschaft nach Nutzbarmachung des Menschen (pp. 121–228). Landsberg: Hüthig Jehle Rehm.

Körner, S. (2005). Landschaft und Raum im Heimat—und Naturschutz. In M. Weingarten (Ed.), Strukturierung von Raum und Landschaft. Konzepte in Ökologie und der Theorie gesellschaftlicher Naturverhältnisse (pp. 107–117). Münster: Westfälisches Dampfboot.

Körner, S. (2006a). Die neue Debatte über Kulturlandschaft in Naturschutz und Stadtplanung. http://www.bfn.de/fileadmin/MDB/documents/service/perspektivekultur_koerner.pdf. Accessed 21 March 2012.

Körner, S. (2006b). Eine neue Landschaftstheorie? Eine Kritik am Begriff "Landschaft Drei". *Stadt und Grün, 55,* 18–25.

Körner, S. (2006c). Heimatschutz, Naturschutz und Landschaftsplanung. Technische Universität Berlin (Ed.), Perspektive Landschaft (pp. 131–142). Berlin: Institut für Landschaftsarchitektur und Umweltplanung.

Körner, S., & Eisel, U. (2006). Nachhaltige Landschaftsentwicklung. In D. Genseke, M. Huch & B. Müller (Eds.), Fläche—Zukunft—Raum. Strategien und Instrumente für Regionen im Umbruch. Schriftenreihe der Deutschen Gesellschaft für Geowissenschaften 37 (pp. 45–60). Hannover: Deutsche Gesellschaft für Geowissenschaften.

Kortländer, B. (1977). Die Landschaft der Literatur des ausgehenden 18. und beginnenden 19. Jahrhunderts. In A. v. Wallthor & H. Quirin (Eds.), Landschaft als interdisziplinäres Forschungsproblem. Vorträge und Diskussion des Kolloquiums am 7./8. November 1975 in Münster (pp. 26–44). Münster: Aschendorff Verlag.

Kost, S. (2013). Landschaftsgenese und Mentalität als kulturelles Muster. Das Landschaftsverständnis in den Niederlanden. In D. Bruns & O. Kühne (Eds.), Landschaften: Theorie, Praxis und internationale Bezüge (pp. 55–70). Schwerin: Oceano.

Kotkin, J. (2006). *The city: A global history.* New York: Modern Library.

Krabbe, W. (1989). Die deutsche Stadt im 19. und 20. Jahrhundert: Eine Einführung. Göttingen: Vandenhoeck & Ruprecht.

Küchler, J., & Wang, X. (2009). Vielfältig und vieldeutig. Natur und Landschaft im Chinesischen. In T. Kirchhoff & L. Trepl (Eds.), Vieldeutige Natur. Landschaft, Wildnis und Ökosystem als kulturgeschichtliche Phänomene (pp. 201–220). Bielefeld: transcript.

Kühne, O. (2006). Akzeptanz von regenerativen Energien—Überlegungen zur sozialen Definition von Landschaft und Ästhetik. *Stadt + Grün, 60*(8), 9–13.

Kühne, O. (2007). Soziale Akzeptanz und Perspektiven der Altindustrielandschaft. *Das Beispiel des Saarlandes. RaumPlanung, 132*(133), 156–160.

Kühne, O. (2008). *Distinktion, Macht: Landschaft. Zur sozialen Definition von Landschaft.* Springer VS: Wiesbaden.

Kühne, O. (2011). Akzeptanz von regenerativen Energien—Überlegungen zur sozialen Definition von Landschaft und Ästhetik. *Stadt + Grün, 60*(8), 9–13.

Kühne, O. (2012). *Stadt—Landschaft—Hybridität: Ästhetische Bezüge im postmodernen Los Angeles mit seinen modernen Persistenzen.* Springer VS: Wiesbaden.

Kühne, O. (2013). *Landschaftstheorie und Landschaftspraxis. Eine Einführung aus sozialkonstruktivistischer Perspektive.* Wiesbaden: Springer VS.

Kühne, O. (2014). Die intergenerationell differenzierte Konstruktion von Landschaft. *Naturschutz und Landschaftsplanung, 46*(10), 297–302.

Kühne, O. (2015a). Wasser in der Landschaft—Deutungen und symbolische Aufladungen. In C. Kressin (Ed.), *Oberwasser—Kulturlandschaft mit Kieswirtschaft* (pp. 35–48). Rees: PIUS.

Kühne, O. (2015b). Historical developments: the evolution of the concept of landscape in German linguistic areas. In D. Bruns, O. Kühne, A. Schönwald & S. Theile (Eds.), Landscape culture—culturing landscapes. The differentiated construction of landscapes (pp. 43–52). Wiesbaden: Springer VS.

Kühne, O. (2015c). Results and perspectives of the conference "landscapes theory, practice and international context". In D. Bruns, O. Kühne, A. Schönwald & S. Theile (Eds.), Landscape culture—culturing landscapes: The differentiated construction of landscapes (pp. 33–40). Wiesbaden: Springer VS.

Kühne, O., & Schönwald, A. (2015). *San Diego – Eigenlogiken, Widersprüche und Entwicklungen in und von, America´s finest city'.* Wiesbaden: Springer VS.

Küster, H. (1999). *Geschichte der Landschaft in Mitteleuropa. Von der Eiszeit zur Gegenwart.* München: C. H. Beck.

# References

Lehmann, H. (1968). Formen landschaftlicher Raumerfahrung im Spiegel der bildenden Kunst. Erlangen: in Kommission bei Palm & Enke.

Lehmann, A. (1996). Wald als ‚Lebensstichwort'. Zur biographischen Bedeutung der Landschaft, des Naturerlebnisses und des Naturbewußtseins. BIOS: Zeitschrift für Biographieforschung, oral history und Lebensverlaufsanalysen, 9(2), 143–154.

Lehmann, A. (2001). Landschaftsbewusstsein. Zur gegenwärtigen Wahrnehmung natürlicher Ensembles. In R. Brednich, A. Schneider & U. Werner (Eds.), Natur—Kultur. Volkskundliche Perspektiven auf Mensch und Umwelt (pp. 147–154). Münster et al.: Waxmann.

Lekan, T., & Zeller, T. (2005). The landscapes of German environmental history. In T. Lekan & T. Zeller (Eds.), Germany's Nature. Cultural Landscapes and Environmental History (pp. 1–16). New Brunswick, New Jersey, London: Rutgers University Press.

Lipietz, A. (1991). Zur Zukunft der städtischen Ökologie. Ein regulationstheoretischer Beitrag. In M. Wentz (Ed.), Stadt-Räume (pp. 129–136). Frankfurt a. M., New York: WW Norton.

Maase, K. (2001). Krisenbewusstsein und Reformorientierung. Zum Deutungsmuster der Gegner der modernen Populärkünste 1880–1918. In K. Maase & W. Kaschuba (Eds.), Schund und Schönheit. Populäre Kultur um 1900 (pp. 290–342). Köln, Weimar, Wien: Böhlau.

Makhzoumi, J. M. (2002). Landscape in the middle east: An inquiry. *Landscape Research, 27*(3), 213–228.

Megerle, H. (2015). Landschaftswandel in den Savoyer Alpen als Resultat geo- und wirtschaftspolitischer Machtstrukturen. In S. Kost & A. Schönwald (Eds.), *Landschaftswandel-Wandel von Machtstrukturen* (pp. 141–163). Wiesbaden: Springer VS.

Mels, T. (2013). Emplacing landscape in Sweden. In D. Bruns & O. Kühne (Eds.), *Landschaften: Theorie, Praxis und internationale Bezüge* (pp. 71–82). Schwerin: Oceano.

Mills, C. (1997a). Myths and meanings of gentrification. In J. S. Duncan & D. Ley (Eds.), *Place, culture, representation* (pp. 149–170). London: Routledge.

Mills, S. (1997b). *The American Landscape*. Edinburgh: Keele.

Mitchell, W. (2002). Imperial landscape. In W. J. T. Mitchell (Ed.), *Landscape and power* (pp. 5–34). Chicago, London: University of Chicago Press.

Mitchell, D. (2008). New axioms for reading the landscape: Paying attention to political economy and social justice. In J. Wescoat & D. Johnston (Eds.), *Political economies of landscape change: Places of integrative power* (Vol. 89, pp. 29–50). Dordrecht: Springer Science + Business Media.

Mitchell, D. (2009). Work, struggle, death, and geographies of justice: The transformation of landscape in and beyond California's Imperial Valley. In K. Olwig & D. Mitchell (Eds.), Justice, Power ant the Political Landscape (pp. 177–195). London, New York: Routledge.

Moulaert, F., & Swyngedouw, E. (1989). A regulation approach to the geography of flexible production systems. *Environment and Planning D: Society and Space, 7*, 327–345.

Müller, G. (1977). Zur Geschichte des Wortes Landschaft. In A. H. v. Wallthor & H. Quirin (Eds.), ‚Landschaft' als interdisziplinäres Forschungsproblem (pp. 3–13). Münster: Aschendorff.

Münker, S. (2009). Emergenz digitaler Öffentlichkeiten. Die Sozialen Medien im Web 2.0. Frankfurt a. M.: Suhrkamp.

Nassauer, J., & Wascher, D. (2008). The globalized landscape: Rural landcape change and policy in the United States and European Union. In J. Wescoat & D. Johnston (Eds.), *Political economies of landscape change: Places of integrative power* (Vol. 89, pp. 169–194). Dordrecht: Springer Science + Business Media.

Olwig, K. (1996). Recovering the substance nature of landscape. *Annals of the Association of American Geographers, 86*(4), 630–653.

Olwig, K. (2002). *Landscape, nature, and the body politic*. From Britain's Renaissance to America's New World. London: University of Wisconsin Press.

Olwig, K. (2008a). The 'actual landscape', or actual landscapes? In R. Z. DeLue & J. Elkins (Eds.), *Landscape theory* (pp. 158–177). New York, London: Piepmeier.

Olwig, K. (2008b). The Jutland ciper: Unlocking the meaning and power of a contested landscape. In M. Jones & K. Olwig (Eds.), *Nordic Landscapes. Region and Belonging on the Northern Edge of Europe* (pp. 12–52). Minneapolis, London: University of Minnesota Press.

Oppel, A (1884). Landschaftskunde. Versuch einer Physiognomik der gesamten Erdoberfläche in Skizzen, Charakteristiken und Schilderungen. Breslau: Hirt.

Osborne, B. (1988). The iconography of nationhood in Canadian art. In D. Cosgrove & S. Daniels (Eds.), The iconography of landscape: Essays on the symbolic representation, design and use of environments (pp. 162–178). Cambridge et al.: Cambridge University Press.

Özgüner, H. (2011). Cultural differences in attitudes towards urban parks and green spaces. *Landscape Research, 36*(5), 599–620.

Paasi, A. (2002). Place and region: Regional worlds and words. *Progress in Human Geography, 26*(6), 802–811.

Pazaurek, G. (2007 [1912]). Guter und schlechter Geschmack im Kunstgewerbe. In U. Dettmar & T. Küpper (Eds.), Kitsch. Texte und Theorien (pp. 116–128). Stuttgart: Reclam.

Piepmeier, R. (1980). Das Ende der ästhetischen Kategorie 'Landschaft'. Zu einem Aspekt neuzeitlichen Naturverhältnisses. *Westfälische Forschungen, 30,* 8–46.

Pregill, P., & Volkman, N. (1999). *Landscapes in history: Design and planning in the Eastern and Western traditions.* New York et al.: Wiley.

Radkau, J., & Schäfer, I. (1987). *Holz: Ein Naturstoff in der Technikgeschichte.* München: Rowohlt.

Riedel, W. (1989). *"Der Spaziergang" Ästhetik der Landschaft und Geschichtsphilosophie der Natur bei Schiller.* Würzburg: Königshausen & Neumann.

Riehl, W. (1854). Die Naturgeschichte des Volkes als Grundlage einer deutschen Social-Politik. Land und Leute. Volume 1. Stuttgart: JG Cotta.

Rifkin, J. (2007). *Access—das Verschwinden des Eigentums: Warum wir weniger besitzen und mehr ausgeben werden.* Campus: Frankfurt a. M.

Ritter, J. (1996 [1962]). Landschaft. Zur Funktion des Ästhetischen in der modernen Gesellschaft. In G. Gröning & U. Herlyn (Eds.), Landschaftswahrnehmung und Landschaftserfahrung (pp. 28–68). Münster: Lit-Verlag.

Rodewald, R. (2001). *Sehnsucht Landschaft: Landschaftsgestaltung unter ästhetischen geschichtspunkten.* Zürich: Chronos.

Roters, E. (1995). *Jenseits von Arkadien. Die romantische Landschaft.* Köln: DuMont.

Rothfuß, E., & Winterer, A. (2008). Eine Natur—Zwei Kulturen? Schutzphilosophien im transnationalen Kontext der benachbarten Nationalparke Bayerischer Wald und Šumava. *Standort—Zeitschrift für Angewandte Geographie, 32,* 147–151.

Rudorff, E. (1994 [1897]). Heimatschutz. St. Goar: Schaper.

Safranski, R. (2007). *Romantik. Eine deutsche Affäre.* München: Carl Hanser.

Sauer, C. (1963 [1926]). The morphology of landscape. In J. Leighly (Ed.), Land and life: Selections from the writings of Carl Ortwin Sauer. Berkeley, Los Angeles: University of California Press.

Schein, R. (1993). Representing urban America: 19th-century views of landscape, space, and power. *Environment and Planning. D: Society and space, 11*(1), 7–21.

Schenk, W. (2005). Historische Geographie. In W. Schenk & K. Schliephake (Eds.), *Allgemeine Anthropogeographie* (pp. 215–264). Gotha: Klett-Perthes.

Schenk, W. (2006). Der Terminus ‚gewachsene Kulturlandschaft' im Kontext öffentlicher und raumwissenschaftlicher Diskurse zu ‚Landschaft' und ‚Kulturlandschaft'. In U. Matthiesen, R. Danielzyk, St. Heiland & S. Tzschaschel (Eds.), Kulturlandschaften als Herausforderung für die Raumplanung. Verständnisse—Erfahrungen—Perspektiven (pp. 9–21). Hannover: Verlag der ARL.

Schenk, W. (2011). *Historische geographie.* Darmstadt: Wissenschaftliche Buchgesellschaft.

Schenk, W. (2013). Landschaft als zweifache sekundäre Bildung—historische Aspekte im aktuellen Gebrauch von Landschaft im deutschsprachigen Raum, namentlich in der Geographie. In D. Bruns & O. Kühne (Eds.), *Landschaften: Theorie, Praxis und internationale Bezüge* (pp. 23–34). Schwerin: Oceano.

# References

Schönwald, A. (2013). Die soziale Konstruktion ‚besonderer' Landschaften. Überlegungen zu Stadt und Wildnis. In D. Bruns & O. Kühne (Eds.), Landschaften. Theorie, Praxis und internationale Bezüge (pp. 195–207). Schwerin: Oceano.

Schönwald, A. (2015). Bedeutungsveränderungen der Symboliken von Landschaften als Zeichen eines veränderten Verständnisses von Macht über Natur. In S. Kost & A. Schönwald (Eds.), *Landschaftswandel—Wandel von Machtstrukturen* (pp. 127–138). Wiesbaden: Springer VS.

Schott, D. (2008). Die europäische Stadt und ihre Umwelt: Einleitende Bemerkungen. In D. Schott & M. Toyka-Seid (Eds.), Die europäische Stadt und ihre Umwelt (pp. 7–26). Darmstadt: section.

Soja, E. (1989). *Postmodern geographies: The reassertion of space in critical social theory*. London, New York: Verso.

Spanier, H. (2006). Pathos der Nachhaltigkeit. Von der Schwierigkeit, „Nachhaltigkeit" zu kommunizieren. *Stadt + Grün, 55*, 26–33.

Spanier, H. (2008). Mensch und Natur – Reflexionen über unseren Platz in der Natur. In K.-H. Erdmann, J. Löffler & S. Roscher (Eds.), Naturschutz im Kontext einer nachhaltigen Entwicklung. Ansätze, Konzepte, Strategien (pp. 269–292). Bonn-Bad Godesberg: Bundesamt für Naturschutz.

Taylor, K. (2009). Cultural landscapes and Asia: Reconciling international and Southeast Asian regional values. *Landscape Research, 34*(1), 7–31.

Tolia-Kelly, D. (2004). Landscape, race and memory: Biographical mapping of the routes of British Asian landscape values. *Landscape Research, 29*(3), 277–292.

Trepl, L. (2012). Die Idee der Landschaft. Eine Kulturgeschichte von der Aufklärung bis zur Ökologiebewegung. Bielefeld: transcript.

Tuan, Y.-F. (1979). *Landscapes of Fear*. New York: Pantheon Books.

Türer-Baskaya, F. (2012). Landscape concepts in Turkey. In D. Bruns & O. Kühne (Eds.), *Landschaften: Theorie, Praxis und internationale Bezüge* (pp. 101–113). Schwerin: Oceano.

Ueda, H. (2010). A Study on Residential Landscape Perception through Landscape Image. Four Case Studies in German and Japanese Rural Communities. Kassel: Doctoral Dissertation, Universität Kassel.

Ueda, H. (2013). The concept of landscape in Japan. In D. Bruns & O. Kühne (Eds.), *Landschaften: Theorie, Praxis und internationale Bezüge* (pp. 115–132). Schwerin: Oceano.

Uekötter, F. (2007). *Umweltgeschichte im 19. und 20. Jahrhundert*. München: Oldenbourg Verlag.

Urmersbach, V. (2009). Im Wald, da sind die Räuber. Berlin: Vergangenheitsverlag.

Vicenzotti, V. (2005). Kulturlandschaft und Stadt-Wildnis. In I. Kazal, A. Voigt, A. Weil & A. Zutz (Eds.), Kulturen der Landschaft. Ideen von Kulturlandschaft zwischen Tradition und Modernisierung (pp. 221–236). Berlin: Universitätsverlag der TU Berlin.

Vicenzotti, V., & Trepl, L. (2009). City as wilderness: The wilderness metaphor from Wilhelm Heinrich Riehl to contemporary urban designers. *Landscape Research, 34*(4), 379–396.

Weber, I. (2007). *Die Natur des Naturschutzes. Wie Naturkonzepte und Geschlechtskodierungen das Schützenswerte bestimmen*. München: Oekom-Verlag.

Weber, F. (2015). Diskurs—Macht—Landschaft. Potenziale der Diskurs—und Hegemonietheorie von Ernesto Laclau und Chantal Mouffe für die Landschaftsforschung. In S. Kost & A. Schönwald (Eds.), Landschaftswandel—Wandel von Machtstrukturen (pp. 97–112). Wiesbaden: Springer VS.

Winchester, H., Kong, L., & Dunn, K. (2003). *Landscapes: Ways of imaging the world*. Harlow: Routledge.

Winiwarter, V., & Knoll, M. (2007). Umweltgeschichte. Köln: UTB.

Wojtkiewicz, W., & Heiland, S. (2012). Landschaftsverständnisse in der Landschaftsplanung. Eine semantische Analyse der Verwendung des Wortes ‚Landschaft' in kommunalen Landschaftsplänen. *Raumforschung und Raumordnung, 70*(2), 133–145.

Wypijewski, J. (1999). *Painting by numbers. Komar and Melamid's scientific guide to art*. Berkeley: University of California Press.

Zapatka, C. (1995). *The American landscape*. New York: Princeton Architectural Press.

Zhang, K., Zhao, J., & Bruns, D. (2013). Landschaftsbegriffe in China. In D. Bruns & O. Kühne (Eds.), *Landschaften: Theorie, Praxis und internationale Bezüge* (pp. 133–150). Schwerin: Oceano.

Zink, M. (2006). Von den Elementen zur Landschaft. In H.-G. Spieß (Ed.), *Landschaften im mittelalter* (pp. 199–206). Stuttgart: Franz Steiner.

Zukin, S. (1993). *Landscapes of power: From detroit to disney world*. Berkeley, Los Angeles, London: University of California Press.

Zukin, S., Lash, S., & Friedman, J. (1992). Postmodern urban landscapes: mapping culture and power. In S. Lash & J. Friedman (Eds.), *Modernity and Identity* (pp. 221–247). Oxford et al.: Blackwell Publishers.

Zutz, A. (2015). Von der Ohnmacht über die Macht zur demokratischen Neuaushandlung. Die geschichtliche Herausbildung der Position des Planers zur Gewährleistung ‚Landschaftlicher Daseinsvorsorge'. In S. Kost & A. Schönwald (Eds.), Landschaftswandel—Wandel von Machtstrukturen (pp. 65–94). Wiesbaden: Springer VS.

# Chapter 4
# The Social Genesis of the Definition of Landscape

Previous chapters have established that the concept of landscape is a product of the social mechanisms by which meanings are defined in language—in this case, as Wescoat (2008) has noted, especially (but not exclusively) in the language of politics and economics; for other organizational systems—administrative, educational, scientific, and medial—are also interested in the interpretation, conservation, and change of the spaces in which we live. All these systems exercise power—including the power of hierarchical social distinction—over the linguistic mechanisms of definition. The entire process is an aspect of socialization.

## 4.1 Socialization of Landscape Constructs

The paucity of instincts and patterns of congenital behavior in the human has as its converse a marked openness toward the world (see e.g. Plessner 1924; Gehlen 1956; Berger and Luckmann 1966) and a natural adaptativeness to life in society. "The individual, however, is not born a member of society. He is born with a predisposition toward sociality, and he becomes a member of society" (Berger and Luckmann 1966, p. 129). The socialization process in which this occurs is one of learning which involves experience, thinking and doing. According to Fend (1981) the process has a twofold function: on the one hand to reproduce society (through the transmission of roles, values and norms), and on the other to equip the individual to act within society—a self-programming (Mead and Morris 1967) of consciousness that entails the acquisition and internalization of the roles, values, customs, and patterns of action operative in a particular society (see also Durkheim 2013 [1912]). Far from making the individual a simple "victim of circumstances"

(Nissen 1998, p. 12), however, the socialization process is one in which "the personality germinates and grows in mutual exchange with its mediated social and material environment" (Geulen and Hurrelmann 1980, p. 51).

It is in this everyday environment that meaning—including spatial meaning—is constructed in accordance with material and cultural objects, patterns, and structures mediated by others (see Geulen 1991, 2005; Piaget 1937). Here the individual develops the moral and ethical, cognitive and emotional, social and aesthetic competencies that underlie "a productive interaction with the environment centered on individual needs and interests" (Nissen 1998, p. 32; see also Peatross and Peponis 1995; Grundmann 2006). Berger and Luckmann (1966, p. 130) distinguish between primary and secondary socialization: "Primary socialization is the first socialization an individual undergoes in childhood, through which he becomes a member of society. Secondary socialization is any subsequent process that inducts an already socialized individual into new sectors of the objective world of his society." Primary socialization largely determines an individual's chances in life, for it is here that the foundations of a normal 'everyday' access to world and self are laid.

Far from being neutral, the patterns of meaning mediated in the socialization process are infused with power. As Prengel (1994, p. 64) puts it: "The situations in which people are socialized are generally hierarchical, reflecting the hierarchies of culture, social class, and gender." Successful socialization therefore entails both "adaptation and resistance to hierarchies, the art (and artfulness) of living within them and seeking profitable aspects for oneself [...]" (Prengel 1994, p. 64; and see Monk 1992). Here the early experience of social inequality, especially in the distribution of life's chances, is crucial, and with it the awareness of the existence of different interpretations of the social and material environment (Mansel and Hurrelmann 2003)—an environment whose objective 'reality' can either be accepted or critically questioned.

Nor is the socialization process completed with entry into adulthood: it is a lifelong process. The idea that a young adult has achieved relatively stable maturity no longer matches the conditions in which we live. These require spatial and social flexibility, lifelong learning, and the fragmentation of 'normal' biographies. Every phase of life today imposes its own specific developmental tasks (Elder 2000), if only at the level of daily routines that must be practiced in a socially accepted way; and with age, prior personal experience plays an increasingly important role (Hurrelmann 2006). Given the shift in life conditions, and the multiplicity of competing patterns of social action—from problem solving to conflict regulation—with which we are continuously confronted, analysts speak of a new type of 'patchwork identity' peculiar to the postmodern era (Keupp 1992, p. 176) that has turned socialization into a permanent reflective process (see Veith 2008; Böhnisch et al. 2009; for a more detailed analysis see Kühne 2008a, b).

## 4.1.1 General Socialization of Landscape in Childhood, Youth, and Early Adulthood

One of the material elements that falls within the complex cultural scope of socialization is landscape. The step in which the indefinite 'other' confronting my childhood self becomes a named and recognizable 'landscape' is rooted in cognitive, emotional, and aesthetic competencies exercised not only in relation to that 'other', but also to my reactions to it, whether socially desired, tolerated or proscribed (Proshansky et al. 1983; and see Thrift 1983; Nissen 1998; De Visscher and Bouverne-De Bie 2008; Kühne 2008a, 2013; Somerville et al. 2009). Socialization makes of these competencies a blueprint that successively modifies, and is modified by, individual experience of the spatial phenomena we call landscape. The process can be thought of as an aspect of the acquisition of everyday knowledge on which the social reality of our world as a system of manageable—because habitual and unquestioned—certainties is founded (Schutz and Luckmann 1973). As already explained earlier, landscape can be understood as a 'special case' of space (each understood as social constructions). Accordingly, landscape socialization can also be understood as a special case of the socialization of space, an this "spatial socialization may be defined as the process through which individual actors and collectivities are socialized as members of specific, territorially bounded spatial entities, and through which they more or less actively internalize territorial identities and shared traditions" (Paasi 1999, p. 4).

In Berger and Luckmann's (1966) sense, the socialization of everyday knowledge, emotional relations, symbolic representation and others is part of primary socialization, and as such based on "one's own experiences and the knowledge mediated by parents and friends, books and films about prescriptions and prohibitions, beginning with the simple labeling of things as good and ugly, edible and inedible" (Kruse-Graumann 1996, p. 172; and see e.g. Somerville et al. 2009). Even the extent to which, and in what circumstances, emotion is an appropriate reaction is subject to this kind of regulation (Eisenberg et al. 1998).

Apart from knowledge gained in school and other educational institutions, acquaintanceship with landscape and its attributes is generally unsystematic. In childhood—or more precisely in the concrete operational stage of cognitive development between six or seven and eleven or twelve years of age (Piaget and Inhelder 1948; and see Ahrend 1997)—such knowledge is generally mediated through 'significant others' (Mead and Morris 1967), especially the father. This statement is backed by evidence empirically gathered by Kühne (2006a, p. 182) in a survey with a quantitative sample of n = 455 and a qualitative analysis of 31 interviewees, of which the following example is typical:

I first became conscious of landscape as a 6–7 year-old when I walked with my father on Sunday mornings from [place name] to [place name] and we looked across from [place name] to [place name].

Characteristic of this early construction of landscape is its detachment, its lack of immediacy, although it still takes place against a backdrop of concrete events and perceptions (see Piaget and Inhelder 1948; Peatross and Peponis 1995). Landscape is construed as a concrete physical object whose content is gradually assimilated to received interpretive patterns (Kook 2008). In later childhood and youth, these patterns are complemented with independent ascriptions, on which the peer group exercises increasing pressure, as the following example (see Kühne 2006a, p. 183) illustrates:

So I first really became aware of it [landscape] when I was 6 or 7, during the war or shortly afterward, when we here in [place name] began to play further away from the house. We would say 'We're going into the fields', or 'We're going into the woods'. Then sometime later—we would have been about 9 or 10—we would say to each other 'Let's play landscape games'—meaning cops and robbers.

Within its interactive horizon, the peer group develops its own world of understanding in which individual activities and attitudes are mutually attuned (Veith 2008) and meanings defined—for example in the agreement (whether implicit or explicit) about what is and is not 'landscape' as a physical setting for a young boys' game. Especially important in this youthful appropriation of space is what Burckhardt (1980, p. 140) calls "no-man's-lands": spaces left unmarked by society or individual ownership that are available without any particular sanction for the group (or individual) as secret meeting places, hideouts, dens etc., if needs be furnished with the insignia of proprietorship. In this respect, the acquisition of space reflects "everyday practice with its forms of behavior and action, its cognitive and affective processes" (Chambart de Lowe 1977, p. 26; and see Graham 1998).

Among the landscape constructs whose foundations are laid in childhood and adolescence, pride of place must be accorded to the normative home landscape (Kühne 2006a, b, 2008a); the emotionally charged surroundings in which one grows up and which generally pass into adulthood as a stable and unquestioned standard untouched by aesthetic or cognitive criteria. One's home landscape is "filled with early memories of regional speech and sounds, with smells, colors, gestures and moods—things that speak to one and remain deeply anchored in the memory (Hüppauf 2007, p. 112; and see Proshansky et al. 1983; Stremlow 2001; De Visscher and Bouverne-De Bie 2008; Micheel 2012): molding consciousness, the contours of this landscape "offer an enduring maternal shelter and home" (Hard 1969, p. 11). In adulthood this is reinforced by the experience of other landscapes, but even here the home landscape will remain subliminally dominant until external events impinge with sufficient force to question its normative status—either by a change in its physical shape or by the re-rooting of the individual in another, different landscape, whether through travel, change of domicile, or (generally to a lesser extent) the influence of the media (Kühne 2008b; Kühne and Schönwald 2015a).

A second major influence is the stereotypical landscape (Kühne 2006a, b, 2008a; and see Bacher et al. 2016). As with stereotypes in general, the simplifying impact of such landscapes is not related to objective criteria of truth. But they, too, underpin the production and development of individual matrices of social action. Here, however, the operative factor is not so much the direct confrontation with physical space as the secondary presentation of such space in the media (Kühne 2008a, 2015b). In Central Europe this tends to focus on pre-industrial signifiers and meanings: woods, meadows, streams, mountains etc. (see e.g. Hoisl et al. 1987, 1989; Ulrich 1977; Konold 1996; Hunziker and Kienast 1999; Ipsen et al. 2003). The attractiveness of such landscapes lies in the cognitive simplicity of their symbolism (Jackson 1984), the relative universality of their aesthetic standards, and their emotional 'homeliness' (Fig. 4.1).

The internalized interpretive patterns of normative home landscape and its stereotypical counterpart underlie and form our perceptions as an active, evaluative force: "Whenever we are confronted with new experiences, we involuntarily compare our perceptions and feelings of landscape with the images stored in our memory" (Nohl 2004, p. 37)—'the mountains are higher than ours', 'there is less industry than at home', 'that high-rise block spoils the landscape', 'the power station is ugly' etc. (see Said 2000; for a more detailed analysis see Kühne 2006a, 2008a, b; Fig. 4.2).

It follows that the normative home landscape cannot itself function as a stereotype: it is too familiar and emotionally charged for that (see Kühne 2006a, 2007a, b, 2008b; Kühne and Schönwald 2014, 2015c). The immunity of the native landscape to stereotypical ascriptions preserves it from subjection, either mental or real, to abstract, alienated standards of beauty, whether at the individual level (in the sense of loss of value of the home landscape in comparison with a stereotype) or at the social level (in the reality of dysfunctional attempts to 'improve' its physical appearance).

**Fig. 4.1** A stereotypical Central European landscape from the Bliesgau (Saarland; *Photo* Olaf Kühne)

**Fig. 4.2** A physical complex, which, in its misty, romantic dawn atmosphere, expresses a Central European stereotype. When the mist lifts, the sun will shine, however, on high-rise blocks in the valley and a line of pylons carrying electricity across it (*Photo* Olaf Kühne)

## 4.1.2 Influences on Landscape Awareness: Gender, Mobility, Mass Media

The social landscape is no more than a body of meanings and values onto which an individually perceived landscape is, so to speak, plotted. These two are not identical. However, individually traced patterns also have a social dimension, which can be briefly described in the impact of three variables: gender, mobility, and mass media.

According to Ipsen (2002; more fully in Kühne 2015c), the cognitive, functional, emotional and aesthetic components of landscape awareness reveal gender-specific differences both in their socialization processes and in their adult structures. In the quantitative survey referred to above, Kühne (2006a; n = 455) established a significantly higher level of cognitive knowledge of landscape in male than in female respondents, while in the qualitative survey women on the whole perceived landscape aesthetically and men more cognitively. These results can be ascribed to polarized gender role types and socialization processes, which for girls and women lay emphasis on the emotional and aesthetic, for boys and men on the cognitive, functional, and exploratory-proprietary (Hagemann-White 1993; and see Saugeres 2002). Typifications of this sort also marked earlier geography teaching in

Germany, whose gender image Schultz (2006, p. 114) pithily categorized as *"soul and feeling are womanly, understanding and action manly"* (original emphasis). To the present day, the socialization of girls has, in fact, often tended to passivity and —informed by an implicit fear of rape—centered on house and home, whereas that of boys has taken greater account of the socializing force of active spatial exploration (Monk 1992; Radding 2005; Nissen 1998; Kühne 2008a, b).

Another important factor in the construction of landscape—in this case an intergenerational difference—is mobility. In 1955, Pfeil noted that children explored the space around their home in more or less concentric circles, but this has now changed. Developing means and patterns of transportation (automobile, bus, streetcar, subway, airplane etc.) have long since linked locations with different specific functions, with the result that landscape—even visually—is increasingly perceived as a set of spatial islands (Bertels 1997).

A third factor is the virtual landscapes of media and cyberspace, which stand in competition with the constructs of immediate perception. Their informative-interpretive models frequently have recourse to stereotypes which themselves draw on individual stereotypes. One sees this, for example, in films and computer games, which regularly choose scenic elements that reflect common associations of outdoor space with action (Asmuth 2005; and see Sects. 4.3.4 and 5.6 below). However, "the actual impact of mass media" on social perceptions of landscape is difficult to determine, as it "depends closely […] on individual usage" (Veith 2008, p. 29). Nevertheless, it is clear from both qualitative interviews and quantitative surveys that the Internet in particular has increasingly influenced individual constructions of landscape (Kühne 2014b; Fig. 4.3).

## *4.1.3 Socialization of the Concept of Landscape Among Specialists*

While the non-professional socialization of landscape, whether in childhood, adolescence, or adult life, takes place—apart from school geography classes—in a largely unsystematic fashion, its conceptual development among adult professionals is systematic. This 'secondary socialization' takes place in what Berger and Luckmann (1966, p. 138) call the "institution-based 'subworlds'" of tertiary education: college and university departments of landscape planning, architecture, ecology and gardening, as well as (to some extent) geography and biology. Glaser and Chi (1988) have demonstrated that landscape specialists possess outstanding knowledge in their field: rather than simply experiencing landscape, they 'read' it quickly and meaningfully on both large and smaller scales, use their memory effectively, and perceive and address problems at a deeper level than laypeople. They also spend a lot of time on the qualitative analysis of landscape issues, and are more self-reflective in their procedures than laypeople, so they are generally more sensitive to problems and errors.

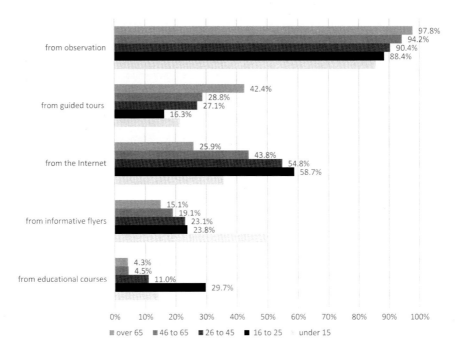

**Fig. 4.3** Frequency of answers by age group (n = 1546) to the question 'How do you acquire knowledge of landscape?'—multiple answers possible. The pale bar for the 'under-15' age group, and the lack of a percentage figure, reflect the statistical unreliability of answers in this cohort (diagram after Kühne 2014b)

The differences in socialization between specialists and non-professionals are clearly expressed in Kühne's interview with the physical geographer K.U. (Kühne 2006a, p. 189):

> My sense of landscape was greatly reinforced at university. In fact, I only became really conscious of it then […], thought about why landscapes are as they are. Before that time I just accepted it, saw landscapes as impressive. It was only at university that I began to look analytically at that question: why they have that particular shape. Moreover, that's where I still am. I think: Just look at those strata!

University—as Hilbig (2014, p. 98) comments of architectural students—turns the world-view of young people upside down: "By the time they graduate they can *only* see buildings in that way: as beautiful, ugly, honest, inauthentic etc." (original emphasis). The same is evidently true of landscape. Today the training of landscape specialists has created a class of "organized professionals who are entrusted with the solution of problems" (Tänzler 2007, p. 125), and a corresponding acceptance of the distinction between "those who are competent [in this field] and those who are not" (Bourdieu 1977, p. 13). The former, in our functionally differentiated society, possess a problem-solving monopoly (see e.g. Larson 1977; Freidson 1986; Luhmann 1990; Stichweh 1997; Weingart 2003; Tänzler 2007; Hilbig 2014; Kühne 2014c; Matheis

2016). In Ipsen's (2002) terms, their approach to landscape is cognitive rather than emotional, while in the lay awareness of landscape the dominance of the latter dimension is evident in the normative appeal of the native landscape and later aesthetic stereotypes.

Empirical research (Kühne 2006a, 2014b; Hunziker et al. 2008; Hunziker 2010; Hokema 2015; Kühne and Schönwald 2015a; Bacher et al. 2016) suggests in this context that Ipsen's triad of cognition, aesthetics and emotion should be complemented with a functional dimension—although experts and non-experts tend to interpret 'functional' in widely differing senses. For experts it means the function of a 'landscape' (e.g. rough grazing or recreation), whereas for laypeople it tends to refer more narrowly to their personal free-time requirements (e.g. walking, cycling, paragliding, or photography), the typical question here being 'Is it good for *me*?' (Kühne 2015c). Among experts, the aesthetic dimension of socialization also tends to differ according to training and specialty. This is evident above all in the difference between planners, who give less, and designers, who give more weight to aesthetics. For Debes (2005, p. 117), landscape design means "intentionally altering landscape in a symbiosis of the functional and the artistic", while planning seeks "convincing solutions that combine rational consideration of conflicting interests and objective evaluation [of alternatives]." Above all in the planning disciplines, the socialization of the concept of landscape is, with few exceptions, positivist, with constructivist positions treated, if anything, cursorily (Jessel 2000; Kühne 2008a). Accordingly, typifications such as 'green belt', 'saprobic system', 'landscape element', 'climate classification' etc. are applied to physical space—for the most part uncritically and to the point of ritual repetition—creating unreflected stereotypes (see Kühne 2008a and Sect. 4.2 below) that differ from those of laypeople only in being based on scientific method and a corresponding canon rather than on tradition and emotion. In both cases the constitutive mechanism is what Berger and Luckmann (1969) call 'objectification' (see also Hard 1991, 2002c [1970], 2002 [1971]; Schneider 1989; Weingart 2003; Hilbig 2014).

The scholarly approach to landscape in the relevant disciplines also reveals considerable dependence on schematic interpretations in Bourdieu's (1979, 1985) sense, which, for all their ostensible neutrality with regard to the physical genesis of landscapes, mediate an implicit or explicit scale of values that elevates 'historical cultural landscape' to a primacy verging on the sacral. Wöbse (1999, p. 271), for example, declares that "the conservation of a historical cultural landscape is an obligation, its development a task" (for similar views see Zimmermann 1982; Wagner 1997; Quasten 1997; Denecke 2000; for further detail see Sect. 4.2 below). In this sense physical change is "experienced as 'loss', 'decline' or 'destruction' of the real or supposed former qualities" of a landscape (Hokema 2009, p. 241). Stereotypes of this kind reduce the complexities of a world or situation to a simple disjunction (see Luhmann 1984), subjecting multiple sense impressions to a rapid (and often rigorous) evaluation that categorizes them as either 'fitting' or 'unfitting'. That experts are, in fact, more liable than laypeople to make judgments of this kind is noted by Tessin (2008, p. 136; and see Hilbig 2014), who comments: "Urbanization is largely seen in professional landscape aesthetics as [...] a 'disfigurement' or 'devouring' of the landscape," whereas laypeople see it as "also having something to do with 'living' comfortably."

## 4.1.4 Relation Between Lay and Expert Views of Landscape

Lay and expert attitudes should not, however, be thought of as polar opposites: they represent, rather, the opposite ends of a continuum (Kühne 2008b) whose intervening space has become increasingly important in a world where education and information are widely available; the informed newspaper reader or scientifically literate member of a citizens' initiative are cases in point. Nevertheless, this does not nullify the fundamental distinction between professionals and laypeople (see Weingart 2003, 2012; Kühne 2008b).

Some of these differences are illustrated in Fig. 4.4, a photograph shown to a total of 399 people who were asked to characterize it with three concepts. Laypeople almost unanimously described the scene as beautiful: a peaceful, balanced natural landscape. Experts, on the other hand, immediately noted its shortcomings, which differed (sometimes widely) according to their discipline. Landscape conservationists remarked on the sparseness of fruit trees and hedges in the foreground and the encroaching woodland on the slopes in the middle ground;

**Fig. 4.4** A landscape open to varying professional and lay constructions (*Photo* Olaf Kühne)

## 4.1 Socialization of Landscape Constructs

agriculturalists considered the hay bales on the field in the foreground as too numerous for balanced husbandry; forestry scientists objected to the monocultural spruce plantations in the background, and urban planners to the irregular roof shapes of the settlement in the middle ground (Kühne 2008a, b).

Laypeople have frequently, in varying intensities, acquired unsystematic bits and pieces of expert knowledge and interpretive models from their schooling, or from books, newspapers, excursions etc., and these may often be in conflict (Korff 2008, p. 103) with the aesthetic and emotional dimensions of their landscape awareness. Tourists in the Black Forest, for example, may be prepared to accept what they might otherwise regard as ugly aspects of the landscape "once they recognize these as regionally specific" (Korff 2008, p. 103). And this type of conflict may also beset experts (Kühne 2008b). Thus a biogeographer interviewed by Kühne (2006a, p. 192), answered the question 'What makes a piece of the earth's surface or a specific area into a landscape?' as follows:

> OK – on the one hand closed ecological systems, on the other hand whatever I see. In addition, that may be the totality of all the various aspects of the landscape – of the ecosystems. That, of course, contradicts what I just said. On the one hand I defined landscapes as ecosystems; but on the other hand … when I look out of the window I see an urban landscape, and behind it the Bliesgau, for example, or Saarkohlenwald Forest … that's a contradiction, I admit.

The intense selectivity of specialist training—an ecologist sees landscape primarily as an ecosystem, a geomorphologist as a contoured surface)—results in structures that transcend the categories of the particular discipline being relegated to general knowledge. Thus, the biologist may have no more differentiated access to the beauty of landscape than a layperson. Experts develop various strategies to cope with this disparity (Kühne 2006a):

(a) bypassing primarily socialized landscape awareness or treating it as irrelevant
(b) consternation at the apparent contradiction of their own concept of landscape
(c) equating the beauty of landscape with ecosystem factors
(d) acceptance of the polyvalence of their own concept of landscape
(e) creation of a synthesis of primarily and secondarily socialized landscape awareness.

These five strategies reveal different evaluations of lay and expert attitudes: (a)–(c) set the expert approach to landscape hierarchically above the lay; (d) sees them as hierarchically equal; (e) sees lay awareness as complementing and extending that of the experts.

Table 4.1 presents a comparison between various perspectives on landscape.

Differences in awareness of landscape deriving from different levels of professionalization give rise to social distinctions. This will be treated in the following section.

**Table 4.1** Contextually variant evaluations of appropriated physical landscape (after Hunziker 1995, 2010)

| Context | Meaning of landscape | Ideal state of landscape |
|---|---|---|
| Tradition | Cultural heritage | Traditional forms |
| Nature conservation | Nature | Biodiversity, rare species |
| Revenue | Productivity | Economically sound husbandry |
| Emotion | Recreation | Wealth of colors, forms, and symbols |

## 4.2 Social Distinction and Landscape

The attractiveness of physical landscape as a social construct cannot be gauged in terms of spatial complexity (see e.g. Ulrich 1979; Kaplan and Kaplan 1989; Prominski 2004; Ipsen 2006; Weis 2008; Ode et al. 2010; Papadimitriou 2010), or accessibility to individual or communal appropriation (see Burckhardt e.g. 1978c; for greater detail Kühne 2013a); it is an inherent function of processes of social distinction which are reproduced in landscape not only as a social construct, but also in the individual and even physical actualizations of that construct—or, to put it another way, as the symbolic communication of physical order (see Duncan 1999). In this context, Rotenberg's words, albeit primarily referring to urban landscapes, have a wider application: "The forms that impose upon their environment represent their social selves. Because they can build on a monumental scale, institutional leaders produce elaborate and complete representations of their vision of the metropolis. As groups of institutional leaders succeed each other in power, they appropriate a specific set of public landscape design possibilities to represent their vision" (Rotenberg 1995, p. 4). Seeking sustainability and 'normalization', power tends to inscribe itself permanently on the physical environment—not least in its claim to exercise social control over life's opportunities.

### 4.2.1 Social Distinction and the Aesthetics of Landscape

#### 4.2.1.1 Landscape as a Class Signifier in the Wake of Social Modernization

As a synoptic construct of symbols and physical objects, landscape is largely indebted to the code of landscape painting (see Chap. 3). Art—and with it landscape both *in* and *as* art—is particularly adapted to express social distinctions of taste and the legitimacy accorded them by the ruling class (Bourdieu 1979; Kühne 2008a). After all, art-and-landscape discourse employs a special register and code (Resch 1999), whose nuances must be learned from experts. Access to the stylistic critique of Classicism and Romanticism, of the Enlightenment and Biedermeier, is confined to the educated classes. Only they have the necessary susceptibility to "the

## 4.2 Social Distinction and Landscape

landscape ideals" implicit in the "parallel manifestations of the beautiful and the sublime" (Riedel 1989, p. 45) specific to each of these movements. In the German-speaking countries of Central Europe, Schiller and Eichendorff, Droste-Hülshoff, Lorrain, Friedrich and Runge all contributed to the genesis of a stereotypical social ideal of landscape, whose evolution can be seen as both accompanying and countering that of the normative home landscape (Manwaring 1965; Kühne 2008a).

The inscription on physical space of the landscape visions of painting—and to a lesser extent poetry—is graphically illustrated in the phenomenon of the so-called 'Claude glass', a tinted mirror used by artists and travelers in the 18th and early 19th centuries to emulate the tonal gradations of Claude Lorrain: "the glass was moved to and fro until it captured, as it were, a 'Lorrain landscape', on which, significantly, the observer had turned his or her back" (Kortländer 1977, p. 37; and see Cosgrove 1985; Groth and Wilson 2003). The very possession of such an instrument—all the more so its use and the description of the impressions it conveyed—demonstrated not only the constructed nature of landscape but also the distinctiveness of those aware of it.

When, in the wake of the Enlightenment, landscape was rediscovered, few physical spaces in the open countryside could actually represent the social ideal. Their place was taken by the landscaped gardens and parks of the 18th and 19th century ruling class—precincts, like the French Garden, with highly restricted rights of access (see Maier-Solgk and Greuter 1997). The Romantic aesthetic that infused and appropriated both wild and agriculturally productive landscape also followed the functional logic of landscape as a universal (in the sense proposed by Veblen 1899), elevating the non-useful into the aesthetic code and applying it—at a symbolic rather than physical level—as an instrument of social distinction (see Kühne 2006a, 2008a). For with the Industrial Revolution, the social as well as economic significance of agriculture had diminished rapidly, and the aesthetic vision of nature henceforth found expression in emotional contemplation and enjoyment (Haber 1992; Fuhrer 1997; Hoeres 2004). Oppositional pairings derived from the Enlightenment, such as "nature and God, nature and history, nature and science, and even nature and culture" (Dethloff 1995a, p. 7) became the standard by which educated people viewed the world. Nevertheless, by implication, these polarities affirmed that nature qua nature was also constructed—and not merely by "alien, imposed connotations or religious and moral claims on meaning" (Warnke 1992, p. 137): the Romantic answer to this development was the wholesale re-enchantment of nature (see Chap. 3).

The aesthetic of the sublime was predicated of nature as a new world-view steeped in symbolism (Hammerschmidt and Wilke 1990), borne by the new urban elites of the 19th century (Graham et al. 2000). The enlightened, with their technical prowess, could furnish a designer landscape with a ruin as a symbol of bygone days, of the transience of fame, and of a yearning for the imagined authenticity of a former age (Burckhardt 1979b; Hartmann 1981; Sieferle 1984; Thacker 1995; Hauser 2001). In this sense, the Romantics developed a sentimental relation to the Middle Ages in sharp opposition to the Enlightenment (Hegel 1970 [1835–1838];

Illing 2006). The medieval castle became a symbol of the "particularity of feudal resistance" (Warnke 1992, p. 54), and—especially after Novalis—of the "quest for a long-lost era, mirrored in individual childhood and that of the human race, [...] when faith and love had [not yet been] supplanted by knowledge and possession" (Safranski 2007, p. 129). The irony of romanticizing castles, which until the Early Modern Age had stood for "the more aggressive forms of local violence" (Liddiard 2005, p. 2), went largely unnoticed.

According to Dethloff, the Romantic concept of landscape projected a dominant social construct—that of the elevated individual of the 18th and 19th centuries—onto the image of divine creation. Gripped in the "high frenzy of the *promeneur solitaire*" (Dethloff 1995b, p. 27; original emphasis), the enlightened wanderer read into both wild and hand-hewn landscapes the "ideal contours of a fusion between society and culture" (Jessel 2004, p. 22), personified in the figure of the peasant as unspoilt 'natural man' (Bray 1995; see Haupt 1998; Pollard 1998; Greenberg 2007 [1939]). The motif of the *promeneur solitaire* has persisted in Western society in the form of the lonesome rider of film (whether on horseback or motor-cycle), the *flaneur* of essays and journalism, and the city 'night-hawk' fraught with the burden of impending decision or threatening fate.

The landscape aesthetic was maintained by the aristocratic and bourgeois classes: the affluent industrialists, merchants, and intellectuals whom Bourdieu (1979) collectively defines as the 'ruling class'. Not immediately dependent on the physical produce of the countryside, they enjoyed the cultural competence and necessary economic distance to the wild—as well as peasant-farmed—landscape to treat it as an aesthetic complex and, as such, a discursive marker of social distinction. They were in possession not only of the codes of landscape art handed down "in the literary forms of the 18th century: in essays, geographical and natural history travelogues, and fictional (for example idyllic) writing" (Hard 1977, p. 14; and see Hard 2002a [1969]; Cosgrove 1984), but also of the complex religious and mystical patterns of Romantic landscape painting. Natural beauty was accordingly something that could "only be upheld in the silence of educated contemplation" (Zeller 2002, p. 25); landscape became a "cultural *weltanschauung*" (see Manwaring 1965; Hard 1977, 2002a [1969a]; Cosgrove 1984; Hugill 1986; Haber 2000), or as Piepmeier (1980, p. 32) forcefully put it: "Natural beauty only exists in elitist seclusion." But, for all their desire to "flee the social shackles and physical confinement of the city" (Kaufmann 2005, p. 59), the educated urban classes were inevitably subject to the forces of "power and money—the tribute of taxes, tenancies, interest, and produce flowing into the city" from the surrounding countryside (Burckhardt 1995b, p. 272; and see Cosgrove 1984). In connection with the economic-aesthetic distancing of the educated classes from nature, Burckhardt (1977, p. 19) summarizes the reciprocity between the social appropriation of landscape and its physical planning in the following terms: landscape "is a perceptual construct used by a society that no longer lives directly from the soil: a way of looking that can shape or disfigure the outside world when that society begins to implement its perspective in the form of planning."

## 4.2 Social Distinction and Landscape

Achleitner illustrates the obduracy with which the farmers, who are responsible for the stereotypical beauty of many landscapes, resisted (and in places still resist) the implications of this perspective. He comments: "in many Alpine areas it was not until the 1950s that the farmers finally succumbed to the hard drug of sentimental regional films and began to look on their working environment as beautiful, or even worth preserving" (Achleitner 1997, pp. 54–55). In this latter aspect they were aided and abetted by the realization that "there are people who find the [Alpine] environment desirable even for a few weeks of the year" (ibid.) and are willing to pay real money for the pleasure of experiencing it as tourists (see Bätzing 2000; on the understanding of physical structures as 'heritage' see Graham et al. 2000; Matless 2005).

In the course of the late 19th and 20th centuries, the established landscape code was popularized and inevitably thereby trivialized (see Burckhardt 1977; Hard 1977). The lower middle class, aspiring (unsuccessfully) to imitate the 'legitimate' taste of the day (see Bourdieu 1979), began to look on the physical phenomena of the countryside as 'beautiful' without, however, enjoying access to the powerfully metaphorical language that underlay such an epithet—a language confined to those whose aesthetic and scientific discourse had (especially in the wake of Romanticism) molded and continued to sustain that perspective (see Hard 1965; and more generally Maasen and Weingart 2013). In this light, the adulteration of taste can be seen as a futile attempt to gain access to the social opportunities of the ruling class by mimicking their aesthetic of physical space. This aesthetic did, however, spread into the language of science. Hard observed how many landscape researchers of the 19th and earlier 20th centuries (e.g. Rüstow, Hehn, Ratzel, Vischer, Ritter, Winckelmann) employed what he called a "singular geographical prose"—a language marked by the "concepts and vocabulary of classical aesthetics" (Hard 1965, p. 16), distinctive in style as well as content. And he further remarked that while landscape still played an important role in poetry, novels, stories and essays between 1920 and 1940, by the late 1960s "it occurred, in writing of quality, almost exclusively in contexts of ironic parody, or at least in deeply disguised variants; [...] taken at face value, landscape was now characteristic only of sub-literature" (Hard 2002b [1969b], p. 114; and see: Schönenborn 2005; Fig. 4.5).

Another key factor in the trivialization of landscape is its global availability in terms of the changing ratio of cost (and ease) of travel to distance traveled. In earlier times, when journeys could only be made on foot, on horseback, or by stagecoach, travel was slow, uncomfortable, and (especially if one counts incidental outlay) expensive. The railway, and later the private car and airplane, changed all that: travel became more comfortable, and today even flying has become cheaper. Landscape of all sorts "is now so close, so easily attainable that it has become a virtual byproduct of transportation technology, a consumer article: landscape has become homeless" (Vöckler 1998, p. 278). Nevertheless, mass taste still lacks access to the complex codes of legitimacy, from the Romantic sublimity of the mountains to the tranquil slopes of Arcadia (see Sachs 2007 [1932]). Up to and including the long-haul flight to Australia, landscape is today accessible to almost

**Fig. 4.5** The mass reproduction of the appropriated physical landscape of Cracow and its environs has reduced it to a stereotype bereft of any distinguishing characteristic: a 'sunken cultural gem' (*Photo* Olaf Kühne)

everyone (Lippard 1999), from the 'stunning view' with its mood of exaltation rendered consumable by the AAA roadmap (Vöckler 1998; see Lippard 1999 and Fig. 4.5) to the industrially produced posters of roaring stags before an Alpine panorama, or (even more topically) palm trees silhouetted against the sunset on a South Sea beach. Bourdieu (1979) saw the spread of *kitsch* motifs of this kind as the expression of a working class aesthetic typical of the mid third of the 20th century, honed—according to Greenberg—on the functional mechanization of factory routine: "second-hand experience with pre-injected feeling" (Greenberg 2007 [1939], p. 206). Once this point is reached, landscape—robbed of any distinguishing mark —has become impervious to a more differentiated aesthetics.

#### 4.2.1.2 A New Distinctiveness—Landscape in the Age of Deindustrialization and the Growth of Ecological Consciousness

The contemporary transition from an industrial to a post-industrial society (Bell 1973) reflects, in its structural premises and aesthetic implications, the earlier transition from an agricultural to an industrial society (and see Kühne 2006c, d). On the other hand, to put it another way, the significance of the secondary economic sector has dwindled, just as that of the primary economic sector did in the wake of the Industrial Revolution.

With the shift from a modern (Fordist) to a postmodern (post-Fordist) regime of accumulation (see also Sects. 3.1.6 and 3.1.7 above)—above all in the old industrial areas of Western Europe and North America—a semiotic transformation has also occurred: from the signs and symbols of industrial space to those of post-industrial space (Lash and Urry 1994, p. 193). The derelict monuments of industry "have been infused with new meaning" (Hoppmann 2000, p. 159), their distinctive value has been restored in terms of a new self-reflecting perspective: that of media aesthetics. The design of post-industrial urban landscapes "associates rusting blast furnaces with a Baroque aesthetic of ruins, and the scars of opencast lignite mining with the picturesque eighteenth century garden; industrial plant is now a 'cultural landscape' in its own right" (Hauser 2004, p. 154). Post-industrial landscape architecture sees ruins of this sort, according to Weilacher (2008, pp. 94–95), "as a witness to the past and a source of information that makes landscape legible". But ruins are "not only preserved, they are, on occasion, newly built," as in the classical English garden—a fact Burckhardt (1995a, p. 104) sees as "paradoxical confirmation […] that the lost claim to significance remains; its *raison d'être*, however, is no longer in the realm of logic but in that of remembrance and etymology"—i.e. in the names and signs buried in layers of history. The aesthetic accessibility of old industrial landscapes derives especially from their contrast with stereotypically beautiful landscapes. Thus Burckhardt (1979, p. 36) observes that "the pleasure one takes in these objects is proportional to the effort one must put into integrating them into the [traditional] scheme of the attractive"—an integration that has a lot to do with the rehabilitation of the sublime noted by Lyotard (1991): applied to anthropogenic objects of the industrial past, the sense of sublimity may be rooted either in their sheer size ('the mathematical sublime') or in their symbolic force as a direct counter to conventional denotations of beauty ('the dynamic sublime').

At the social level, the aesthetic of ruins can again be interpreted as an expression of a 'ruling' taste that almost arbitrarily singles out a profane complex from the wasteland of dysfunctionality and elevates it to relevance and value (see Bourdieu 1979). For the ruling class, the sacralization of the profane (Durkheim 1984) fills such a landscape with a distinctive and distinguishing function, for middle class and popular taste—i.e. the standard of the ruled—has little access (as yet) to the Schillerian pathos (see Schiller 1970) that can perceive sublimity in industrial ruins (Kühne 2006a, c, d). As a symbolic record of the "simple, hard life of the workers" (Vicenotti 2005, p. 231), the monuments of industrial archaeology reproduce the scheme of values applied to the simple, hard life of the agricultural community at the time of industrialization (see Höfer 2001)—a model brought up to date by Bachtin in his influential book *Rabelais and His World* (1987; see Illing 2006). To conventional and popular taste these relics of a lost industrial past speak, however, more of economic—and even social—decline and failure, and as such were better removed from the landscape altogether (Kühne 2006a, c, d, 2007, 2008a).

Alongside the social distinction inherent in the aesthetic perspective of industrial archaeology, a further important distinguishing factor of recent decades is the ecological approach to landscape. And the two tendencies frequently coincide. Burckhardt points out in this context that ecology, as a scientific construct, is also a social sign system: after all, "ecology itself […] is invisible" (1995b, p. 278). Like all sign systems, it is, moreover, a product of social actions and subject to changing social forces (see Sect. 4.3), which in turn reflect dominant aesthetic perspectives. Rooted in the Romantic Movement but increasingly informed by scientific ecology, these have raised what was once valueless into a dimension of planetary value. However, without the infusion of a biologically based semiotics, the aesthetics of landscape would have lacked the energy required for a sustained and socially distinctive conservation of the natural environment. Significantly, dammed valleys, for example, are often seen today as enhancements of nature rather than condemned as "offenses against the beauty of the landscape" (Blackbourn 2007, p. 288). Indeed one of the biggest German breweries advertises its products nationwide in press and television (especially in connection with soccer games) with a thoroughly romanticized image of just such an artificial lake.

At the same trivial level, but with a clearly contrasting social message, Zillich's characterization of the 'typical German yard' (in the magazine of 'BUND', the German Association for the Environment and Nature Conservation) sets the diversity of the native cultural landscape, which "has acclimatized itself over centuries to its habitat" (Zillich 2004, p. 21), against a Fordist uniformity of popular taste: "Row upon row of sterile gardens with close-cropped lawns and ugly conifers. It's not enough that our forests—now that spruce and fir promise quick profits—are carpeted with pine needles rather than rustling leaves. No! At home, too, we must surround ourselves with the products of the garden center, from globular 'teddy Thuja' cypresses to the mussel-shaped variety that resemble pointed cabbages (Zillich 2004, p. 21). For his environmentally conscious readership, the gain in definition (and social distinction) is unmistakable. That (in this example) it is at the cost of a latent xenophobia—'our trees are better than theirs'—may well go unnoticed (see Körner 2005c; Eissing and Franke 2015).

Heinrich Spanier points up parallels in this context between the communicative modes of nature and landscape conservation and the history of the perception of Romantic landscape painting. Both are concerned with the mediation of powerful feelings, which he sees as approximating "in language and gesture the pathos of suffering and religion" that was characteristic of the Romantic world-view (Spanier 2006, p. 26). He concretizes this in the figure of Caspar David Friedrich (1774–1840), whose work, he notes, undergoes a remarkable renaissance whenever environmental and nature conservation discourse peaks (see also Kühne 2006c). Hauser (2001, p. 198) takes this argument a step further, adding that nature, "given the wealth of publications from many disciplines and perspectives, […] has in the course of the environmental debate become a simulacrum"—a term derived from Baudrillard (1978). A simulacrum points to something real but non-existent, and in so doing, becomes a reality itself. Nevertheless, this reality is not a fake: far from lying, it tells its own truth, for there is no other—no 'reality' to serve as a benchmark.

## 4.2.2 Social Distinction and the Aesthetics of Planning

The internalized norms of taste are already evident in the opportuneness of the descriptions presented by planners—a fact noted more than a century ago by Ratzel (1904, p. 231): "The tasteless is inevitably wrong. Where description is called for, images are required, and the choice of images immediately reveals taste or its opposite. Where taste is absent, description cannot succeed." Among landscape professionals, then—especially those of different disciplines and convictions—and between landscape experts and their lay audiences (see Sect. 4.2 passim), language is again a medium of social distinction (see also Cosgrove and Domosh 1997; Maasen and Weingart 2013). Crucial here is the modernist assumption that a good environment produces a good society—i.e. one ordered in accordance with the normative taste of the day—and that a good society will in turn produce a good environment.

The argument is in this sense future-oriented and as such "elevated above both present and historical standards of judgment, indeed dismissive of the present altogether" (Schneider 1989, p. 19). To plan is "to provide for the future" (Spitzer 1996, p. 14), and this provision often follows the modernist axiom 'form follows function'. In this sense landscape planning has always seen its task, in Schneider's words (1989, p. 101), as establishing "purity, beauty and order," a perspective that inevitably marks it as "a profoundly ideological discipline, set on imposing governance." And (as will later be described in greater detail) the governance in question, at both political and administrative levels, is a tool of power—a power of definition, reserved to professionals and experts, which equates the functional with the pure, the pure with the normative, and the normative with the aesthetic (see Haug 1986; Stevens 2002; Paris 2005). The force of social distinction could hardly be more explicit; and it met with early opposition from the empiricist Gustav Theodor Fechner, who in his 1876 *Vorschule der Ästhetik* (Handbook of Elementary Aesthetics) condemned 'an aesthetics imposed from on high' based only on the conceptual deductions of experts from unproven and unprovable postulates. Against it he set an empirically established spectrum of aesthetic norms and values—thereby risking the accusation of having merely supplanted an old authority with a new, albeit on different principles.

Goodman (1971) highlights the unintended social side effects of compliance, in the era of functionalist urban planning, with the norms and interests of the ruling class. Highways were routed through neighborhoods whose residents lacked sufficient symbolic capital to make their voices heard within the existing political system; whole areas were refurbished with apartment blocks built so cheaply that they failed to meet the standards the planners themselves had laid down; large municipally funded estates segregated groups with differing social structures—all of this "aggravated the suffering of an already disadvantaged urban populace" (Burckhardt 1978b, p. 175). Planners and municipalities alike took the path of least resistance—social as well as orographic—implicitly measured on the availability of symbolic capital per unit of area. And, as Bourdieu (1997, p. 164) has remarked, the lack of economic, social, and cultural capital "reinforces the sense of confinement: it chains you to a specific place"—a place whose (real and symbolic) burden of acoustic, olfactory, and visual disturbance increases with every realization of infrastructural plans.

### 4.2.3 Social Distinction and Landscape Experts—The Aesthetics of the Urban-Rural Hybrid

In its aesthetic potential for distinctiveness, the postmodern pluralization of landscape does not confine itself, however, to old industrial environments; it affords room for the systematic appropriation of all the elements of the landscape patchwork, especially those that contravene modern paradigms of purity. Expert discussion in recent years has proposed many different terms for describing the increasing hybridization of the city and its surroundings: among them are 'suburbia', 'urban sprawl', 'intermediate city' (Sieverts 2001), 'exopolis' (Soja 1995), 'city-land' (Holzner 1996), 'city-landscape' (e.g. Hofmeister and Kühne 2016), and 'urban-rural hybrid' (Kühne 2012; and see Kropp 2015; Hofmeister and Kühne 2016; Kühne 2016a, b). With varying emphases, all these terms encapsulate patterns of development that have superseded the obsolete dichotomy of town and village. Thus 'intermediate city' (in German 'Zwischenstadt') refers broadly to the urbanization of "a [rurally conceived] landscape and the concomitant 'ruralizing' of the city" (Vicenzotti 2011, p. 15; and see Vicenzotti 2008, 2012). The expression has given rise in German-speaking debate of the past two decades to considerable controversy.

According to Sieverts (2001, p. 7) the intermediate city is "the urban area between the old historical city core and the open countryside, between places where people live and non-places transcending spatial definition, between the small cycles of a localized economy and dependence on the global market." As such it lies between individual places as the definable centers of historical life and the "uniform apparatus of a global division of labor, between space as an inhabited environment and a non-space that is measurable only in terms of time used, between the enduring myth of the old city and the dream of a once vital cultural landscape" (Sieverts 2001, p. 14; and at a more general level see Thabe 2002; Massey 2006). In Kühn's (2002, p. 95) words, the idea of the intermediate city represents the "antithesis of the traditional conception of the European city with its inherent opposition to 'landscape'." Commenting on the intense debate between supporters and opponents of this position, Bodenschatz (2001) advises against unqualified support for either side: "The alternative either 'European city' or 'intermediate city' is a dead-end: it leads to confusion and the stultification of scientific discourse [...], implying in many cases a rejection of the concept of urban center and a strategic neglect of suburbia." This opens up the field for a third position, that of the qualified supporters of the idea of the intermediate city. The individual positions can be outlined—following Vicenzotti (2008, 2011, 2012) and Schultheiss (2007)—as follows:

## 4.2 Social Distinction and Landscape

Opponents of the intermediate city tend to contrast it with an 'organic historical-cultural landscape' on the one hand, and the 'old city' or 'European city' on the other. In contrast to these, the intermediate city, given its ahistorical nature, lacks identity—and identity is essential for a place to become 'home'. The chief characteristic of the 'unorganically developed' intermediate city is its fragmentation and heterogeneity: "instead of legible structures it offers only a disorderly mess of settlements" (Vicenzotti 2011, p. 85), which in turn prescribes severely, curtailed lifestyles. Such arguments are visibly conservative in their world-view (see Kühne 2015d).

Rather than seeking to establish a specific local-historical identity, euphoric proponents of the intermediate city emphasize its indefinite, fragmentary openness to any and every mode of living. Behind this lies an emphatically urban notion of lifestyle (Vicenzotti 2011, p. 87): "The city means [...] unlimited freedom, unlimited possibilities, the mixing and layering of every conceivable social interest." In its world-view, this is unadulterated liberalism (see Kühne 2015d).

Qualified supporters of the intermediate city, like its opponents, consider the 'identity' of settlements as central, but (unlike them) maintain that the intermediate city possesses its own identity. For this group, too, history is important, but not exclusively so. Historical elements of the appropriated physical landscape—village center, pathways, streams etc.—should be integrated into the planning of the intermediate city; or as Hauser and Kamleithner (2006, p. 213) put it: "Meaningful strategies [...] presuppose [definable] goals," which in turn rest on an aesthetic of "thoughtful observation and conscious perception" (Boczek 2006, p. 230). Again here, the guiding principle is order (Schultheiss 2007). Thus it is a meaningful strategy, for example, to transform the fragmented elements of a spatial patchwork into perceptibly related units, while another, contrary strategy is to "stage their fragmentation and heterogeneity [...] on the premise that this is their specific character" (Vicenzotti 2011, p. 89). The urban is understood here as the possibility of individual self-determination, on the one hand within the framework of communally negotiated limits—which indicates a fundamentally democratic outlook. On the other hand, the very qualification of the concept of the intermediate city depends on expert philosophies of order and design that are difficult to reconcile with lay participation (Schultheiss 2007).

In the final analysis this view is imbued with a twofold potential for distinction, for on the one hand it opposes the old urban-rural dichotomy, with its inherently conservationist demands, while on the other it relies for implementation on both the science of planners and the cooperation of laypeople, thus creating a hierarchical system of meta-experts, experts, and laity. It is in this latter respect, as Hahn (2014, p. 83) observes, that the entire strategy of qualification fails, for "the inhabitants of the so-called 'urban region' were [...] not asked whether they wanted to have their spatial 'identity bestowed' on them by planners." As a rule they do not have the impression of living in a faceless void, nor have they ever expressed a desire for 'identity'—at least not one that could be fulfilled by planners and architects, with their inalienable *déformation professionelle* (Kühne 2008a; Hahn 2014; Hilbig 2014; and see Stevens 2002). The construction of identities, however, is not entirely devoid of side-effects: "Regional identities and affiliations with region are not always rosy visions of solidarity or unity but may coexist with internal oppositions based on cultural, economic and political conflict and processes of Othering" (Paasi 2011, p. 15).

### 4.2.4 Appropriated Physical Landscape as an Embodiment of Social Distinction

The immediately preceding sections have shown that the social distinction inherent in planning is a recursive feature of physical landscape as appropriated by society and its individual members. The following sections will differentiate and concretize this argument.

#### 4.2.4.1 Basic Considerations

The need of the ruling class to be seen and felt as such has left its imprint on physical space in many different ways (see Fig. 4.6 and Sects. 5.1 and 5.3 below). Aided and abetted by the prevailing system of social distinctions, the concepts and images currently favored by the more powerful are imposed on the less powerful and endowed with the permanence of law (see Cosgrove 1993; Higley 1995; Duncan 1999; Duncan and Duncan 2004; Kühne 2008a). Especially in the postmodern age, the physical landscape has been subjected to the designations and symbolic meanings of its hierarchical masters, with all the milieu-specific staging and social distancing this involves (see Mitchell 2001; Krämer-Badoni 2003; Seidman 2012). Fine distinctions in the availability of symbolic capital (Bourdieu 1979) express themselves in rights of access, sojourn, and disposal. The contours of a subtly differentiated private-public allocation of such rights are visible, for instance, in shopping malls and gated communities: two examples of private spaces with regulated, socially selective public access (see Goss 1993; Selle 2002, 2004; Kühne 2015a; for greater detail see Sect. 5.2).

A further manifestation of social distinction and its spatial expression is the distribution, occupation and design of sites with appropriate landscape views. Historically this has often meant that those with a higher level of symbolic capital congregate on the city's higher ground, whose environmental quality and wider views are preferable not just for health reasons (Kühne 2012). The 'better' residential districts offer the prospect of 'suitable' company and contacts, with all the events of the social round that go with them, from hunting and cruising, through balls and receptions, to sporting occasions, cultural ceremonies and the like. Thus Bourdieu (1983, p. 67): "As if by chance [such spatial arrangements] occasion meetings between individuals who in all that bears on the life and survival of the group are as homogeneous as possible."

## 4.2 Social Distinction and Landscape

**Fig. 4.6** Visible from afar, the radio transmitter on the Puy de Dôme in the French Massif Central is a physical symbol of technological mastery over nature (*Photo* Olaf Kühne)

Already in the 17th and 18th centuries "city planners, military engineers, architects, constructional theoreticians, and landscape gardeners" (Markowitz 1995, p. 121)—all of them intellectuals of one kind or another—concerned themselves, from the standpoint of the city, "with different aspects of the view into the landscape." One such way of appropriating—or staging—the surrounding countryside for the city-dweller was to lay down broad avenues as "[visual] axes, affording masterfully architectonic landscape perspectives" (Markowitz 1995, p. 122). Another way was to open the city ramparts—in places where their martial function could safely be consigned to the past—either demolishing them completely or converting them into gardens, as exemplified in the late 18th century cities of Düsseldorf, Wolfenbüttel, Celle, Brunswick, Göttingen, Oldenburg, and Hanover. A third modality was the creation of viewing terraces in the gardens of the former princely residences—or adjacent citadels—of such cities, and connecting them with rampart promenades enjoying views into the surrounding landscape, as in Mannheim, Würzburg and Münster.

Yet another opportunity for urban connoisseurs of landscape developed in 16th century Rome and other Italian cities, whose girdle of villas boasted terraces, landings, loggias, galleries, roof gardens and verandas, and whose gardens were furnished with spiraling walkways and hillocks, all of these features affording aspects of neighboring parks and gardens, as well as of the open country—the appropriated physical landscape. Architectonic forms of this kind spread first throughout Europe, and then worldwide; today they are an established item of the global architectural canon (Markowitz 1995). One thing all such constructs had in common was their privileged status: access was reserved to the master, with his family and companions, and definitively closed to those who dwelt beneath their gaze—as closed as was the aesthetic governing that perspective. It was the unprivileged class, nevertheless, that bore the burden of physical change to the environment which that arrangement entailed. Life's opportunities were indeed unequal (for greater detail see Kühne 2012; Lenski 2013).

An interesting aspect of more recent urban development is the differences in symbolic (and with it economic) structure occurring in areas of very similar architectonic substance according to the social background of their residents. Drawing on Michael Thompson's 1979 book *Rubbish Theory*, Burckhardt (1991) examines this phenomenon in some London streets of 18th century brick houses, where, depending on the background of their current purchasers, three different stories can be told:

- If the houses are bought by Pakistanis, their monetary value and potential as social symbols drop steeply without any change being made (initially) to the fabric (brickwork and window frames may later be painted in the brighter colors favored by the new owners). Nevertheless, lived in by people with low symbolic capital, the house and row will have lost their snob appeal (Burckhardt 1991, p. 224).
- If the houses are bought by skilled workers—printers, gas station proprietors, electricians etc.—devaluation will be less sudden (Burckhardt 1991, p. 224). The fabric will be well maintained, but in such a way that the symbols of an earlier heritage—wooden doors with brass fittings, wood-framed paned or sash windows—will be gradually replaced with hygienic mass products from the DIY store that meet the taste and symbolic capital of the lower middle class.
- If the houses are bought by the educated middle class—a group with higher cultural and symbolic, and at least medium economic capital—they will be preserved (or restored) in a way that sustains their fabric and boosts their economic value in line with their symbolic social distinctiveness: a new residential cycle will begin.

Independently of utility value there are, then, "opinion makers that create value" (Burckhardt 1991, p. 225) in economic terms, and non-opinion-makers that destroy it (see Kühne and Schönwald 2015a). Zukin (2009, p. 544) puts the matter succinctly: "Properly speaking [...] gentrification is an individual action, involving the preservation, restoration and re-use of old houses of some certified architectural quality, which—when broad in scale—produces both a demographic change and a change in a space's social character."

### 4.2.4.2 The Multi-sensory Atmosphere of Appropriated Physical Landscape as a Social Definer

The social construction of landscape is on the whole—and even more so in the case of experts—confined to the visual: "Implicit in the landscape idea is a visual ideology which was extended from painting to our relationship with the real world" (Cosgrove 1985, p. 55). Non-visual aspects are often relegated by professionals to a sub-discourse, which is understandable, given the absence of an adequately developed vocabulary to express them (Brady 2005). This is above all the case in the spatially oriented sciences, whose insights tend to be expressed in maps, models and other visual media (Tuan 1979b). But the modern mind in general—including modern science—is heavily reliant on visual perception, as even our metaphors reveal (see Latour 2002 [1999]): we 'see' the world through a distinct 'perspective'; our 'vision' is disturbed by prejudice; our 'world-view' is clear or blurred (and see Maasen and Weingart 2013).

The non-visual dimensions of an appropriated physical landscape, however, are crucially important for its atmosphere. This is not something that can be physically located with any accuracy, nor is it strictly measurable. On the contrary, atmosphere, according to Fuchs, "is a holistic phenomenon of spatial expression, indefinably diffused over the length and breadth of a place or situation, as when one feels the cool darkness of an alleyway in a Romanesque church, or plunges into the noisy gaiety of a county fair, or senses the oppressiveness of a thunderstorm closing over the landscape" (Fuchs 2000, p. 213).

Atmosphere is a fleeting thing, neither spatially nor temporally sustainable (Kazig 2007). Rooted in the sensory relation of the self to its surroundings, it is a medium of subjective wellbeing—or its opposite (Thibaud 2003; Kazig 2007, 2008). Böhme (1995) endows atmosphere with independent reality, but it is a reality that is unutterable—and hence cannot be confined—by any sign system; for atmosphere is a matter less of cognition than of affect and emotion (Hasse 1993, 2000; Seel 1996; Kazig 2008; Forkel and Grimm 2014). Alongside emotion, alertness is a defining factor; not that "the senses are directed like a spotlight on a particular segment of surrounding space", but rather that "according to the situation, senses, mind and body act in a specific way together" (Kazig 2008, p. 150)—alertness, then, in the sense of an acute bodily wakefulness and sensitivity. In addition, a third dimension of this complex, according to Kazig, is the motor functions, above all inasmuch as "atmosphere is perceived in connection with (specific styles of) movement" (Kazig 2008, p. 150). Raab (1998) relates the systematic disregard, especially of the sense of smell, with the demands of Western science (universal validity, provability, and lack of subjectivity). For "while, for example, visual perception is informed by optical qualities (colors) that are definable by physical measurement (wavelengths of light) and can be classified in a readily available system of subjective categories (basic colors), the olfactory realm enjoys no such consistent relations between the chemical and physical characteristics of scents and their impact, nor any aspects under which their subjectively perceived qualities might be classified" (Raab 1998, p. 16). The social implications of smells was already noted by Simmel in his sociology of the senses (1908), where he pointed out that the social question has an olfactory as well as an ethical dimension: "That we can smell the atmosphere a person radiates is the most intimate of perceptions, one in which he penetrates our inmost being, as it were, in aerial form. How obvious, then, that a heightened sensibility to scent must lead of itself to the selection—and, if needs be, distancing—that is among the sensory foundations for the sociological reserve of the modern individual" (Simmel 1999, pp. 734–735). To say, for example, that you 'don't like the smell of him', or that a particular gathering 'stinks of corruption', is an expression of social stigma as well as profound antipathy (see Payer n.d.), and a sign of the way in which the olfactory functions as a medium of social distinction. But this is a far broader issue, for particular scents immediately express the spatial workings of social power: the powerful—i.e. those endowed with greater symbolic capital—can determine where those who 'smell unpleasant' shall reside, and can themselves avoid the noisome odors of subways, pedestrian tunnels, crowded sidewalks, and public transportation by using their private automobile for every journey, and the stench of factory chimneys and other sites and products of cheap labor by withdrawing to the higher ground of their expensive villas (see also Wyckoff 1990).

The elimination of society's olfactory burden obeyed (and continues to obey) the logic of social differentiation. Payer (op. cit.) illustrates this with the example of late 19th century Vienna, where fecal odors had been effectively quenched by the construction of flushable sewers, but other, less omnipresent sources of repugnance, from factory chimneys to the recently invented automobile, were dealt with a great deal less energetically, for the simple reason that wealthier citizens could withdraw from the nuisance—and in doing so underline their statement of social distinction. Only with the onset of the environmental movement in the late 1970s and 80s, at least in Germany and Austria—in other parts of Europe and North America ecological awareness developed earlier (or in some cases later)—was massive criticism of these causes of pollution voiced, whereupon politics intervened relatively quickly to reduce their intensity (see Payer op. cit.).

Applied to ecological communication, Niklas Luhmann's systems theory (Luhmann 1986) explains the harnessing of politics to the environmental movement as a matter of 'resonance' between the two systems (for further analysis see Sect. 4.3.1.4): in other words, the uptake of ecology enhanced the scope and power of politics. Subsequent legislation had an immediate acoustic and olfactory impact on the landscape. On the one hand, admissible levels of pollution were lowered, controlled, and sanctioned, but on the other hand this very process largely flattened the sensory profile of the landscape. Where the presence of industrial plants had long been obvious to nose and ears, modern production facilities (where they exist at all in Western countries) are acoustically and olfactorily sealed from their surroundings, and only the omnipresent motor vehicle still pours its sound and stench into the environment. This regulatory process, however, is yet another example of the workings of power: politicians and experts determine the limiting values, experts monitor them, and experts propose new steps to meet new situations and materials (for greater detail see Sect. 4.3).

Noise and stench currently count as environmental pollutants to be neutralized with mufflers, chimneys, effluent disposal systems, and air conditioning plant (Bischoff 2005a); they are not accepted as a dimension of landscape in their own right. In this context, the olfactory-acoustic dimension of social segregation has a twofold impact on the landscape: on the one hand the imposition of target levels represents an element of social standardization; but on the other the intentional or unintentional neglect of these factors by landscape experts (especially planners) contributes to the burdening and eventual stigmatization of whole segments of landscape and their inhabitants (see Bischoff 2005b). The imbalance between the de facto irreducibility of the sense of smell—for "the scents that […] correspond with our wellbeing" do so through the very activity of breathing (Bischoff 2003, p. 45; and see Bernat and Hernik 2015)—and its dismissive treatment by those responsible for the design of the appropriated physical landscape (or of physical space as such) is an indication of the latent or manifest refusal of the more powerful to concern themselves with the load borne by the less: after all, they are not themselves affected.

### 4.2.5 Contingent Paradigms: The Conservation of 'Historical Cultural Landscape' and Its Alternatives

The appropriate shape of the landscape in the wake of postmodern change is a controversial issue between both experts and, to a lesser extent, laypeople. Whether in specialist literature or interviews, four paradigmatic positions are discernible—positions whose protagonists as a rule stand in direct competition for interpretational sovereignty (Kühne 2006a, 2008a, 2013; on environmental paradigms in general see Hannigan 2014). These will be described in the following subsections.

#### 4.2.5.1 The Conservationist Paradigm

Supporters of the conservationist paradigm of 'historical cultural landscape' distinguish between 'cultural' and 'natural' landscapes, and between these and landscapes that have not 'developed historically'—even to the extent of denying these latter the predicate 'landscape' altogether. The concept of cultural landscape "expresses the idea of 'cultivation'" (Ewald 1996, p. 100): these are landscapes that have grown under human hands (see e.g. Heiland 2006; Wöbse 2006). Wöbse (1999) characterizes them in the following six propositions:

- cultural landscapes have positive value
- not every natural landscape changed by human hands is a cultural landscape
- cultural landscapes are materialized spirituality
- cultural landscapes are multifunctional and guarantee diversity

- cultural landscapes maintain a balance between economic, ecological, aesthetic and cultural aspects
- cultural landscapes provide us with a long-term home.

He summarizes the concept of historical cultural landscape as follows: "A historical cultural landscape is one formed by men and women of an earlier age. It tells of their commerce with nature and the landscape, and allows conclusions to be drawn about their relation with nature. It tells, too, of lifestyles, needs and opportunities. Historical cultural landscapes contribute much to the individuality and beauty of a region" (Wöbse 2002, p. 186).

The terminological distinction between natural and cultural landscapes reflects, for Siekmann (2004, p. 32), a dichotomy between nature and culture that runs through Western thought: a philosophy that "delimits human activity from external natural events" and, for Holzinger (2004) can be seen as a project of the Modern Age (and see Zierhofer 2002, 2003; Groß 2006; in greater detail Kühne 2012; in creative literature D'hulst 2007). Zutz (2005, p. 39) perceives a close link between the concept of cultural landscape and conservationist argumentation: "Today, when a landscape planner mentions cultural landscape, it is generally in the context of its maintenance and preservation." And Wöbse (1994, p. 37) sees it as an unchallenged truth that "cultural landscape is a thing of value, worth preserving." Even more vigorously, Wagner (1999, p. 36) judges the leveling of the appropriated physical landscape in the wake of globalization and its concomitant "abolition of regional differences" as a cultural decline that "cannot be condemned in strong enough terms." In the same tenor, "the preservation of the regional differences between diverse cultural landscapes" (Quasten 1997, p. 19; and see Henkel 1997; Weber 2007) is the stated aim of organizations dedicated to landscape conservation.

The concept of cultural landscape is, then, not merely descriptive: it implies positive values (see Heiland 2006) which are often borne on a current of melancholy as if faced with loss or threat. Butler (2001, p. 177) describes melancholy as "a suppressed rebellion" which, far from remaining passive, evolves "into a sort of ongoing labor of distraction." In the case of loss of the inherited conditions of a particular landscape, melancholy of this sort involves acknowledgment and sublimation of the prevailing social power structures, as opposed to open confrontation with them (Kühne 2008a). In this context, Burckhardt points up an inherent fault in the paradigm of historical cultural landscape: its innate connection with the 'old'. Cultural landscape conservation—like that of historic monuments—is inseparably "connected with old things, and 'old' has a double connotation: it is both what we throw away and what we cherish" (Burckhardt 1991, p. 222).

The conservation of historical cultural landscapes—especially when this is connected with the home environment—often reveals a dogmatic undercurrent: the perspective reduces to a core of mutually defining self-referential propositions (see Paris 2005). Thus Thieleking (2006, p. 51) argues that action in support of a cultural landscape depends on people's sense of its being their home, and the sense of home is generated precisely in a "historical cultural landscape" (see e.g. Born 1995; Wagner 1997, 1999; Güth 2004; Heringer 2005 for similar arguments). In the

## 4.2 Social Distinction and Landscape

same tenor, Wöbse's axiomatic statement, "the conservation of a historical cultural landscape is an obligation, its development a task" (Wöbse 1999, p. 271), elevates the inherited landscape to sacral status—an attitude of semi-belief that allows for neither counter-argument nor qualification (see Paris 2005). From this derives a moral imperative which, as Bogner (2005, p. 172) comments, "today has both a private and a public aspect," for nature protection, sustainability, and the conservation of "historical cultural landscape" are issues which, in Spanier's words (2006, p. 31), "regularly excite the profoundest emotions. Given the scale of the task, only these emotions seem appropriate. Whether that is really the case, however, is a matter that calls for reflection: excessive emotion and pathos can be off-putting." An appraisal published in 1999 by the Scientific Advisory Board to the German Federal Government exemplifies this attitude. Entitled *Conservation and Sustainable Use of the Biosphere*, it describes "the biosphere crisis" in downright apocalyptic terms: "We are currently experiencing the sixth annihilation of genetic and species diversity. This could exceed in scope the last great crisis, 65 million years ago, when the dinosaurs (among other species) became extinct" (WGBU 1999, p. 3). Against such a conservative mindset (see Vicenzotti 2012; Kühne 2013a, 2015d), change to the physical structure of landscape takes on a quasi-religious aura. As Zimmermann pointedly remarks, the intellectual development that led through the Enlightenment and industrialization to modern individualism "gave birth in due course to the ecological crisis"—from which he concludes that "the ecological crisis was preceded by a religious one [...]" (Zimmermann 1982, p. 92).

The 'profoundest emotions' excited by issues of landscape and sustainability have infused an amalgam of ecologically, aesthetically, economically, and politically motivated changes to the landscape with a moral hue characteristic of current discourse (see Illing 2006). In this context, Luhmann (1993, p. 332) observes that the "moral level of public communication" rises in proportion to the risks, uncertainties, and lack of knowledge inherent in a situation, and that the transformation of a perceived change into a moral problem may facilitate its public communication by reducing it to the universally accessible systemic code of 'good/bad'; but the inevitable side-effect of this reduction is at the very least to make adequate communication of the problem more difficult (and see Bogner 2005). Moreover, an equally inevitable quality of moral commitment within "a community, with its ingrained habits and conventions, traditions, norms and values" (Berr 2014, p. 31) is that it can only with difficulty be rescinded; for the moral code applies not just to a particular role, but also to the whole person. Moreover, moral communication tends to look at breaches of the code rather than compliance with it, to discredit rather than approve a person's actions and intentions, and frequently, therefore, to discredit the person as such (Luhmann 1993). Because of its universality, the moral code also has a leveling bias that is not open to compensation by appeal to a higher instance: the playing field, here at least, is level. Moreover, as the whole person is involved, rather than just a specific role, the game tends to be played with greater vigor. Thus Luhmann (1989, p. 370): "Morality is a risky business: to moralize is to accept that risk. Where resistance arises, more forceful means may be required if one is not to lose face"—which explains the innate tendency of moral positions "to engender strife, to flow from strife, and to exacerbate strife" (Luhmann 1989, p. 370). For the 'good/bad' code may well be understood in different ways: one person's moral tenets positively invite scrutiny through the lens of another's (Luhmann 1993; and see Kneer and Nassehi 1997; Kühne 2008c, 2014c). Moreover, when semi-belief requires the continuous underpinning of dogma, what Bourdieu (2000; also Bourdieu and Eagleton 1992) calls the 'doxa' underlying moral judgments will be applied to others in all its scope and rigor.

The paradigmatic defense of the 'historical cultural landscape' has a number of facets, which have arisen in opposition to other established approaches to landscape conservation, such as:

(a) the preservation of the physical landscape, especially the restriction of activities to the 'natural landscape' in the sense initially promulgated in the National Parks concept, whereas here the focus is on landscapes modified by human hands;
(b) the protection of species and biotopes, which focuses on individual endangered species and their habitats, whereas here the focus is on the aesthetic dimension and wider physical spaces;
(c) the conservation of cultural monuments, which can cover landscape features as well as buildings, but not (or only minimally) biotic elements of the physical landscape;
(d) most of all, perhaps, recent planning initiatives, as exemplified in Schroeder's (1994, p. 79) remarks on the perceived dichotomy between the historical cultural landscape and actual encroachments: "Towns and villages used not to flow out into the landscape, but held together in mutual protection like a herd of animals. Orchards and meadows encircled a settlement, guarding it and forming a transitional space into the landscape"—which goes far to explain why today's protagonists of conservation, are skeptical, if not openly hostile, toward contemporary demands for urban space (Kühne 2008a; and see Graham et al. 2000).

Once recognized and accepted, interpretations of landscape and its human import tend to be uncritically perpetuated, stereotyped and moralized as solid elements of the natural and social world. However, the appropriated physical landscape is a consequence and by-product of social action, and "culture is action, invention, progress" (Burckhardt 1994, p. 92): an open-ended process of transformation. For Burckhardt, the historical cultural landscape is, therefore, "a momentary historical view [which] could be topical, current, and progressive—only that is no longer allowed today" (Burckhardt 1994, pp. 92–93; see also Jackson 1984; Trepl 2012). Consistently, he defines cultural landscape as landscape "into which one comes late in the day, whose charm lies in the fact that one can still (just) read it as it once was; and as it once was is, for us, how it should always be—as in the days when gentlemen rode out from town to hunt and looked on the peasants and said to one another 'Happy folk of field and meadow, not yet awoken to

## 4.2 Social Distinction and Landscape

liberty'" (Burckhardt 1994, p. 92). A cultural landscape is not just a product of description and normativity: it is infused with ideology, yearning, social distinction and other vested power interests, as well as with the intellectual forces of analysis, ethics, and morality (Kühne 2006c). A telling instance of this composite in modern dress is the golf course. Designed for the most part on the model of the English park, golf courses approximate the stereotypical ideal of beautiful landscape. Yet many cultural landscape conservationists and home environment protection groups reject them as "Americanizations […] of the landscape, destructive and pernicious" (Kaufmann 2004, p. 90). The drive for social distinction predicated of the mostly urban golfers is countered here by the distinction of superior knowledge of the landscape in its historical and cultural development, accompanied by the relevant attributes of power, rights of access, and the will to assert them (see Kaufmann 2004; Kühne 2006c; Zutz 2015).

In this sense the conservation of the historical cultural landscape can, with Simmel, be described as a triumph of 'fake authenticity'. Seen through the lens of normative aesthetic projections representing bygone social conditions, the physical landscape loses its de facto authenticity for society as a whole (in the sense of a correspondence between form and function). A partial system—as a rule the political—imposes the official stamp of expert opinion on other societal systems, with the aim, according to Bourdieu (1976, p. 90), "of transforming 'egoistic', private, individual motives and interests […] into disinterested, publicly presentable, collective—in short 'legitimate'—motives and interests." The imposition is generally conducted through legal directives and regulations, but it is also implicit in the socialization of the concept of 'beautiful landscape' itself (Kühne 2008a).

In a qualitative study of landscape awareness among both experts and laypeople, Kühne (2006a) determined that the paradigm of conserving and restoring appropriated physical landscape was upheld above all by those respondents whose approach to landscape, whether functional or aesthetic, excluded other perspectives. The functional approach evaluates landscape according to its individual, social and/or ecological functionality (Kühne 2006a), while the aesthetic approach takes correspondence with classical—or frequently stereotypical—concepts of beauty as its benchmark. In either case the goal is the preservation (or restoration) of landscape in accordance with exclusive principles. The implementation of these principles is as a rule viewed as a moral imperative subject to conventional norms of right and duty (on which see Kohlberg 1974; Colby and Kohlberg 1978). In Kühne's investigation, the functional approach revealed itself to be based almost wholly on secondary socialization. Elements of landscape consciousness derived from primary socialization were either denied or suppressed; moreover, personal convictions were proposed as exclusive and absolute (Kühne 2006a; on exclusivism in general see Sloterdijk 1987). The aesthetic approach was similarly absolute, and categorically rejected any alternative. Exclusivism of this kind is found among experts and non-experts alike, the difference being that experts often stress the distinctive quality of their knowledge. Thus Wöbse: "The loss [of cultural landscape], albeit occurring less from ill will or intention than from inadequate knowledge or untrained awareness, is nothing less than a destruction of culture" (Wöbse 1994, p. 40).

#### 4.2.5.2 Alternative Paradigms

The paradigm of the conservation and preservation of 'historical cultural landscape' dominates both specialist and public discourse, but—according to Kühne (2006a)—the current debate on landscape development is also marked by three other interpretive paradigms: those of successional development, reflective design, and reinterpretation.

The paradigm of successional development ascribes a fundamentally passive role to the appropriated physical landscape in the face of social transformations (see e.g. Vervloet 1999; Weber 2007), restricting the concept of an aspired target state to the results and by-products of socioeconomic development. This may mean leaving the landscape to natural succession, or it may entail a change (e.g. intensification) of usage. Adherents of this paradigm—and they include both laypeople and experts (see Kühne 2006a)—generally share an inclusive, tolerant perspective on landscape: they either express no preference about target states, or do not regard their own preferences as privileged. The position is criticized on the one hand for ignoring the impact of social (and especially economic) forces on overly stretched ecological systems, and on the other for failing to do justice to the importance of familiar cultural landscape in the sustenance of a 'felt home environment' (see e.g. Schenk 1997; Dosch and Beckmann 1999; Härle 2004).

The paradigm of reflective design aims to consciously endow the appropriated physical landscape with new symbolism as a step toward a new aesthetic based on defamiliarization (in Mukarovsky's (1970) sense of that term). The emphasis on intrusive, transformational design distinguishes this approach from the conservationist/restorationist paradigm, and a distinguishing feature is its predilection for historical forms and artifacts—especially as undisguised simulations (Hartz 2003; Hartz and Kühne 2006). Polyvalence is not only admissible, it is encouraged—to the extent that local inhabitants are prepared to accept it, for this paradigm sets great store on lay participation (see Brown 1989; Michert 2000; Bezzenberger et al. 2003). Protagonists of this position—in Kühne's study (2006a) these were all experts—generally share an inclusive perspective on landscape, characterized by the desire for a synthesis of aspects of primary and secondary socialization of landscape awareness with reflection on its onward development (Kühne 2006a). Changes in the physical foundations of landscape are often explained with reference to semiotic interpretive patterns. The position is criticized not only for its radical polyvalence (or lack of unambiguous principle)—e.g. in undermining the ability to read the historical development of the landscape with any clarity, as well as blurring the meaning of historical objects by subjecting them to frequent redesign (see e.g. Güth 2004)—but also for its superficial attitude and practice as regards lay consultation.

The immediate aim of the paradigm of reinterpretation is not to intrude into or redesign the physical landscape but to achieve a reflective change in the way it is received and appropriated by society—and by the individual through society. The overriding aim is to render redundant any major modification of the physical landscape demanded in the name of wellbeing (see Lacoste 1990). To this end, evaluative categories like functional/non-functional and beautiful/ugly, as well as positively and negatively charged symbolisms, are modified or radically transformed, and normative and/or stereotypical landscape concepts neutralized (see e.g. Piepmeier 1980; Höfer 2004; Prominski 2004a, 2006b; Kühne 2008a). Höfer comments: "A landscape in the process of conversion is a stimulus to free oneself from clichés—but also a reminder not to throw out the baby with the bathwater. Landscape is in constant motion, and although we don't yet know where the journey will end, it is our task to explore that end" (Höfer 2004, p. 33). However, liberation from clichés is one thing; the worry that, with a change of paradigm, the very object of the landscape expert's science (and the source of his or her self-definition) will be lost is quite another. Hence Debes' (2005, p. 124) plea for a differentiated and multi-layered approach to landscape that "goes beyond traditional thought patterns." Supporters of this paradigm generally share an affinity with semiotic interpretations of landscape development and a basically inclusive and synthesizing perspective. Criticism comes on two scores (see Kühne 2006a): first, the paradigm is an expert construct that is entirely foreign to the world of those primarily concerned—the inhabitants of the landscape—most of whom are neither prepared nor able to give up their stereotypes; secondly, it largely obscures the motivation of those who propose it. This motivation, it is objected, lies in the perceived impossibility, in the face of restricted national budgets and the requirements of free trade agreements, of leaving the landscape in a state that will meet generally held (i.e. stereotypical) criteria of beauty.

In comparison with the conservation and restoration paradigm (see Sect. 4.2.5.1), the three alternative positions presented in this section are less normative; all of them allow for what Colby and Kohlberg (1978) called a 'post-conventional' ability to form (moral) judgments about the matter in question —here the physical landscape. The equilibration strategies of the four paradigms— in Piaget and Inhelder's sense (1948)—also differ. While the conservation/ restoration paradigm and that of successional development are based on a strategy of assimilation—i.e. the adaptation of the environment to one's own needs—the paradigm of reinterpretation pursues the opposite line of accommodation, developing new, flexible and contingent patterns of perception and interpretation that fulfill Ingarden's definition (1992) of 'points of indeterminacy'. The paradigm of reflective design falls in this context somewhere between these two poles, seeking accommodation through a path of reflective assimilation.

The four paradigms possess disparate potential for social distinction. Their superior knowledge of—and consequent ability to identify and evaluate—elements of the appropriated physical landscape in their historical development provides advocates of the conservation/restoration paradigm (especially experts) with little inherently conceptual distinctiveness, as this knowledge in any case largely coincides with stereotypical notions. The paradigm of successional development enjoys higher potential in this respect for the simple reason that it departs from the conceptual status quo; even more so the paradigm of reflective design, which explicitly breaks with traditional parameters of perception and evaluation (see Kühne 2008a). Maximum distinction potential at the conceptual level is attained by the reinterpretation paradigm, whose constructivist perspective represents a radical break with existing (stereotypical) approaches, which it seeks to replace with new patterns of perception and interpretation.

The foregoing discussion of the paradigms that will determine future attitudes to landscape (in particular appropriated physical landscape) illustrates the close relation between scientific and non-scientific (mainly everyday) knowledge, like the influences of primary socialization, aesthetic traditions, personal claims and preferences in the use of landscape etc. Despite the scientific claims of the various disciplines operating in this field, their knowledge, as Ziman (2000) aptly observes for science as a whole, remains—like other systems of belief, from religion to the football club—based on faith. As such, it is engaged rather in a process of negotiation and political implementation than with the discovery of new laws. In Zimen's terms, this means that scientific knowledge is reflective rather than objective.

## 4.3 Landscape and Power

The genesis of landscape as the physical manifestation of social power is, according to DeMarrais, Castillo and Earle, marked by two central processes: "First, an elite with the resources to extend its ideology through materialization promotes its objectives and legitimacy at the expense of competing groups who lack those resources. [...] Second, materialization makes ideology a significant element of political strategy. Because ideas and meaning are difficult to control, it is impossible to prevent individuals who oppose the dominant group from generating their own ideas about the world and then attempting to convince others of their validity" (DeMarrais et al. 1996, p. 17). The second point in particular highlights the close link between power and knowledge—a connection with spatial implications, which, as social products, are subject to societal change. Hence Stehr's comment that "knowledge and power are allies, [for] knowledge and control of the conditions of action are allies when it is a matter of setting something in motion with the aid of knowledge" (Stehr 2006, p. 39). If, in the modern era, power was concentrated in the political, economic, social and cultural centers, it has, with the onset of postmodernism, become decentralized: its topology no longer assigns it a single privileged source (Deleuze 1975). According to Foucault (2006 [1976]) this decentralization is a key aspect of the transition from an absolutist to a discipline-focused society, while for Lefèbvre (1972) it is rooted topographically in the city: the power of social organization no longer derives in principle from industry, but from the everyday urban pattern of consumption, planning, and public spectacle (and see Prigge 1991).

Society generates, controls, and appropriates space through a range of recursive mechanisms and processes (see Lefèbvre 1974). The contours of physical space reveal a society's patterns of production, reproduction and usage, as well as its conceptions of physical space, all of which vary in reciprocal relation with its specific temporal, cultural, stratigraphical, and functional structures. Harvey (1991, p. 158) characterizes the complex as follows: "Control of spatial organization, and authority over its usage, are central instruments for the reproduction of social power relationships." Accordingly, appropriated physical landscape is not simply a product of fields of power: it is also a cause, an "agent of social, economic, and political processes" (Groth and Wilson 2003, p. 74; Clarke 2008): it bears recursively on society by structuring its patterns of interpretation and action (see Harvey 1996; Kühne 2008a). Bound up in this process, the various organizers of physical space—the state, municipalities, real estate agents, property owners—carefully conceal "their impact on social reproduction behind an ostensibly neutral power over the organization of space" (Harvey 1991, p. 158): the social outreach of their activities is simply taken for granted.

The development of appropriated physical landscape in Central Europe may be seen very largely as a by-product of social activity, an embodiment of society (Kühne 2005, 2006a, 2008a). Landscapes arose through economic necessity, modified by communal norms and values. Social order, Paris observes (2005, p. 28), "channels the contingencies of freedom." Where it relates to space, it channels the contingencies of its physical development, including those of what we call landscape. Landscape in this sense is, then, the product of an everyday geography of authoritative control within the normative remit of political regionalization (Werlen 1995, 1998; see also Saar 2010).

## 4.3.1 Landscape, Power, and Economics

### 4.3.1.1 Appropriated Physical Landscape and the Staging of Economic Power

The establishment of rights of ownership over space represents the transformation of space into a commodity (see e.g. Smith 1984; Wescoat 2008); but in its impersonal formalism it is at the same time a simplification of per se contingent patterns of appropriation, integrating these in a system that promises greater sustainability and stability. In addition, this system is accepted unquestioningly as the basis for rights of usage and disposal—or as Popitz puts it: "Power consolidates, and is ever more consolidated" (Popitz 1992, p. 234). Moreover, power has an inalienably spatial dimension: "every general and every politician knows that control of space is of crucial strategic importance. [...] Indeed, every supermarket manager is well aware that power over a strategically important location within the entirety of social space is worth its weight in gold" (Harvey 1991, p. 158). The placing of goods in a supermarket—to stay with this example—is governed by economic interests in the same way as the organization of constructed space. In addition, property tax itself is calculated on the basis of the economic capital value assigned by society to the limited temporal rights of usage and disposal vested in a specific parcel of real estate—rights that in turn represent limited territorial control over society's reproductive capacity (see Lefèbvre 1974; Lacoste 1990; Wyckoff 1990; Kost and Schönwald 2015; Megerle 2015; Harvey 1991, p. 166): "The acquisition of private ownership rights forms the basis for exclusive dominance over a spatial entity" and the profits that derive from it (Bourdieu 1991a). Accordingly, property ownership not only bestows social prestige, but "allows one to repel undesired trespassers […] and to keep at a distance persons or things with which one wishes to have no contact" (Schroer 2006, p. 94).

The economic productivity of physical space is as differentiated in terms of goods and services as it is in terms of societal reproduction, and both gradients (measured in tax per square meter) follow relative proximity to the city center (see Fig. 4.7), as this is also the basis on which returns can be calculated. Accordingly, usages that bring the highest returns are as a rule found closest to the city center—although this principle has been modified by the construction of suburban centers, shopping malls, and consumer markets in peripheral, non-integrated locations such as satellite and edge cities, where property taxes are also correspondingly high (see Heineberg 1989; Bathelt and Glückler 2003).

Despite high construction costs, the vertical organization of usage relativizes the economic valuation of the parcel of land on which it is built by multiplying usable floor space and corresponding returns (Alonso 1964; Heineberg 1989; Zukin et al. 1992; Zukin 1993; Krätke 1995). Moreover, vertical organization also has a symbolic component in terms of social distinction: as in the biblical Tower of Babel, the higher you are, the more godlike your power. Skyscrapers are "marvels of nature cast in stone by human hand, able to withstand such natural forces as wind and weather and even the movements of the earth" (Bischoff 2002, p. 120). They express the technological and economic power of a society (Fig. 4.8).

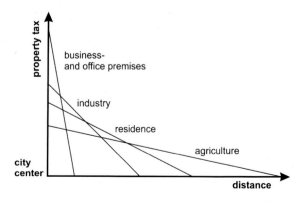

**Fig. 4.7** Urban land usage based on differential property tax zones (Heineberg 1989; Bathelt and Glückler 2003)

## 4.3 Landscape and Power

**Fig. 4.8** Verticality as a symbol of economic power—particularly in a region of predominantly horizontal structures, as here in downtown San Francisco (*Photo* Olaf Kühne)

For within a business skyscraper, an office with vistas across the city embodies the unquestioned power of a military commander situated on a hill above the battlefield—a position of high social status and potential for consolidating stability. Business enterprises follow the same logic, investing their economic capital (in alliance with the political-administrative establishment) to gain the cultural capital of a top address (see Bourdieu 1991a; Schroer 2006). For the "shaping and domestication of space" (Drepper 2003, p. 118) as a staging of organizational and economic rank is not confined to the height and design of a building: it is also crucially a question of address. An example is the so-called 'power tower' of RWE (a major German energy provider) in Essen, whose site on a busy intersection was renamed 'Opernplatz' (Opera Square), giving the corporation's head office the address 'Opernplatz 1', with all the enhanced cultural connotations that accompany such a location (Drepper 2003).

The downside of skyscrapers is on the one hand the displacement of the less powerful by the more—the demolition, in order to build them, of a district generally inhabited by groups with less symbolic capital—and on the other the diminution of the quality of usage and even casual sojourn in their immediate vicinity. For, as Rodenstein (2000, p. 66) observes: "The centralization of functions in high-rise office towers paradoxically detracts from the vitality and diversity of the city, because it boosts land values to speculative heights, excluding the multifarious small traders who might react flexibly to new demands." The result is an "absence of attractive places to meet after work, and a rush to leave the inner city for one's more or less distant home" (ibid.; see also Kühne 2012).

However, the inscription of economic structures on physical space is inevitably subject to rapid fluctuation. Not only for this reason, it constitutes an inadequate index of the city as such. For, as Zukin has observed (1993, p. 186): "Downtown is in fact, as well as image, a collective memory of objective achievement and sentimental attachment to place" (see also Mitchell and Staeheli 2009). It is here, crucially, that the conflicts surrounding change in the physical order are rooted. For although the city center is a magnet for the forces of global finance as the prime symbol of unbounded economic freedom (see e.g. Ohmae 1999), it is precisely its global horizon that destroys the familiarity of a home environment grown on the particularity of the historical inner city.

### 4.3.1.2 Suburbanization—The Making of a Landscape Between Power, Social Distinction, and Economic Analysis

Suburbanization has been the subject of intense—and as a rule controversial, at times even markedly ideological—discussion in specialist circles for some decades (see e.g. Friedrichs 1995; Häussermann and Siebel 2004; Kühne 2007b, 2012), although the intensity of debate has somewhat diminished in the wake of more recent reurbanizing tendencies. For Bourdieu, suburbia is (in Western societies) a result of the expansion of home ownership; indeed, research into this phenomenon is central to the understanding of these societies: "The massive turn to home ownership is a crucial phenomenon; it must be grasped if one wants to understand what is going on in modern society and political life" (Bourdieu 1991b, p. 144). Suburbanization, therefore, does not just denote the expansion of the city into the surrounding countryside. More significantly, it is a process of deconcentration of the population, and with it of the production of goods, and administrative as well as other services. Key parameters are the ready availability of building land at lower prices, and the increasing motorization of the population. An immediate result is the infrastructural development of the city environs, especially for transportation and traffic.

Suburbanization permanently changes the character of the appropriated physical landscape from a predominantly agricultural (more rarely forestry-dominated) mode with smallish settlements to one that can no longer be adequately contained within the classical dichotomies of conventional landscape description and evaluation (see also Sect. 4.2.3). Suburban space is hybrid, defined by radical mixing, ambiguity, and transition (see Kühne 2012). Its focal point is "the house as real estate, locus of ownership, and site of intimate private life" (Hahn 2001, p. 230). Nevertheless, the quality of the periphery is also measured in terms of its accessibility: it is "a springboard for the conquest of the world beyond the intimate and private, a gateway offering entry to urban and rural alike" (ibid.).

As a social—as well as physical, landscape-impacting—process, suburbanization is a by-product both of socially transmitted norms of behavior and action, and of "barriers, measures of exclusion and selective inclusion" (Hauser and Kamleithner 2006, p. 62). Given a mobile, mass-consumer society, it is in essence the fruit of a rational cost-benefit analysis elevated by *homo oeconomicus* to the level of a social norm—a model which for Bourdieu rests on two postulates held by their supporters to be unshakable axioms: first that "the economy is a separate entity governed by universally valid natural laws not to be breached by misplaced political intervention; [and second that] the market is the optimum means toward just and efficient organization of the production and exchange of goods in a democratic society" (Bourdieu 2002, p. 32). On this basis the lower costs of suburban property, whether owned or rented, are offset against higher costs and longer times of travel to and from the workplace (unless this is also suburban). But suburban living also has other, broader incentives: property ownership is socially valued as a type of old-age insurance, as well as a status symbol, and as a simple multiplication (relative to inner city life) of physical distance to one's neighbors. Another major perceived advantage is the ecologically more relaxed atmosphere of the suburbs, which—especially in the awareness of the many young families—approaches that of the stereotypical rural landscape (see Kühne 2006a, d; Bucher et al. 1982; Palen 1995). Finally, suburbanization can be seen as another form of symbolic social dominance, described by Bourdieu in terms of "the programmatic notion, propagated by women's magazines and fueled by the legitimate fears of parents, that— especially with regard to the schooling and later success of their offspring—they must have a pleasant house with a separate room for each child"—a conviction, he adds, that "bears in its train weighty economic decisions" (Bourdieu 1991b, p. 144).

Not just the economic, also the immediate social consequences of buying a house in the suburbs are—especially for many women in Central Europe—considerable, given the gender-specific role differentiation that tends to mark everyday life on the urban periphery (Menzl 2006). Young mothers in particular often experience a multiple break in their lifestyle, bearing the financial burden of a mortgage and the work of rearing a family, while at the same time giving up their former career and network of relations, and having to put down new social roots in unfamiliar surroundings. The resultant everyday pattern—above all if a woman's role is overwhelmingly domestic—may well be "largely dependent on local facilities" (Menzl 2006, p. 2; and see Monk 1992; Duncan and Duncan 2004). The male role, on the other hand, is characteristically dual, with the working day retaining its dominant position and evenings and weekends spent in the 'counter-world' of the suburban family home. Hence, according to Menzl (op. cit.), men generally experience the change in lifestyle associated with a move to the suburbs as less abrupt and far-reaching than it is for many women.

An essential feature of suburban life is the garden: an expression of the yearning for a perceived rural idyll within the convenient confines of the city (see Mitscherlich 1980; Hard 1985; Schneider 1989; Sieverts 2001; for the UK and USA Brunce 1994; Palen 1995; Schein 1997; Knox and Pinch 2010). Conceived, designed, and cared for as an epiphany of nature, the suburban garden, like a luxury garment (Schneider 1989), is there to be looked at and enjoyed—a unique expression of the principle of monetary exchange of goods and services, as even a rudimentary cultivation of crops is for the most part excluded in favor of a mass aesthetic of pure adornment (Veblen 1899). Again, this represents the busy ambition of the middle class to imitate the 'legitimate taste' of an old leisured society. Hence "the purity of an after-work atmosphere" (Schäfer 1981, p. 258) should be undisturbed by any physical sign of labor, or indeed any other unseemly usage such as children's games—a principle extended to the exclusion of all activities reminiscent of a former urban, rural, or village lifestyle, like "living above the workplace, or playing on the street or in the (uncultivated) fields" (Sieverts 2004, p. 87).

A major by-product of suburbanization is the passive segregation described by Bourdieu (1991a) as arising historically from the socioeconomic ability of individuals to select their relationships, excluding the undesired, embracing the desired (see e.g. Mekdjian 2008; Knox and Pinch 2010; Pietila 2010; Kühne 2012; Kühne and Schönwald 2015a). The upshot, to oversimplify somewhat, is on the one hand a homogeneous suburban population of young relatively affluent middle class families, and on the other an inner city (or indeed also countryside) housing those segments of the population that lack sufficient economic or social capital to leave their inherited environment. Häussermann and Siebel speak in this context of "a concentration [of households], based on a combination of inadequate economic resources and social discrimination, in similarly marginalized situations in derelict areas (Häußermann and Siebel 2004, p. 159; and see also Dangschat 1997, 2000; Schroer 2006; Knox and Pinch 2010). Selective mobility brings with it collective decline and 'no-go' districts of poverty and exclusion (Häussermann and Kapphan 2000; Knox and Pinch 2010; Kühne 2012), whose residents themselves—unless gentrification has set in—all become tarred with the same brush, looked on as thieves and drug dealers (Belina 2006). An inevitable result of this polarization is the glaring spatial patchwork of the appropriated physical landscape (Harvey 1990; Cosgrove 1993; Rose 1995; Soja 1993; Duncan and Duncan 2004; Knox and Pinch 2010; Pietila 2010; Zukin et al. 1992; Zukin 1993), where dereliction and neglect of material substance and public space (see Schnittger and Schubert 2005) contrast with the glossy façade of "www.suburbia" (Kunzmann 2001, p. 218): a landscape defined by spatial borders that embody not only socioeconomic but also aesthetic power (Duncan 1973; Harvey 1996; Duncan and Duncan 2004; Kühne 2012; Kühne and Schönwald 2015a).

Bourdieu (1990, 2002) sees the spread of home ownership as an index of the extent to which the working class has assimilated to the bourgeois, integrated and stabilized within a system of stakeholders disciplined by the constraints of mortgage companies and hence less likely to rebel or go on strike. Ostensibly profiting from their new bourgeois identity, house owners are in fact "tied to a house that is frequently no longer sellable" (Bourdieu 2002, p. 41), thereby losing spatial mobility. The ideal customer in this market is "neither one with high economic and cultural capital, who knows his rights and restrictions, nor one lacking either economic or cultural capital, who will sign anything in order to realize his dreams, but the low or middle ranking government employee who on the one hand has *just* enough financial resources to be able to offer the necessary collateral and enough job security to face the future with confidence, but is not sufficiently wealthy to dispense with a loan altogether, and on the other has the cultural resources to understand and accept the requirements of the mortgage company without being in a position to offer organized resistance to their maneuverings" (Bourdieu 2002, p. 140, original emphasis).

Moreover, between the vendor and purchaser of a newly built house—or of the mortgage that mostly goes with it—there is an asymmetry of both knowledge and power: "Strengthened by the experience of thousands of similar cases, often codified in sales handbooks that contain appropriate answers to every foreseeable question and situation [...], and armed with the information that every purchaser involuntarily provides and that enables him to classify them and to predict their every expectation and preference—even their banal defense mechanisms, catch questions and pretended competencies—the functionary within such a system can treat the customer, for whom this is a one-off experience (and one that is all the more daunting the higher the stakes and the thinner the real information) like a mere number in a series, arbitrary and exchangeable with any other" (ibid., p. 128).

Yet suburbia also reflects the increasing individualism of postmodern society. After all, "living in these newly urbanized landscapes [requires] a high level of personal initiative and management skills, given the widespread lack in the immediate vicinity of public or other collective facilities" (Hauser and Kamleithner 2006, p. 113; see also Pred and Watts 1992). Conversely, it also requires the need to compensate this lack in the form of the family automobile, the symbolic 'machine of mobile privacy' (Kühne 2012) that furnishes contact with the outside world (Jackson 1990). In all these aspects, suburbia is the emblem and symbol of a society based on individual opportunity.

The wish for a suburban home, one may conclude, is derived not from objective economic conditions but from social perceptions and values (Kühne 2007b), above all from a symbolic sense of mastery. But there is a paradox here, for the suburb and its inhabitants in fact destroy the very stereotype of 'beautiful landscape' to which they seek access (see Kühne 2008a). The logical answer is to move still further from the city center and in doing so to stabilize still more effectively the system of social dominance established in the bespoke mansions of the surrounding countryside. More recently, however, to escape this logic, a counter-movement has set in, albeit one that is no less unquestioningly subject to the existing hierarchies of socioeconomic and aesthetic power. The new quest for residential distinction finds expression in a return to the inner city and its immediate ring of housing, and the creation of a new type of settlement, the URFSURB.

#### 4.3.1.3 Reurbanization and Its Physical Consequences: The URFSURBS of Southern California

The contemporary development of urban-rural hybrids in the USA has departed from the quasi-linear modernist path of de- and exurbanization: "The sparse technical and social infrastructure of many suburban areas, combined with a possibly permanent rise in energy prices, puts a question mark on the whole concept of suburbia" (Hesse 2008, p. 230). Especially in older suburbs with low property values, municipalities often lack the income from real estate taxes to reverse the remorseless deterioration of the technical infrastructure (for an overview see Hesse 2008, 2010; Hanlon 2012; Gallagher 2013). The problem has been further aggravated by the financial crisis, leading to an above average depreciation in property values in areas dwelt in by people with little capital (Kühne 2012).

Demographic factors have contributed to a growing preference for locations closer to the urban center. Where in 1960 three-quarters of all US-Americans were married, today the figure is 50%; and where half the households then had children, current forecasts for 2025 indicate a drop to 25% (Gallagher 2013). More than three-quarters of younger Americans prefer urban living, and the number of driving license holders—the prerequisite for suburban living—is correspondingly falling: in 1980, 66% of sixteen year-olds possessed a driving license; in 2010 the figure was 47% (ibid.). These demographic shifts are reflected by the house-building industry: the new millennium has seen a continuing increase in activity in urban, and a decrease in suburban districts. Big supermarket chains like *Walmart* have launched inner city stores tailored to the requirements of a new breed of customer, whereas only one shopping mall—the symbol par excellence of the suburban lifestyle—has opened in the entire United States since 2006 (ibid.). Modern ICT also plays a major role in decoupling social and emotional proximities: "our neighbors may be total strangers, while our closest friend can live on the other side of the world" (Rosa 2013, p. 62). The quest for a community of the likeminded that underlay the suburban ideal has yielded to the appeal of a location offering above all flexibility.

Depending on accessibility and symbolic valuation, the urban-rural hybrid has witnessed a veritable pastiche of historical developments, with the classical streetcar suburbs built for the rising middle classes in the initial phase of suburbanization gradually being taken over, for lack of investment, by people with low capital assets, before a new wave of gentrification started in the 1980s (see Zukin

## 4.3 Landscape and Power

et al. 1992; Zukin 1993; Palen 1995; Mills 1997a, b; Hanlon 2012; Gallagher 2013; Kühne and Schönwald 2015a; Kühne 2016b). The Los Angeles conurbation, for example, contains old suburbs like Glendale or Garden Grove whose population, thanks to immigration, is now once again growing (Hanlon 2012, p. 68). So long as they are incorporated as independent municipalities, other suburbs can maintain their socioeconomic status with the help of the—by European standards restrained—planning regulations that keep parcels of building land large but the permissible area for building small (Palen 1995, pp. 103–105), and at the same time prevent the construction of socially subsidized apartment blocks (Davis 2004). Characterized by architectonic simulacra and panoramic landscapes reflecting a socially stereotypical aesthetic, Bel Air and Beverly Hills, for example, retain their virtually universal attraction, while other suburbs draw residents with specific spatial requirements (e.g. stabling or similar; see Hanlon 2012).

Thus, while some conditioning factors (like rising energy prices) are global constants, and others (like demographic change) characterize many Western societies, the historical development of urban-rural hybrids in the United States remains subject to regional and local specifics. Growth regions like Southern California undergo developments that to some extent differ significantly from the general U.S. pattern, particularly as regards the frequently chronicled decline of the 'inner suburban ring'. Alongside a gentrification that maintains continuity of (generally residential) usage while undertaking widespread refurbishment—exemplified in San Diego's South Park (see Kühne and Schönwald 2015a)—other developments reveal a new quality in the reshaping of Californian cities: more specifically of those parts of the cities built in the first phase of suburban expansion. In recent years these have in some cases undergone a process of functional as well as structural reurbanization that justifies their classification as 'urbanized former suburbs' (URFSURBS for short: Kühne and Schönwald 2015a, b; Kühne 2016b). The development may consist in the extension of an existing urban center (as in East Village and Barrio Logan in San Diego, or Skid Row in Los Angeles), or in a clear structural-functional separation from historical downtown (as in West Hollywood or San Diego's Hillcrest,: Kühne 2012; Kühne and Schönwald 2015a, b). Here reurbanization takes the form of new apartment blocks with shopping and gastronomic infrastructure constructed on the generous areas of land vacated by industry (as in San Diego's Barrio Logan and East Village). While in Los Angeles' Skid Row former industrial buildings that fulfill urban socio-aesthetic criteria have simply been converted into loft apartments (Füller and Marquardt 2010), other areas further from the center have attracted an urbanophile population that has rededicated empty buildings as stores or restaurants and stimulated the construction of new ones with typically urban functions. With their new settlement patterns, especially by the gay community these URFSURBS have generated a characteristically urban style evident both in San Diego's Hillcrest and in West Hollywood (see Fig. 4.9).

**Fig. 4.9** URFSURB variants: (left) loft conversion of former industrial buildings, Skid Row, Los Angeles; (upper right) new apartment block, East Village, San Diego—both locations are linked structurally and functionally to the respective urban center. In contrast, Hillcrest (lower right), some 5 km north of downtown San Diego, reveals a typically urban pattern of public and private service provision (*Photos* Olaf Kühne)

#### 4.3.1.4 Ecosystem Services—Subjecting Nature to Economics

The translation of various aspects of the natural environment into social action has in recent years received a new conceptual framework in the shape of 'ecosystem services' (ESS). Three central documents, the *Millennium Ecosystem Assessment* (MEA 2005), *The Economics of Ecosystems and Biodiversity* study (TEEB 2009), and the *Ecosystem Services and Biodiversity in Europe* policy report (EASAC 2009), set out to show how "ecological aspects in general, and freely available natural forces in particular, can be better integrated into decision making processes in order to safeguard sustainable land usage and to counter the exploitation and degradation of natural life conditions" (Grunewald and Bastian 2013, p. 2).

The ESS concept systematizes these aspects and forces, focusing on "technological access to the self-generating and regenerating systems of dynamic relations among living organisms and their environment" (Voigt 2015, p. 204), and thereby seeking to make society aware of what nature offers in economic terms. ESS enhances "public awareness of the natural consequences of human decisions about consumption and investment and their impact on our wellbeing" (Schröter-Schlaack 2012, p. 10), transforming these consequences, in terms of pollution, into economically quantifiable units that can serve as a measure of the obligation on the polluter to compensate the polluted (Knorring 1995, p. 2). This can then take the form of a withdrawal of alternative goods of exchange—calculable unit for unit—against the loss inflicted on the good constituted by the natural environment: an economically rational solution intelligible to any business person. With its inherently economic perspective on biodiversity and ecosystems, ESS therefore represents "nothing less than the attempt to elicit action—economic action—in the face of the threat to humanity of diminishing natural resources" (Hansjürgens and Schröter-Schlaack 2012, p. 16), and its converse, the imperative to conserve natural resources on purely economic grounds (Jessel et al. 2009). This perspective entails aspects of power and power relations that will be addressed in the following pages with reference to Niklas Luhmann's systems theory (for greater detail see Kühne 2014a).

Luhmann's theory of social systems sees modern societies as a complex of subsystems tasked with the management of specific social sectors and problems (Luhmann 1984, 1986). Each of these systems "introduces its own categories and distinctions, with which it grasps the situations and events that serve it as information" (Luhmann 1990 [1986], p. 45; original emphasis). The extent to which it grasps this information at all depends on whether the data impinges on—or in Luhmann's terms 'disturbs' or 'resonates with'—the system. In addition, even where it does so, "the system does not react to 'the environment', but to its own concept of environment" (Luhmann 1990 [1986], p. 47). A central societal factor here is communication: climate change, loss of biodiversity, the formation of surface-level ozone and other anthropogenic impacts on the natural environment will elicit no resonance within society as chemical, physical or biotic facts "so long as this is not communicated" (Luhmann 1990 [1986], p. 63); and non-communication can have dire consequences for the system. For one must "at least envisage the possibility that a system so impacts its environment that it can at some point no longer exist in that environment" (Luhmann 1990 [1986], p. 38).

Communication within the different subsystems of a society follows their inherently individual logic (Van Assche and Verschraegen 2008): for example, landscape is interesting for the economic sector when it is connected with money ('have/lack' code); and it is interesting for science when it can be investigated and yields new knowledge through the application of scientific methods ('true/untrue' code)—although what counts as true and what as untrue is again discipline-specific (with reference to landscape see e.g. Kühne 2008a, b, c); the natural environment is politically relevant when it concerns questions of power ('power/powerlessness' code). No subsystem—not even that of science—can, on the basis of its own code, grasp the environment objectively; nor is societal resonance to environmental change simply "the sum of the resonances of its subsystems" (Luhmann 1990 [1986], p. 98). For each subsystem sees the other subsystems as environmental factors conditioning and disturbing it: the process is mutual (one need only think of the sometimes conflict-laden relations between politics and industry).

Given that ESS implies the convertibility of the common good of the natural environment into an economic good, which can as such be subjected to the rational parameters of the economic system (see Heiland 1999; Schneider 2016), it follows from a systems theory perspective that economic actors face the alternative of "either being owners or not" (Luhmann 1990 [1986], p. 102). For them, the gap in the social communication of the natural environment arising from the logic of functional differentiation within society is thereby closed; for "no functional system can replace or even relieve another" (Luhmann (1990 [1986], p. 207). In this way ESS offers the economic system the opportunity of at least partially escaping from the moral and political judgments (and condemnations) toward which it has in the past necessarily been passive.

The economic code can, however, only be applied restrictively to the natural environment, for only known influences and their impacts can be monetarized: what is unknown in terms of either cause or effect cannot, by definition, be converted into monetary terms. Future impacts, for example, are particularly difficult to predict, as future market prices are unknown (Kühne 2004). Nor does the same cause—or what is held to be the same—have everywhere the same effect: global and even regional economic differences preclude any uniform evaluation of local environmental factors. All one can say in principle is that the smaller the area under consideration (e.g. the catchment area of a stream or brook) and the fewer the environmental factors involved (e.g. water), the better the chances of achieving an accurate monetary balance between cause and effect of damage.

The attempt to grasp the natural environment in exclusively monetary terms runs the further risk of blurring the borders between the various subsystems of society. If the economic code is allowed to dominate, it will impose itself on other, non-economic subsystems. And if science begins to perceive changes in the natural environment, especially those engendered by society, in accordance with the economic code of possession rather than with the scientific code of truth—e.g. in light of the availability of third party funding for the department's environmental research—or if politics suddenly decides to dispense with the instruments of law enforcement classically embedded in the political code of power, society will soon cease to function. Kühne (2003) cites Eastern Europe under communism as an example of the results of imposing a systemically alien political logic on economic decisions (see also Sect. 5.3). Moreover, the conversion of aspects of nature into monetary units ignores the ecological function of many less common species (see Kühne 2004; Voigt 2015), with the result that their protection becomes difficult to defend; after all, "ecosystem services are generally rendered by common species that are tolerant of change […] rather than by rare and endangered species" (Voigt 2015, p. 211).

The expansion of the economic code can also be observed as an infra-theoretical phenomenon, inasmuch as current discussions of ESS seem wholly focused on monetary categories; other interpretations are marginalized. But hegemonic discourse of this kind not only endangers the plurality of the key concept, but also propagates the illusion that the complexity of the world—here the natural environment—can be reduced to a sum of money, while at the same time fostering the illusion that the (frequently subjective) ascription of economic values to (often merely hypothetical) damage is actually objective (Kühne 2014a). Moreover, the cultural rootedness of evaluations concerned with nature and landscape (see Sects. 3.1 and 3.2) is ignored. Different cultural contexts can view—and also evaluate in monetary terms—the preservation or loss of a certain object very differently (Voigt 2015), all the more so given the disparities of wealth that pertain among the world's (or even between different regional) economies (see Kühne 2004).

## 4.3.2 Landscape as a Medium of Symbolic Communication at the Interface of Science, Politics, Administration, and Civil Society

### 4.3.2.1 Symbolic Communication—Language, Power, and Landscape

To scrutinize the relation between landscape and power is to inquire into the social function of knowledge, its transformation, and its bearers (see e.g. Weingart 2003, 2012; Stehr 2006; Lynch 2016). Transdisciplinary sociological research of this kind cannot fall back on a closed canon of knowledge: it is involved in a strategic conceptual feedback process with other social subsystems, and individual as well as collective worlds. Moreover, according to Giddens (1990), personal assimilation of expert knowledge is increasing: "Individuals engage in their environment with the aid of specialist information, which they routinely interpret and use as the basis for action" (Knorr Cetina 2006, p. 110; see also Kukla 2000; Lynch 2016). However, this engagement is particularly susceptible to expert interests rooted in power rather than truth.

A particularly significant factor in the symbolic communication of power and landscape is authority over the definition of its codes of communication. Once it is accepted that language is not simply a neutral instrument of communication but a transformative mechanism of power, it becomes necessary to consider the entire range of meanings contained in speech about landscape (see Pred 1990). Words, according to Bourdieu (1982, p. 83), exercise "a typically magical power: they make one see, they make one believe, and they make one act." In the spatial dimension, this often takes the initial form of assigning toponyms; for to name a place is to impose the power of definition on it (Myers 1996; see also Gailing 2014). Moreover—unless the speaker is prepared to appear a novice among initiates—this definition is bound to a particular semiotics: for like all discourse, landscape discourse is "a means not only of expression but also of censorship. Paradoxically a language always consists not only of the things it permits one to think and say, but also of those that it forbids: things that other language systems may allow [...]" (Bourdieu 1977, pp. 19–20). A relevant example of these "other language systems" in the context of landscape is lay discourse.

"All dominance," Burckhardt has said, "is linguistic. Language is the instrument of executive action" (Burckhardt 1982, p. 106). A corollary of this is that those not specifically educated in landscape-related fields are forced to remain silent in the face of professional discourse about the forms and development of the physical landscape and its societal basis—especially in the context of voluntary (or legally prescribed) participation in planning or conservation procedures. They can choose only "between an officious alien jargon and their own colloquial idiom" (Bourdieu 1977, p. 27; see also Jackson 1990). In this respect the silence of the non-professional is often merely a matter of avoiding a put-down in the face of professional self-assurance and precision—although, as Gelfert observes (2000, p. 85), the language of the professional can itself be censured as "browbeating kitsch" shot through with irrational myths and highfalutin expressions (see also Maasen and Weingart 2013).

In its bearing on landscape, the deployment of power can be measured by the resistance to change in its symbolic language. Thus single signs can be more readily changed than whole complexes (Ipsen 2006): it is a simple matter to move a hazel bush in order to make way for a path, but to move an immemorial oak—or any other landscape element protected by law—may ignite the resistance of honorary as well as official nature conservationists, and possibly of an entire local population; in which case the goal can only be achieved by invoking supra-regional forces and the instruments of official authority (Kühne 2008a). Alterations to the symbolism of an urban complex are even more difficult and require a *force majeure* which, in its short-term mobilization of both power of action and instrumental power, as a rule oversteps the limits of democratic order—as with the construction of the Boulevard Haussmann in Paris, or of the analogous central boulevards of socialist states, all of which represent symbols of a manifestly superior modernity. As far as the alteration of complex symbols is concerned, the shift in the forms of power that must be deployed—ranging from (potential) violence with corresponding compliance to unquestioned authority with corresponding technological dominance—requires an extension of the time dimension. A strategy often employed in such circumstances, when the immediate attainment of a planning target is faced with resistance (e.g. from citizens' initiatives), is to exchange this target for a long-term evolution that will likely go unnoticed, for example the marginal gentrification of a district due in any case for wholesale refurbishment (see Popitz 1992; Jordan 1996; Holm 2006; Ipsen 2006; Kühne and Schönwald 2015a).

The relation between power, dominance, and distinction is clear, but not linear. Power and dominance may always be distinctive, but distinction does not always involve power or dominance. Distinction is based on the knowledge and use of signs and symbols, but power is "concerned very significantly" (Ipsen 2006, p. 45) with the production and control of such symbols. In landscape planning, for instance, power takes the form not merely of being able to read a landscape, but of being able to change its symbolic statement (also in its physical manifestation) in line with the decision implicit in the planning proposal as to "how people in these concrete circumstances should live" (Hauser 2001, p. 41; see also Irrgang 2014). Power thus expresses itself as the sovereignty of definition held by a small body of specialists over the codes of communication (see Hugill 1995): a power of authority manifest also in their control of the technical standards governing a physical space —i.e. Popitz's 'technical' or structural power (Popitz 1995; see Sect. 2.3).

### 4.3.2.2 Appropriated Physical Landscape and Power—Expert-Lay Relations

The development of 'knowledge societies' is based on "systems of experts [...] penetrating every aspect of social life" (Knorr Cetina 2002a, p. 11) and developing their characteristic knowledge cultures in the form of "practices, mechanisms, and

## 4.3 Landscape and Power

principles bound by mutual relation, circumstance, and historical coincidence, which determine in a specific field how and what we know" (Knorr Cetina 2002a, p. 11). Among these, in accordance with the differentiation of modern society, are "specialists in [terrestrial] space" (Prigge 1991, p. 105)—or in what we call 'landscape', whose function consists in objectivizing the inherently self-referential tendency of a professional discourse encapsulated and perpetuated in scornful disregard for "the uneducated taste of capitalist society" (Elias 2002, p. 157). In the same vein Paris (2005, p. 116) characterizes experts as acting "as if they could present their credentials at any time if they wanted to, but actually don't need to."

An important aspect of modern differentiated societies is the bureaucracy (Weber 1976 [1922]), which—in combination with expanding technological capabilities (railroads, steamships, excavators etc.) has promoted a more powerful political and administrative grip on the development and shaping of the appropriated physical landscape (see Gregory 1994). Along with this has come the distinction between the professional landscape planner/architect and the nonprofessional—part of the growing (and by now universal) "separation of the competent from the incompetent" (Bourdieu 1977, p. 13). Enhanced by a disparity in the possession of information, this has inevitably resulted in an uneven accumulation of power—or in Theodor Geiger's terms (1947) a divide between the more and the less powerful—in relation to the planning and structuring of the landscape (see also Paris 2005).

The recursive nature of professional landscape discourse is aptly illustrated in the discussion between the landscape architect Prominski (2006a, b) and Stefan Körner (2006) about 'Landscape Three' (broadly speaking what has here been called 'appropriated physical landscape'). As the undisputed field of the landscape architect, landscape is a key aspect of professional identity, guiding both perception and action in accordance with an established rationale and instrumental scope, rather than with communicative principles of action (Habermas 1970). The way in which experts cut through the complexities of landscape planning was already exposed by Burckhardt, when he pointedly asked "what does a designer or architect suggest when faced with a problem? What does an apple tree suggest …? Apples, of course. Moreover, the architect suggests buildings. Every problem leads to a building" (Burckhardt 1967, p. 44)—a pre-eminent example of *déformation professionelle* exacerbated by "the belief that in designing a technology the constructor can determine its use" (Irrgang 2014, p. 12).

Contemporary research in the sociology of science sees experts less as "repositories of competence and knowledge than as representing the [strategic] interests of a scientific or technological community" (Saretzki 2005, p. 359); for, as Bourdieu has remarked, the science construct is the result of a long and arduous process of gathering many different indicators whose consideration is recommended from the point of view of practical knowledge of various positions of power […]: on the one hand of personages regarded as 'powerful' or 'influential', and on the other of the qualities commonly proposed—or pilloried—as hallmarks of power (Bourdieu 1985).

In this sense, scientific work, according to Cetina (2002b, p. 175), "is characterized by an opportunistic rationality embedded in transepistemic contexts of argument" that readily activates the strategies of officialdom in both individual and professional interests (for further details see Sect. 4.2.5). Thus the postulated intrinsic value of a natural resource can serve to secure significant economic, political, communal, and sociocultural power for the landscape and nature conservation professions (Kühne 2006c) and at the same time to satisfy their internalized moral norms, for "to help is to exercise power selflessly—but with uplift for the self" (Paris 2005, p. 25): an uplift that gains social capital (and social capital can, when the time is ripe, be cashed in for economic capital—see Sect. 5.4 for an example from the field of renewably sourced energies).

Based on secondary socialization (see Kühne 2006a), and as such characterized by the incorporated and institutionalized possession of cultural capital, the training of landscape experts is an aspect of the 20th century expansion of higher education (see Bell 1973). The professional authority of such experts is primarily communicated in the definition of ecological standards sanctioned (more or less stringently and negatively rather than positively) by law. However, while scientific approaches focus primarily on ecosystems and quantitative methods, aesthetic approaches are frequently based on (and derive distinction from) a Romantic concept of landscape; this may, indeed, even inform reflective post-industrial concern for the physical landscape (see Kühne 2006a, c). Professional power in this approach is communicated less as a cognitive resource than as a superior taste, a subtle appeal to the middle class instinct to close the gap separating them from their sociocultural masters. In this sense the affirmation of the object of aesthetic distinction—here a landscape deemed worthy of conservation—serves, in fact, to underpin the structures of social power which Bourdieu (1979) sees as an essential foundation for social stratification, and in the final analysis for the inequality that accompanies it (see also Greider and Garkovich 1994).

Their common secondary socialization ensures far-reaching agreement among landscape experts as to the appropriate aims, concepts, and paradigms with which to approach physical—and to a lesser extent social—landscape. Their academic training will as a rule have been positivist and empirical, but it may have had a scientific or aesthetic bias of varying intensity, and in fact natural scientists often expressly exclude aesthetic parameters. Leser (1984, p. 75), for example, remarks that landscape is "certainly not an aesthetic something or other on the borders of art and science." Experts with this background often view landscape-related concerns as definable in terms of "objects and objectivity" (Paris 2005, p. 114; Hilbig 2014; Matheis 2016). To justify their superiority, the 'ruling class' of landscape experts tends to invoke a "special disciplinary competence, scientific method, or sometimes even 'talent' […]" (Bourdieu 1977, p. 14) which, clad in impressive jargon, will testify to their claimed scientific authority (Adorno 1977). A favored instrument of this formalized language is the visual plan, about which Burckhardt trenchantly comments: "Planners often laugh about laypeople who seem unable to read a plan; but plans are not a suitable code for describing reality. If an expert cannot express in words what the plan depicts, there must be some deficiency in the information it contains" (Burckhardt 1982, p. 103; see also Kühne 2006c, d).

Despite their projected distinction, professional landscape planners are hardly "intellectuals in the traditional sense" (Bourdieu 1977, p. 15) of reflecting systematically on social developments; they are rather what Bourdieu calls "intellectuals for specific services [...], masters of action rather than reflection" (Bourdieu 1977, p. 15; Hilbig 2014). The strong practical bias in the profession is evident from the description of their work given on the homepage of the Federation of German Landscape Architects, which also clearly circumscribes their competencies: "Landscape architects today are largely responsible for the shape of the natural environment that forms the basis of our lives, and for its interrelations with the social and physically constructed worlds. With their unique combination of ecological knowledge and planning competence they stand for the feasibility of ideas and projects. In this way they play a key role in the development of the landscape and the planning of both urban and rural open spaces" (BDLA 2006a). These activities, Schneider (1987) observes, constitute for the landscape architect a rather one-sided relationship, a passion that is as fulfilling, as it is intolerant of critique; for critique risks loss of the very source of the artist's recognition, questioning the paradise she or he has created. Expert activity can at times—and by no means only here—approximate all too closely the everyday banalities of a "monistic materialism" (Lerf 2016, p. 247).

### 4.3.2.3 On the Changing Relation Between Science and Politics

Although the power of the individual state to "maintain established structures and control of relations between opposed societal interests" (Belina 2006, p. 13) may be receding—among other things as a result of globalization—it still has many channels through which to exercise its authority. The state, for Foucault, can be described as "a superstructure reaching across a whole series of power networks including the body, sexuality, the family, forms of behavior, knowledge, technology etc." Nevertheless, it can only exercise this "superordinate function" (Foucault 1978, p. 116) because it is itself rooted in a series of multi-faceted and undefined power relations that "constitute the indispensable foundation for these major forms of negative power" (ibid.). Among these are the relations between politics and science. Interest will focus in the following section on the mechanisms within the general context of landscape with which the state maintains its power and that of its servants, along with the "micro-powers" (Foucault 1977, p. 39) of its external intellectual resources.

The political system has recourse both internally and externally to scientific research; not only the relations between science and politics, however, but the system of science itself has fundamentally changed. The separation of fundamental from applied research ('mode 1' knowledge production) has given way to systematically mixed forms of applied fundamental research ('mode 2' knowledge production; see Gibbons et al. 1994; Latour 1999; Nowotny et al. 2001; Bender 2004; Nowotny 2005; Berr 2013). This modal shift foregrounds the "ad hoc, context-driven nature" (Berr 2013, p. 130) of many modern research projects. Latour, in fact, distinguishes 'science' from 'research': "While Science had certainty, coldness, aloofness, objectivity, distance, and necessity, Research appears to have all the opposite characteristics: it is uncertain; open-ended; immersed in many lowly problems of money, instruments, and know-how; unable to differentiate as yet between hot and cold, subjective and objective, human and nonhuman" (Latour 1999, p. 20)

According to Gibbons et al. (1994) and Nowotny (1995), the transition from mode 1 to mode 2 entails an epistemological break: science no longer investigates the basic laws of nature, but produces 'socially robust knowledge' in applied interdisciplinary contexts (see Kukla 2000; Viehöver 2005). This also bears on the relation between science and non-science: science is seen as a node within a real-world-focused feedback system of autonomies, alliances, and public presences (Latour 1999):

- The real-world focus implies the incorporation into the scientific of a range of instruments from technical appliances, through surveys and data collection, to expeditions (and the various sites associated with these activities and objects), in which information about the 'non-scientific' world is gathered.
- The autonomies in question are those associated with the innate drive of every discipline, profession and clique to establish its own reference and value system —i.e. to gain disciplinary independence, interpretive sovereignty and distinction.
- An essential feature of mode 2 science is the alliances it promotes between researchers and non-scientific (e.g. industrial) organizations. Thus physicists must interest the military, geographers (or more specifically cartographers) must interest the governors of states, chemists must interest the captains of industry, political scientists must interest parliamentarians, and educationalists must interest teachers and educational administrators for their concerns. Only in this way can they mobilize the necessary (in particular financial) resources for their projects.
- Science must have a public presence: it needs relations with the press in order to have relevance for the public at large. So it is in many cases desirable to publicize research projects pointedly and programmatically (see also Weingart 2012).

On these feedback loops depend the material and immaterial (symbolic capital) resources required for the effectiveness of individual research projects. Without alliances, science can develop sophisticated and autonomous theories with large empirical databases, and may even achieve high public awareness, but its results will not be put into practice in the real world.

The alliance of the political with the external scientific system (Luhmann 1997) in the context of the transition from mode 1 to mode 2 implies a widening of the outreach and responsibilities of science (Nowotny 2005) that bears within it, however, the seeds of an acute problem. For, Bourdieu argues, given the major influence of science and scientists on political decisions, "even if only to underpin their legitimacy with rational, 'objective' arguments" (Weingart 2003, p. 92), "science gains, in the battle of ideas [...] that are accepted by society as true and valid, a unique power, which furnishes its representatives"—or whoever possesses or appears to possess scientific knowledge about the socially available world—with a monopolistic legitimacy in the form of a self-fulfilling prophecy (Bourdieu 1985). But a prophecy about the future is always an act of hubris with regard to those whose future it will (or may) be—the more so since every prognosis "paves the way for a development in just that direction" (Bourdieu 1977, p. 26).

A corollary is that the laws of science, according to Bloor (1982) are not established and propagated on the basis of scientific evidence but because of their perceived public impact, justification and legitimacy. But such a process runs the risk of governments and parliaments being "colonized" (Weingart 2003, p. 98) by a single perspective: one school of thought (e.g. natural succession of the landscape) will exclude others (e.g. preservation of the physical landscape), and the consultation between politicians and scientists—which is in any case subject to other social influences than the pure quest for knowledge—will be exposed to "a constant threat to its primary basis of legitimacy […], the will of the people expressed via the ballot box" (Weingart 2003, p. 92). In any case the very specialization and differentiation of the sciences, each with its own methods and language, already creates a dangerous gap between science and society (see Weingart 2003; Berr 2014; Matheis 2016).

This applies to landscape research in all its subdisciplines, whether in the natural sciences, the humanities, the social sciences, or in architecture and design. Here, too, communication is frustrated—or at least considerably hindered—by the multiplicity of intellectual traditions and specialist languages involved. The result, when it comes to the actual ordering of physical objects in a 'landscape', is that the (often stereotypical) concepts of local citizens are either overridden in favor of ideas proposed by committees of politicians and their expert advisers (whether internal or external to the respective administration), or a third way is taken, a compromise between expert recommendations, what a decision maker sees as personally power-enhancing, and what is perceived, on the basis of personal observation, as a successful model (Schimank 2012, p. 385; Forsyth 2004; Blum et al. 2014).

The consultation process in the multifarious alliances between science and politics is not linear. Science communicates—sometimes circuitously via presentation and public opinion—issues that are taken up by politicians, for example climate change, demographic change, biodiversity, or soil erosion, in order "to forestall either danger or loss of legitimacy" (Weingart 2003, p. 94). In this process, scientists (or at least their leading representatives) may well find themselves in a position to put specific problems on the political agenda—problems for which they themselves will then be invited to suggest solutions. This requires two-way translation between the 'true/untrue' code of science and the 'power/powerlessness' code of politics (see Luhmann 1990, 1997; Weingart 2003; Weingart et al. 2008; Kühne 2013a, 2014a)—a situation in which expertise is necessarily transgressive, for "all experts must overstep their scientific competencies, because all are faced with questions from other fields than their own" (Nowotny 2005, p. 37). This has considerable consequences for scientific statements, for "despite an official face of neutrality flowing from scientific expertise, members of expert panels regularly make moral and political claims and choices" (Hannigan 2014, p. 96; see also Forsyth 2004; Holzinger 2004; Levidow 2005; Hilbig 2014; Matheis 2016).

Scientific opinion—given the usual asymmetrical power relationship between consulter and consulted—is generally sought on condition that evidence be presented in a summary form intelligible to the layperson, preferably with explicit recommendations for action; the detailed evidence will then serve as source material for an eventual decision (see Weingart et al. 2008; Berr 2014; Hannigan 2014). This forces experts to fall back on inexact prescientific language that renders them vulnerable to criticism. The comprehensive lists of plant and animal species often found under their scientific names (along with other technical vocabulary) in landscape planning proposals, will be relegated (if anything) to an annex; the recommendations themselves will contain only general terms like 'mown meadow' or 'mixed woodland'. Moreover, the predilection of the politics-administration-science triangle for recursive structures will often mean that new studies are repeatedly commissioned on the basis of old—frequently as subordinate inquiries in ever greater detail—with the result that they excite ever-diminishing public interest, acquire ever-diminishing funds, and generate ever-increasing competition for those funds, to say nothing of the associated personal animosities among politicians, civil servants, and scientists, which endure in some cases until retirement, dismissal, or even death.

As problems become more complex and their outreach widens, the dilemma of 'expert scientific advisers' becomes more acute (see Beck 1986). For as complexity grows, the scientific position loses its firmness. Funtowicz and Ramirez (1990) illustrate this with the example of anthropogenic climate change: despite global research, the growing complexity of the relations of individuals and societies to the environment has increased the scope of scientific uncertainty and ignorance, and with it the "hypothetical risks" (Fischer 2005, p. 111) to all concerned. The climate-stabilizing mechanisms of the earth's atmospheric temperature are not only complex, they are often embedded in feedback systems; and with a standard measurement period (the gap between readings) of thirty years, it is highly unlikely that prognoses, however expert, will be verified. The consequence is a public loss of scientific credibility: if the danger of global warming is emphasized and no short-term change is noticed, scientists will be blamed for raising unnecessary alarm; if they refrain from giving advice, they will be blamed for neglecting their duty (Weingart 2001; see also Cosgrove and Domosh 1997). One way or the other, the experts' dilemma—in its essence a conflict between opinions (see Nennen and Garbe 1996)—is understood in scientific circles neither as an unintended questioning of one's own position in the sense proposed by Beck (1986) nor as a sign of the growing significance of not-knowing in the sense proposed by Nowotny (2005) and Lerf (2016), but as the failure of decision makers and administrators to correctly implement the 'scientific facts'. For Nowotny the argumentation of climatologists reveals the "typical pattern of the enlightened scientist, along the lines: 'We know what we're talking about, but you'll have to acquire a minimum of scientific knowledge before we can even begin to talk to you!'" (Nowotny 2005, p. 40).

### 4.3.2.4 Landscape, Power, and Administration

Far from basing their decisions in matters relating to landscape on their own knowledge and mature consideration of circumstances and consequences, the legislative and executive as a rule delegate their authority to (qualified but not elected) bodies of experts and/or to organizations working directly in the field (Burckhardt 1978a; Beck 1997; Michelsen and Walter 2013). Until about the mid 20th century this took the form of recruiting academically qualified staff into the various administrations, which could then build up a body of specialist departmental knowledge—with all the conflicts of power and loyalty that went with it. However, as science grew more specialized and society—both internally and in its external relations—more complex, this structure revealed itself as inadequate, and "since that point, appeal to specialist scientific knowledge became necessary" (Weingart 2003, p. 90). The classical model that still largely governs the interface of politics with science reflects this division of labor, viewing planning and decision making as separate functions: "The government commissions specialists to conduct research, feasibility studies, and/or projects, and the specialists present the results of their inquiries and studies, and their various alternative plans, to the government, which then decides what shall be done—that's the received doctrine" (Burckhardt 1974, pp. 72–73).

It is, perhaps, inevitable that the drive for discursive dominance over the physical landscape among the representatives of different social subsystems should entail disallowing the legitimacy of competing arguments and perspectives. The following excerpt from the *Fundamentals of Regional Planning and Development* illustrates this attitude: "The professional rationality and long-term perspective [of the Regional Planning Department], with its carefully balanced consideration of communal interests, is restricted in its decision making not only by self-interested resistance on the part of enterprises and other administrative departments, but also by ineradicable characteristics of the political system and its processes: coalition agreements, questionable political bundling of issues, concerns of individual local politicians due for re-election [...]" (ARL-Regional and State Planning Academy 2011, p. 18).

Far-sighted, professional planners dedicated to the common good are, in other words, frustrated in the pursuit of their legitimate goals by the profit interests of industry and the sectarian interests of (e.g.) departments of the natural or cultural environment, whose members are implicitly accused of lacking rational long-term vision. The political system is disavowed in even stronger terms: its decisions are questionable, its alliances remote from actual needs, and its interests focused on the retention of power—accusations corroborated by von Arnim, who censures politicians for their overwhelming self-interest. Not, he concedes, that they break the law: "they don't even have to—they make the law, and make it to suit themselves" (von Arnim 2007, p. 271). However, in the context of democratically organized communal structures, the demand that planning decisions be left to planners is, to say the least, not altogether without problems, for it amounts to promoting the interests of one subsystem, which enjoys no direct democratic legitimacy, against another. The increasing appeal to lay participation in planning procedures can be seen as an attempt to compensate for this lack of legitimacy (Kühne 2014c; see also Van Assche and Verschraegen 2008).

Von Beyme (2013, p. 13; Michelsen and Walter 2013) see the waxing importance of experts against the background of social change: "The decline of the social classes and the rise of the experts seems to have decisively weakened democratic parties. Disciplinary competence has displaced the enthusiasm of the amateur." A politics "steeped in science" (Jörke 2010, p. 275) has "banished the classical intellectual [...] in favor of experts and planners" (Michelsen and Walter 2013, p. 365). This is combined with an increasingly detailed articulation of social challenges which "in view of the declining ability of political institutions to solve problems, invokes the apolitical measures of administrators, who enjoy the confidence of a civil society bent on output" (ibid. 109). Max Weber understood the remit of politics and administration very differently. His principle of "governance through knowledge" (Weber 1976 [1922], p. 226) is rooted in the distinction between the politician, who seeks a majority for his or her policies, and the administrator, whose job is to execute those policies. A civil servant tasked with political decisions will, however, apply the logic of the administrator and split these into discrete, procedurally manageable (and politically invulnerable) units (see Van Assche and Verschraegen 2008; Schluchter 2009; Michelsen and Walter 2013; Hahn 2014; Kühne 2014c). Sofsky draws the sobering conclusion that the state "can neither protect from material need, nor create jobs and economic growth, [but] lets the transportational infrastructure and education rot [...], while allowing its civil service and quasi-governmental bodies" to grow unhindered (Sofsky 2007, p. 104).

The distinction between science and politics is sometimes reduced to a questionable "rhetorical dichotomy between facts and values" (Pregernig 2005, p. 272; see also Forsyth 2004). In relation to landscape, however, the process of consultation, delegation and decision in which politicians and experts are engaged shows the extent to which the decision making model of politics has been supplanted by a technocratic (and sometimes even pragmatic) approach to managing the issues concerned. Thus preparations for a (political) decision will often be delegated to tiers of administrators, creating a pressure-relieving mechanism that moves conflict risks—e.g. between agricultural interests—down the line to subordinate administrative bodies entrusted, for example, with the definition of specially protected areas (see Burckhardt 1967; Popitz 1992). In this process, the master is the one that "can afford to let others do the dirty work, or can conjure up new offices to produce new knowledge" (Paris 2005, p. 22).

In their empirical investigation of modes of governance and the deployment of power, Sofsky and Paris (1994) distinguish various structures of formal and personal authority:

(a) Official authority is accorded to a position; thus, the minister of a federal state has greater power in the definition and implementation of a landscape norm than the case officer of a subordinate department tasked with executing it.

## 4.3 Landscape and Power

(b) Organizational authority is "a matter of leadership" Sofsky and Paris (1994, p. 69) entailing the distribution of tasks and the overseeing of their execution. In the landscape context, it is connected with the question of where responsibility for that sector resides—in Germany in departments of town and country (rather than regional) planning.

(c) Specialist authority is vested in a person "whose specialist knowledge is attested by others as indispensable for the task in hand" (Sofsky and Paris 1994, p. 51). Above all in exceptional situations, it provides immediate accessibility of incorporated cultural capital. Virtually independent of official hierarchical structures (Krackhardt 1990), such knowledge (e.g. of local botany or zoology), and the authority that goes with it, can be acquired by a junior case officer just as well as by a head of department.

(d) Functional (or operational) authority is vested in the process manager who can successfully break down overall goals into feasible work units. Generally, a middle management position, its equivalent in a German state Ministry of the Environment is the civil servant who drafts funding programs and regulations based on ministerial directives (see Krackhardt 1990).

(e) Finally, charismatic authority is a matter of personality and character. Of crucial importance in leadership contexts, and a key factor in group coherence, it can in principle exist at any level of an official hierarchy, but will excite envy and suspicion in any superior who lacks it.

It is evident from this overview that in performing the "dirty work" (Paris 2005, p. 22) the middle ranks of the hierarchy enjoy considerable authority—the more so, the further the lines of this hierarchy are extended (see Burckhardt 1967). And their power, so far as landscape is concerned, largely lacks democratic legitimacy, inasmuch as consultation with elected organs is (generally for reasons of time) extremely rare. Thus regulations on landscape conservation via such instruments as protected areas, landmarked buildings, or agricultural subsidies derive their legitimacy neither directly from the electorate nor from its representatives in the executive, legislative and judiciary, but as a rule from the same group of experts that applies them (Kühne 2006c). Mosca (1922) described these 'technocrats' as the 'actual ruling class', for they are needed by both ruler and ruled as the medium of transmission and communication without which the social order would collapse (see also Tamayo 1998). But this inevitably results—if only as a by-product—in pursuit of their own interests. Thus Tänzler (2007, p. 114) observes that "the nomination and appointment of representatives" elevates them above the realm of everyday reality. For Sofsky and Paris (1994, p. 164) the problem lies not so much in an inherent leaning toward betrayal as in "their ambiguous in-between position, which means they can only survive in the long run by asserting all-round independence."

This certainly applies in the context of landscape planning, whose officials tend to construct 'practical constraints' which are then—often by appeal to other experts—transformed into apparently 'external' constraints. They "obscure their de facto levels of freedom," Burckhardt observes (1982, p. 106), "by referring to aspects they cannot themselves decide and lamenting this as 'beyond their control'"—a tactic which, in the absence of any code of governance, effectively preempts evaluation or revision by civil society or its delegates. In his study of the development of landscape and transportation infrastructure between 1930 and 1990, Zeller (2002, p. 411) speaks of the "shift in the ideological current, from the politically charged, mythically staged projects of a dictatorship incorporating nature itself, to an ostensibly apolitical sphere of rapid and unhindered circulation of goods and persons under the dominance of academically trained experts" (for a similar analysis see Illich 1979; Dingler 1998; Michelsen and Walter 2013). The result of this dominance is a definition of targets and measures whose lack of transparency (Popitz 1992) often hinders assent by outsiders—a pertinent example being the concerted discovery of protected lichen species along the course of a projected major road without any further mapping of the species in the adjacent area.

Democratic communication between experts and laity is in general fraught with what are seen as the "middle-class presuppositions of planners, their ill-considered myths of intuition, and their ideological severance of means from ends" (Fezer 2006, p. 13). Walgenbach adds to this list the "myth of rationality," and continues that "myths in this sense depend on a shared belief that exempts them from objective examination" (Walgenbach 1999, p. 66; see also Hahn 2014), making them ideal instruments of exclusion for every eventuality that might disturb the decision making process. Berger and Luckmann (1966, p. 111) speak similarly of "elaborated myths [...] that strive to eliminate inconsistencies and maintain the mythological universe in theoretically integrated terms." Many patterns of judgment and action established through secondary socialization and professional practice, whether work- or status-related—e.g. the imposition of restrictions on public access and use)—are in this sense remarkably (one might even say mythically) self-referential. Nor is any fault perceived in this, for professional recognition is accorded in a "process of *mutual exchange* between the privileged" (Popitz 1992, p. 198, original emphasis)—a body steeped in explicit and implicit self-confirmation that extends to experts both within and beyond the ranks of officialdom.

A corollary of these structures is the undercurrent of advantage to be found in smooth decision making processes, for the greater the discrepancy between expert opinions—whether within a profession (e.g. cultural landscape conservationists vs. environmental successionists) or across professions (e.g. biologists vs. landscape architects)—the more weakly institutionalized is the fiction of rationality (Schimank 2012, p. 385; Saar 2010) and the more isolated the decision maker, who can scarcely delegate anything to disagreeing parties. Hence there is a congruence of interests among politicians, administrative specialists (e.g. heads of landscape planning departments), and external specialists (e.g. professors of geography) to find common ground early in the proceedings (see Dingler 1998; Michelsen and Walter 2013). Various permutations are likely in this context (see Fig. 4.10):

## 4.3 Landscape and Power

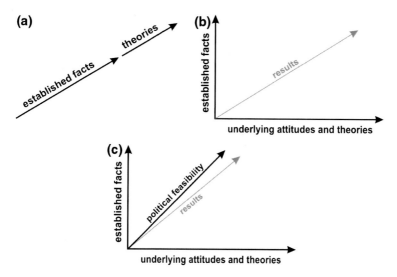

**Fig. 4.10** Traditionally, theory is thought to originate in empirical investigation (**a**). Latour (1999) sees the solution to a scientific question as also depending on the underlying attitude and convictions of the scientist (**b**). Following Horkheimer (1977 [1937]), Habermas (1982), and Latour (1999), Kühne (2008a, b, c) sees applied research as further subject to political (often opportunistic) feasibility

- all round agreement, promising a smooth transition from (scientific) knowledge to power
- agreement between the decision maker and external experts, to which the administrator must submit, but whose implementation he or she can ex officio effectively obstruct
- agreement between administrative officer and expert, who then inform the decision maker, with the risk that their recommendation may be rejected on other (e.g. political) grounds.

The shift from mode 1 to mode 2 knowledge production is reflected in a power shift among experts, administrators, and politicians. In the mode 1 era, experts were existentially and socially—and hence also in their judgments—largely independent of politicians and administrators. In the mode 2 era this has changed: experts depend today, in every respect, on third-party finance, which in the landscape sector means as a rule public sector funding allocated by administrators and politicians. As the research contracts of specialists are invariably for the term of the particular project, these experts are inherently interested in the prolongation of the project or the granting of a new or follow-up project. Moreover, the social status of university professors is now measured not only in terms of research publications and (in the landscape-related sector) by the quality of teaching, but also by success in acquiring third-party funding. (This is a frequent stipulation for professorial appointments, where "relevant research experience including the acquisition of third-party funding" is called for.) Accordingly, the simple requirement of mode 1 research that underlying attitudes and theories concur with empirical results has yielded in mode 2 research to a situation where political feasibility (or indeed opportunism) must also be taken into account if the researcher is not to forego the availability of further funding (see Fig. 4.10). In the concrete case of landscape-related projects, if a certain modification of the appropriated physical landscape seems politically inopportune, it will not be investigated: it will be rejected ab initio either by the provider or by the recipient of the putative funding (see Dingler 1998; Kühne 2008a).

Rooted in the symbolic dominance of the state, the definition and implementation of landscape targets by experts can in the final analysis be described as a corporate discourse revealing every dimension of power defined by Popitz (see Kühne 2006c, 2008a, 2015a; Sect. 2.3):

- Technical (structural) power resides, for instance, not only in the definition of standards (e.g. selection for IUCN Red Lists) but also in the increasing ability, with the aid of modern technology, to create or change landscapes as expressions of symbolic power. Contemporary examples are Palm Jumeirah Island in Dubai and the renatured opencast coalmining landscapes of Germany. Popitz cites the example of residential estates whose "planners and developers take decisions that affect the living conditions and spatial and environmental constraints of many people" (1992, p. 30). Building "worlds for others", they determine spatial functions and implicitly draw lines of segregation not only according to income but also (and increasingly) according to the milieu an area represents (see also Duncan and Duncan 2004).
- Authoritative (directive) power is evidenced in the incorporation of landscape-related concepts in both primary and secondary socialization—e.g. in the virtually sacred construct of an "ecologically intact, beautiful landscape, close to (untouched) nature" (Wöbse 1991, p. 34).
- Instrumental (persuasive) power reveals itself e.g. in the declaration of protected areas with corresponding restricted usage and even expropriation—an act that Schneider (1989) characterizes as "the ultimate step of ideological landscape conservation" securing for the superior end of nature protection what legally and factually belongs to another.
- Active (coercive) power is evident, for example, in police enforcement of restrictions to the use of protected areas.

The targets defined by experts can be implemented administratively by either threat or incentive. A threat, Paris observes, is only effective when "one can perform the threatened action (or have it performed) if the other party continues to resist" (Paris 2005, p. 39). As a conditional act, it binds both parties to a certain course of action. In terms of symbolic capital, it costs little if it is successful but a good deal more if it is unsuccessful (see Schelling 1960). The prohibition to enter a nature reserve, for example, coupled with the imposition of a fine, will only be credible if, upon breach of the prohibition, the fine is actually incurred; but this entails administrative costs both in surveillance of the nature reserve and in processing the fine. Incentives "invert this structure by offering reward and gratification" (Paris 2005, p. 41) to their target group, so they are cheap when rejected but expensive when accepted. In principle, they are more likely to evoke loyalty, but they are not entirely free from difficulty, because once they are accepted as the status quo, any restriction imposed to save resources may cause conflict (see Paris 2005). An example is the EEC agricultural subsidies of the 1950s and 1960s, which aimed to boost crop production in Central and Western Europe in order to provide the population with ample food at affordable prices. But overproduction in and after the 1970s led, within the framework of the so-called Agenda 2000, to medium-term cuts coupled with restructuring of the program—measures which in 1999 and 2003 met with vehement protest from agricultural organizations (see Nassauer and Wascher 2008).

The role of political action outside the constitutional sphere of the legislature and executive has been intensely researched in recent years. Crouch speaks of a "post-democracy" (Crouch 2008, p. 13) in which the formal institutions of democracy persist but the power of the traditional political parties is steadily dwindling. In this situation the political life of a country is increasingly shaped on the one hand by the media, Internet and social networks and on the other by a privileged elite (see Leggewie 1998; von Beyme 2013; Michelsen and Walter 2013; Swyngedouw 2013). The upshot is that "individual citizen-stakeholders lose any say in the proceedings" in favor of NGOs and expert groups who "enhance the aura projected by politicians that their decisions lack any alternative" (Michelsen and Walter 2013, p. 79). The expertise of such groups serves in the final analysis to maximize both output and its acceptance (Michelsen and Walter 2013; Kühne 2008a). The loss of power of the elected representatives in western nation states to experts inside and outside the bureaucracy, as well as to societal actors outside the system of politics and administration, is only one side of the loss of centered national power. The other side of the coin is the strengthening of transnational units (such as the European Union, the United Nations; see for example Beck 1986; Chilla et al. 2016), but also sub-national units (Allmendinger/Haughton 2012; Paasi 2009). The consequence of this is, on the one hand, an increasing complexity of the inscription power in physical spaces and, on the other hand, an equally increasing complexity regarding the sovereignty of interpretation over what is called landscape (see Weber 2015).

### 4.3.2.5 Landscape and the Actions of the (Relatively) Powerless

The mighty stand in polar contrast to their followers, the lord to the vassal, the leader to the led (although the mighty in other contexts may also be powerless—in the dentist's chair or at a frontier, their institutionalized power avails them little). The motives and structures of leadership have been far more extensively researched than those of the follower. An exception is the sixfold typification proposed by

Paris on the basis of a person's "level of internalized compliance, the intensity of their will to follow" a particular leader or program (Paris 2005, p. 104):

(a) The enthusiastic follower is passionate, unconditional, and unreserved—a type often found in more or less autonomous local grass-roots initiatives (see Brand 1999), especially during the initial phase of their constitution, which has been characterized by Kuphal (2006, p. 39) as one of "fiery love and solemn oaths." Large, traditional, environmental and landscape protection groups—also local ones—are less likely to inspire this level of emotion.

(b) The committed follower, motivated more by rational interest than enthusiasm, is commonly found in the lower administrative ranks of landscape-related organizations, whether corporate or governmental (see Brand 1999). The transition from enthusiastic to committed follower is often evident at the point when groups and associations are professionalized, not least because the availability and management of public funds brings with it a need to strengthen structures and clarify lines of competency and responsibility. This tends to dampen overt enthusiasm (Kuphal 2006). Moreover, professional structures offer opportunities for advancement to those whose commitment is firm and rational.

(c) Habitual followers are not really committed to an idea or program, but follow because they are used to doing so: for them, "achievement of the goal is only of mild interest" (Paris 2005, p. 105). The old nature protection associations and the new, ecologically oriented environmental groups—as well as party-political organizations (Brand 1999)—have many such members. They attend major events (like AGMs) and may join in the occasional activity (e.g. landscape conservation), but "not so as to excite notice or confrontation" (Paris 2005, p. 105). This is the default mode for both enthusiastic and committed followers whose motivation has (for whatever reason) flagged, and their numbers are legion in major organizations and in political parties, which continue to run for hardly more reason than that they have been there for so long that "their continuation seems somehow a duty" (Kuphal 2006, p. 40), as well as in public authorities, whose officials perform their tasks out of a sense of duty rather than desire, and certainly no longer from personal ambition.

(d) The automatic follower is characterized by the simple drive to obey without reflection. Found above all in organizations with linear hierarchies (e.g. public authorities), they "function, rather than follow by an act of the will" (Paris 2005, p. 105). An example might be the landscape officer in a local administration who simply 'does her duty' with no real interest in its content, implications, or consequences.

(e) The unwilling follower is one with individual ideas, but who has to follow directives because "not to do so would involve high personal costs" (Paris 2005, p. 106). Such persons are to be found in associations and networks, as well as in public bodies, and especially in this latter case, their disruptive potential is considerable. While they are bound by office, and frequently also by the obligations implicit in the comprehensive security provision they enjoy, they nevertheless often find themselves in disagreement with their superiors on job-related as well as political issues—for example the funding of landscape or village renewal projects, the siting and construction of power stations or transportation infrastructure, the planning of new residential areas etc. Paris remarks, however, that their integration in a rigid hierarchical structure, mitigated by the awareness that a political superior may only last until the next election, means that their "constant grousing and grumbling" is of no consequence (Paris 2005, p. 106).

(f) The protesting follower is one who is vocally unwilling, who considers the plan in question to be wrong, but still follows it, because there is no turning back: too many obligations toward the organization or superior have been accumulated for any other course to be thinkable. In the landscape context, this structure can be found in any type of organization, but in linear hierarchies its impact is greater than in more loosely organized, less institutionalized bodies.

Committed, habitual, unwilling and protesting followers are characteristically people who "half believe" (Paris 2005, p. 110). Half-belief is a compound of belief and its opposite: "the semi-believer both believes and does not believe one and the same thing" (Paris 2005, p. 110), albeit with different intensities. While the enthusiastic follower believes wholeheartedly in every aspect and consequence of the program, the automatic follower neither believes nor disbelieves, but simply follows. In the committed follower, belief outweighs disbelief; but in the unwilling or protesting follower, the opposite is the case. The habitual follower both believes and disbelieves, while ignoring the internal dissonance of that position.

A key characteristic of half-belief—at least in public discourse—is its innate tendency to dogmatism, coupled with highly selective perception and a correlative demonizing of opposing positions. The complexity of landscape as a phenomenon, in its social as well as ecological dimensions, is typically reduced to the semi-believer's professional interest as a representative of farming, hunting, or heritage conservation agencies, or as an official of the forestry, tourist, or nature conservancy board—bodies whose social, cultural, aesthetic, or ecological importance is hammered home in threadbare phrases, while other aspects of landscape are treated with suspicion, granted only ancillary importance (e.g. the role of agriculture in preserving tall oatgrass meadows), or ruled out altogether. Such discourse is ultimately self-referential and self-confirming, underpinning the speaker's own position, lest doubt arise as to its worth (see Paris 2005). Critical scrutiny of one's own position and that of other semi-believers is avoided or over-compensated by externalizing responsibility to distant "centers of evil" (Paris 2005, p. 14) and/or by personally discrediting an opponent as "having no idea of farming, hunting, forestry, or heritage conservation—after all he/she is not a farmer, hunter etc." As if only the farmer had a right to talk about farming, the hunter about hunting, or the politician about politics—a circular argument whose absurdity Latour (1999) has formulated in the postulate that only rats should talk about rats, frogs about frogs, houses about houses, and electrons about electrons. Furthermore, the lack of empathy shown by other semi-believers (oneself always excepted) is compensated by a hyperactivity (Paris 2005) that combines selfless commitment to one's own group, program, or leader with time-consuming projects that leave no room for reflection or self-questioning. Precious leisure hours spent on excursions explaining the significance of the physical landscape or the danger of scrub encroachment fostered by mowing threatened meadows leave no time to ask if one is not, in fact, treading the well-worn path of Don Quixote.

Half-belief is also a typical attribute of the mayors of small to medium-sized municipalities who, in the absence of any distinctive feature in their territory, will indulge in cliché-ridden advertisement, personalized conflict against other municipalities and superordinate bodies (e.g. 'the minister'), emotional moralizing when a neighboring town is praised in a publication that ignores their own, hyperactivity in attending clubs and small-town fêtes, and apathy toward developments they cannot control, like the consequences of demographic change (see Paris 2005; Kühne 2006a). Given the restricted power enjoyed by the heads of such municipalities, landscape constitutes an ideal symbol of energetic leadership: the creation and funding, enabling and conservation of physical landscape structures demonstrates a will-in-action that all can see. As a stage for self-projection, landscape eminently fulfills Bourdieu's axiom that a measure counts only when it is announced, and as soon as it is announced it counts in the public mind as realized (Bourdieu 1985; see also Jain 2000; Michelsen and Walter 2013).

The relative powerlessness of civil society in matters of landscape typically reveals itself not only in the phenomenon of half-belief but also in the frequent rejection by citizens' initiatives of actual and target states as defined (see Sect. 5.4). Resistance of this sort to planning projects was initially "understood as a lack of appropriate explanation: one only had to inform citizens of the benign intentions of the planners, and they would immediately realize that certain incidental shortcomings must be accepted for the greater good" (Burckhardt 1978a, p. 97). Burckhardt points out, however, that the real bone of contention often lies not in the individual measures but in their arbitrary division into means and ends, and the exchangeability of these two logically disparate levels. If the declared end is to widen a street, the demolition of adjacent houses can be presented as a necessary means. However, the real end may be to build stores and office blocks, and the widening of the street is just a means to that end. Burckhardt concludes, "power is the power to define things as means or ends. Resistance arises where the ideological character of this division is revealed" (Burckhardt 1978a, p. 97).

Even if the relatively powerless cannot as a rule prevent the implementation of landscape targets, they can hinder it, especially by drawing public attention to their claims. Here (and sometimes even in contention between the equally powerful) the principle of the double bind is often invoked. Applied in various forms and on various levels, the double bind embodies "the demand for two or more courses of action that are in fact self-contradictory but each of which carries a threat of sanction, so that any reaction to the initial demand can be deemed inappropriate" (Paris 2005, p. 45). The method enables its user to be right at all times—a gratifying perspective in situations laden with high emotion (e.g. recrimination). Politicians, for example, can be labeled incompetent and irresponsible and at the same time called upon to deliver full employment and to radiate altruism, authenticity and credibility (see Paris 2005; Tänzler 2007). The reality constructed by a double bind immunizes its user against arguments from data or experience; its semantic structure is so constructed as to be (on its own logical level) unfalsifiable.

A pertinent example is the plan published by the Saarland to create, in cooperation with local municipalities, a UNESCO biosphere in Bliesgau (the south-eastern part of the state). The plan has met with resistance from farmers who fear restrictions on their use of natural resources (see Hussong 2006; Kühne 2010; Nienaber and Lübke 2012). When it is pointed out that the state government intends no such restrictions on agriculture, the rejoinder is simple: 'The current government may not, but what about a future government?' This is an ecological Cassandra painting in glaring colors a damage, which, she asserts, is only unreal inasmuch as it has not yet happened. A corollary directed at politicians in general is that they all lie, so why should one believe them in this instance? The conundrum is not helped by attempting to be factual, for this is interpreted as talking the danger down, which merely confirms one's lack of credibility (see Paris 2005).

Protest against the workings of symbolic power in the landscape can be physical as well as discursive, ranging from planting one's front garden with xenophytes (in the German context thuja cypresses etc.) in flagrant disregard of a municipal target, to vandalizing protected natural or cultural monuments. A common form of (especially urban) cultural protest is the omnipresent graffiti, which, despite their arguable claim to creativity, are often perceived as a violent confrontation with official culture and majority taste. Paris comments: "While art leaves the public free to approach it or turn away, graffiti occupy the field of perception and leave the passer-by no choice" (Paris 2005, p. 138). Like the soundscapes of rocker subculture, graffiti transgress the received norms that shape public space, symbolizing their slow loss of authority and relativizing what we hold in common (Popitz 1992). Yet at the same time, they represent a 'landscape of resistance' which has so far escaped the attention of scholars (see Palang and Sooväli-Sepping 2011). The graffito shown in Fig. 4.11, for example, expresses the desire for beauty while ironically contradicting—not only by its presence on a public wall, but also (and in particular) by its crude representation of a hemp leaf—the conventional ideal. On the other hand, the passer-by schooled in philosophy may see in it a critique of mainstream Western aesthetics, with its time-worn union of 'the good, the true, and the beautiful'.

### *4.3.3 Landscape, Social Capital, and Power*

As members of the "state intellectual apparatus" (Ó Tuathail 1996, p. 61) and producers of socially appropriated landscape, landscape experts possess an incorporated, institutionalized, and objectified cultural capital that is secured in all its dimensions through stable networks of (more or less) institutionalized mutual awareness and recognition. And as these resources "derive from membership of a

**Fig. 4.11** "The world is good, but is it also beautiful?"—graffito at Kirkel train station, Saarland (*Photo* Olaf Kühne)

group" (Bourdieu 1983, p. 63), their cultural capital also constitutes social capital. The less institutionalized networks include ritualized conversations about colleagues, informal get-togethers, and casual meetings at conferences etc., all of which serve to accumulate social capital and corresponding distinction (Adler and Kwon 2002). Like all such professional networks, these convey a privileged position in the production and distribution of information, so that experts have "a better chance of organizing themselves quickly and effectively" (Popitz 1992, p. 191) than either non-experts or experts outside such networks. Their interests are "not necessarily more intense, but they have *higher organizational potential*" (Popitz 1992, p. 191; original emphasis) than other groups. This process of "networking democracy" (von Beyme 2013, p. 13) is a proven tool for pushing through individual (or in-group) interests. Moreover, in this context Weber's principle of "dominance through knowledge" (Weber 1976 [1922], p. 226) is pervaded by patrimonial elements based not on an official hierarchy but on loyalty to the accepted leader of the clan (see Scott 1986).

### 4.3.3.1 Social Capital in Academic Landscape Discourse

According to Bourdieu, the structure of an academic discipline reflects the forces currently operating between the various protagonists in that field—or more accurately between the forms of power available to them personally and above all by virtue of their institutional affiliation (Bourdieu 1985). And this power derives, as has already been remarked, from the availability of symbolic (especially social) capital. The landscape-related disciplines of geography, landscape planning and landscape architecture (as well as sociology) inhabit a typically practice-oriented, theoretically 'impure' position somewhere between the natural sciences and the humanities. In her feminist psychoanalytic study *Love of Power—Landscape Conservation and Dispossession* (1989), Schneider applies Adorno's (1987) concept of the 'jargon of authenticity' (or uniqueness) to these disciplines, which, she asserts, are inadequately understood by their professional representatives. Because of this intellectual shortfall, their protagonists resort to a "radical projection of landscape conservation [as] the reinstatement of a lost paradise," presenting it in a jargon of "religious promises" with profoundly malignant effects (Schneider 1989, pp. 3–4), above all that of expropriation. Schneider (1989, p. 30) goes on to argue that in garden and landscape design the paradise motif takes on the historico-cultural role of a "redemptive religion"—to which Paris adds that the accumulation of exclusive social capital endows the exponents of that 'surrogate religion' with a "modern version of infallibility" (Paris 2005, p. 114). Landscape is reconstructed, in this view, as the female object of its male exponents (see also Weber 2007; Sect. 4.2.5).

Examples of strongly institutionalized networks are academies and professional associations like the *Bund Deutscher Landschaftsarchitekten* (Federation of German Landscape Architects—BDLA), which formulates its tasks in the following terms: "The BDLA engages in public relations for the profession and represents its views and interests in political, administrative and business circles. As well as providing training and ongoing education opportunities, it lobbies at both federal and state levels for the consolidation and expansion of the professional field and the updating of fee schedules. The Federation offers its members a platform for exchanging views and experiences, for collaborative ventures, and for the enhancement of professional commitment" (BDLA 2006b). In this mission statement, the accumulation of social capital is presented as a clear goal alongside the deployment of corporate influence. That this has not changed in the past ten years is evident from the unaltered wording of the paragraph in the organization's 2016 website.

As well as generating social capital, academies, federations, guilds, institutes, and associations tend to concentrate power in the hands of those appointed to represent them, and hence to control the interface between experts and society (Lahnstein 2000). This power, which is "unrelated to the importance of the individual" (Bourdieu 1983, p. 68), can be used not only to underpin the relevance of the organization's goals, but also to dissipate ad hoc notions of (landscape) development before they even reach the public sphere. In both ways, it secures and enhances the loyalty of its members (Paris 2005). Conversely, deviance from its norms brings with it a withdrawal of social recognition and concomitant distinction (see Adler and Kwon 2002; Kühne 2008a).

With membership limited by election, academies are more highly institutionalized than many other professional networks, and this structural rigor gives them greater leverage with cultural capital. Conversely, it restricts their freedom, for "they are subject to the conditions imposed not only by their own situation and form of organization, but also by the grammar of power itself" (Sofsky and Paris 1994, p. 16; see also Krackhardt 1990). Membership of any of these networks requires a minimum of institutionalized cultural capital and compliance with the organization's statutes. It is conveyed in specific "rites of institutionalization" (Bourdieu 1983, p. 65) which express the acquisition of new social capital: presentation of certificates, invitations to exclusive congresses, and documentary evidence of the new status alongside a mini-biography on the candidate's institutional website. All of this boosts loyalty to the profession and its institutions—a loyalty rewarded by acceptance into the group of 'frequently cited' authors (Hard 2002d [1971]). The way to such distinction has many steps and "to earn that accolade, rigorous tests must be accomplished. Progress is noted and assessed, and the inquisition of knowledge is at the same time a ceremonial demonstration of power" (Sofsky 2007, p. 134). In bodies like academies, with elected membership, the rite of entry implies elevation to an elite status; for, as Paris tells us, elites are "recruited from above, unlike the authorities and leaders that control us politically, whose power comes ultimately from below" (Paris 2005, p. 83). Duty and group loyalty are reinforced by subjective feelings of recognition, respect, and friendship, as well as by the already noted institutional guarantees (Bourdieu 1983; see also Hard 2002d [1971]).

The relation of authority within a network rests on a twofold process: on the one hand "recognition of preeminence" and on the other "the desire to be recognized by the preeminent" (Popitz 1992, p. 29). Crucial to the stability of any hierarchy (Bourdieu 1985; Sofsky and Paris 1994; Hilbig 2014), such relations are reciprocal but asymmetrical, inasmuch as they incorporate "the fixing of our desire for recognition on [...] a superordinate person or group" (Popitz 1992, p. 115). The power of the academic pupil generation, for example, vis à vis their teachers is informed, but at the same time limited, by the (by no means altruistic) desire of the teachers to have pupils appointed to good positions (Bourdieu 1992). "The forces at work in such relationships are ineluctably conformist; for to accept an authority is to accept the *values* it stands for" (Sofsky and Paris 1994, p. 26; original emphasis). Hence, the junior professional will necessarily follow the mainstream, as defined by those of higher status: it is "their perspectives and criteria, [...] their expectations that we take on board. Our sense of self is bound to their recognition and its withdrawal" (Popitz 1992, p. 133; see also Eisenberg et al. 1998). Drawing on the analogous concept of status closure, the process of building an academic and professional elite can in this sense be understood as one of institutional closure (see also Weingart 2003).

## 4.3 Landscape and Power

As a rule, extensive symbolic (especially sociocultural) capital engenders considerable "theme-setting power" (Paris 2005, p. 31)—i.e. power "to determine what may be said in what circumstances by whom and with what consequences" (Paris 2005, p. 31). The placing of an issue on the agenda entails stating a preference for a specific group (see Hard 2002d [1971]; Burckhardt 1974), an action that "requires the aura of necessity" (Burckhardt 1970, p. 48). In a discipline whose representatives are elected on the basis of reputation—and the "spatial [i.e. landscape-related] disciplines" are such (Prigge 1991, p. 105)—it is the supremely reputable who define what can (and cannot) be said. They "give scientific coherence to the spontaneous ideologies of the appropriated spaces of daily life, thereby securing the cohesion of spatial practice. Their mechanisms of exclusion—determining who may legitimately speak about space [and/or landscape (O.K.)]—formulate the dominant mode of such discourse and the power over its object which this entails" (Prigge 1991, p. 105; see also Graham et al. 2000). A particular intensity of the attempt to the sovereignty of definitions can be observed when the specific *déformation professionelle* in relation to an 'object' meets the 'desire for impact': "Regional geographers were deeply involved in power-knowledge relations when creating bounded 'orders' on the earth, fixed in apparently neutral maps and texts that identified separate regions" (Paasi 2003, p. 476). The process of generating and disseminating patterns of interpretation of spatial identity is a process of hierarchical communication: "The construction of identity narratives is a political [and administrative! Note O.K.] action, and, particularly in the case of national identities, this activity is an expression of the distribution of social power in society" (Paasi 1999, p. 11).

Leaving aside the political executive, this power is exercised through the corporate groups that run academic journals, convene conferences, and determine who speaks at them, or what textbooks are 'required reading'. Examples can be seen on the one hand in the recent predilection for topics connected with landscape and the local (home) environment, and on the other in the virtual banishing as 'damaging to the profession' of critical articles on landscape conservation or geography (see Hard 2002d [1971]; Böse et al. 1981 on the Federal Horticultural Show in Frankfurt). Other (especially critical) voices are prevented altogether from gaining a hearing within a profession in which coveted jobs go only to those whose voice is heard (see Bourdieu 1985).

These processes amount to a sort of consecration of the elect, in which university professors endowed with high symbolic (especially social) capital compile disciplinary syntheses whose outreach extends across secondary as well as tertiary education. Of these Bourdieu (1985) remarks that derived from lectures and intended for use in other lectures, they all too often reflect only the received state of knowledge—a condition that may well suit teachers at every level, for whom the thwarting of new notions offers escape from the threat of obsolescence. A clear sign of this strategy is the appeal to obsolete data and literature—as with the stereotypical characterization of the 1960s as a 'social wasteland', which lasted into the 1990s.

Conversely, the "structure of social self-esteem" (Popitz 1992, p. 118) is determined by publications in reputed journals, proceedings, and textbooks, with their institutionalized (and at least partly anonymized) power traps—above all in the form of peer reviews. Weingart sees the danger of abuse of reputation implicit in this procedure as the greatest risk to scientific integrity, for the declared purpose of peer reviewing—to protect the scientific community from "pointless, false, and fraudulent voices" (Weingart 2003, p. 33; see also Fröhlich 2002)—is, according to Hirschauer, frustrated by the sheer unreliability of the procedure: "Most publications in the social sciences (where rejection rates are high) derive from decisions whose divergence from pure chance is so small that peer review procedures could almost as well be decided by a throw of the dice" (Hirschauer 2004, p. 69). On the other hand, as Smith puts it "peer review might be described as a process where the 'establishment' decides what is important. Unsurprisingly, the establishment is poor at recognizing new ideas that overturn the old ideas" (Smith 2010, p. 3; and see Weingart 2001). Proponents of new ideas and paradigms are, then, compelled to launch their own forums, and/or to present at least the appearance of conformity— or indeed, in many cases, to seek alternative employment outside academia (see Kuhn 1973; Cosgrove and Domosh 1997).

An example of paradigmatic change in the landscape-related sciences is the transition that took place in German geography, after the pivotal 1969 Kiel Geographers' Conference, from a classical geography based on countries to a spatially based model of economic and social geography (Hard 2002a [1969b], 2002b [1971]; Bahrenberg 1996; Werlen 1998; Blotevogel 1996, 2000). Exemplifying Bourdieu's (1985, 1998 [1996]) concept of a permanent scientific revolution, this overturned an approach seen as empirically unprovable, methodologically unsound, and overly susceptible to ideology. Based on an "oversimplified realism" (Kaufmann 2005, p. 102), traditional geography had focused almost exclusively on the local and regional, excluding supraregional influences (Mitchell 2005; see also Eisel 1980). Its world-picture "was that of a well-ordered mosaic of spatially segmented natural and social entities […] that bracketed out not only the increasingly important contexts of inter-spatial influence, but also the controversies inherently associated with spatial constructs" (Blotevogel 1996, p. 13). The paradigm of landscape was largely replaced (especially in anthropological geography) with an empirical neo-positivist approach, with the result that "in many mainstream areas of anthropological geography it could damage one's career to speak of landscape at all" (Schenk 2006, p. 17). On the other hand, in physical geography (oriented more on the natural sciences) the landscape construct persisted, and the terms 'geoecology' and 'landscape ecology' were retained in connection with ecosystems approaches.

Any wholesale rejection of critical voices is an abuse of power. Limpet-like adherence to an outdated paradigm, combined with the routine imposition of negative sanctions, leads to the suppression—or belated treatment—of critical issues central to the discipline. An example is the postponement in Germany until the 1980s of serious discussion about landscape and spatial planning under National Socialism. This runs clean counter to the scientific ideal of producing new knowledge (Luhmann 1990). The instinct behind it can be analyzed in line with Sofsky and Paris' classification of structures of authority, according to which technical (structural) power is manifest in the vital economic and emotional dependence of subordinates for future employment, promotion, or simply keeping a job. Following the same classification, the power of official authority is invested, for example, in the president of an academy or member of the advisory board of an international journal; the power of organizational authority in the convener of a congress or editor of a series; the power of functional authority in the person who introduces the results of a working group into wider discourse; the power of specialist authority in the one who possesses profound knowledge in a specific sector; and the power of charismatic authority in the talented and convincing communicator (Sofsky and Paris 1994; and see Sect. 4.3.2.4).

Popitz (1992) distinguishes five facets of the need for recognition—or types of 'social subjectivity'—applicable to relations of esteem and self-esteem in academic landscape (and similar) contexts, and to the self-disciplining, they impose:

(a) The need to be accepted as a member of a specific group—to be "like others, to be 'in'" (Popitz 1992, p. 141)—applies both to educational qualifications (university degrees etc.) and membership of professional organizations (in Germany e.g. the Association of Town, Regional and State Planning; the Federation of German Landscape Architects) or academies (e.g. the German Academy for Urban and Regional Spatial Planning). The sense of belonging to a group with a common world-view (and hence 'reality') is a basic social experience entailing self-confirmation.

(b) The need to be accepted in an assigned role is only indirectly relevant to the type of sociocultural capital generated in such academic contexts, where age, gender, background, and in some cases social standing may determine the allocation of roles. Thus, honorary chairpersons will often be senior members who have already gained high honors in the group, women will be expected to volunteer to take minutes, and men with a certain bearing will be invited to represent the organization.

(c) The need to be accepted in an acquired (especially professional role) depends critically on those of higher status who grant the role and approve (or disapprove) its performance. This applies to freelance workers (who need commissions), to state and municipal employees, and to the elected board members of professional associations or academies. It discourages people from making (or supporting) deviant statements.

(d) The need to be accepted in a public role (e.g. within a professional body or scoping context) requires visible leadership and approbation from the group. Too great a deviation from public expectations will inspire controversy and diminish self-esteem.

(e) The need to be accepted in one's individuality runs potentially (and sometimes openly) counter to other roles and needs. Rooted in the need for recognition in the "singularity of [one's] existence" (Popitz 1992, p. 149), it is limited by the requirements of other roles—or will else face sanctions. Thus, the position of a landscape successionist in an organization dedicated to the preservation of the "historical cultural landscape" may well become de facto untenable.

In conclusion, Weingart's general strictures on academic discourse may also be taken to apply to the landscape-related disciplines: "Success in the production of scientific knowledge is not simply a matter of the 'truth' of this knowledge or of its power to convince. It requires the clever manipulation of the relevant networks—including persons, technologies, and natural objects—in order to gain support for one's ends. Only when such networks can be stabilized will knowledge proposed as true (e.g. an innovative theory) receive social recognition" (Weingart 2003, p. 72).

In the landscape context social recognition is clearly measurable by the extent to which public authorities are prepared to base political action on the 'truths' propounded by academics, and the formal and informal scientific groups and organizations associated with them. However, the academic world presented here should not be thought of simply in the dichotomous terms of progressives versus conservatives. It reveals, rather, what Bourdieu calls 'the coexistence of several independent hierarchical principles' (Bourdieu 1985) relating to different forms of capital and their accumulation: economic capital (salaries), social capital (networks inside and outside the university/profession), and cultural capital (publications, lectures etc.—see also Kühne 2006d).

### 4.3.3.2 Social Capital Within an Informal Hierarchy

Within the relevant formal hierarchies—for example public authorities—the social capital of landscape experts of various professions generates parallel organizational structures that can lead to conflicts of loyalty between corporate/professional and official roles (Krackhardt 1990; Molina 2001; Kühne 2008a). Moreover, it is often the professional group—the association or academic body—that wins: one even hears people say "I think of myself as representing the interests of XY association in the ministry." As already observed (see Sect. 4.3.3), informal professional networks validate this observation inasmuch as only those projects, as a rule, reach the stage of political decision that have already been passed by the relevant networks as not impacting (or in the worst case only minimally impacting) their established cultural capital (see also Krackhardt 1990). The corollary is that success in the public sector is awarded on the basis not of money but of 'what colleagues think'. Hence Burckhardt's admonition: "You can only do what's possible," where possibility in a public authority means what does not interfere with anyone else's plans. For "at least on paper, to cross a colleague's plan causes far more problems than to pull down a few private houses" (Burckhardt 1982, p. 105).

Informal agreements among middle-ranking officials on the one hand minimize the risk of presenting legally questionable or technically faulty concepts to the political decision makers; on the other hand, however, these networks are not subject to any form of democratic control and harbor the risk of bureaucratic sclerosis. In both respects they may well function as stabilizing factors in an inherently hierarchical system, but democratically they nevertheless represent a stumbling block, for they in principle "leave no room for politics in the sense of a normative expression of the will, relegating it to the ancillary level of a stopgap for imperfections in a basically technocratic state system" (Schlesky 1965, p. 457; see also Bourdieu 1992b).

Unofficial networks of status and recognition operate not only in the question whether or not a project reaches the point of decision, but also in how it is then to be realized. The activities of landscape experts are as a rule integrated in official (political) systems, with their positive as well as negative sanctions, and oriented on the perceived needs of the population (or at least on political interests), but the actual execution of the project will follow the accepted patterns and standards of the profession. After all, neither politicians nor people see a landscape with the eyes of the professional or are versed in the differentiated code of the landscape specialist. In this context, Burckhardt cites the example of architecture: "The building itself [and also its plans (O.K.)] is a fount of information, it expresses something, but its message is unclear. The most public of all art forms, architecture, like modern painting, literature, and music, is addressed first and foremost to connoisseurs, to colleagues, to the readers of specialized journals" (Burckhardt 1978a, p. 89).

Depending on its specific area, landscape discourse is, in this view, a more or less hermetically sealed subject, despite the fact that both its output (physical objects, research results) and input (financing) are located in the public sphere. Such systems, Burckhardt continues, tend toward "ruthlessness vis à vis the system as a whole, promoting above all their own development, irrespective of disruption and dereliction. Thus the construction sector is a system bent on erecting new buildings, and this is reason enough for existing buildings to be destroyed" (Burckhardt 1978a, p. 90).

The efficient functioning of the administrative system is constantly threatened by the competition within the unofficial networks of landscape experts—agriculturalists, nature conservationists, game and hunting specialists, landscape planners etc. The "compulsive measuring of the self against others, of the less powerful against those perceived as (perhaps only marginally) more powerful" (Paris 2005, p. 23), is a fertile source of resentment. Again, however, this may work as much for the hierarchical system as against it, inasmuch as the daily wrestling for economic and symbolic capital prevents a more radical questioning of the system itself—whether the competition is for the allocation of competencies, human resources, financing, or EDP within a project, the acceptance of contributions for a publication, or the more existential issue of personal promotion. These conflicts rarely become public; they typically remain within the middle administrative echelons and only enter the political sphere when an issue radically threatens the social, economic or cultural capital of a specific group and no compromise can be reached on the lower administrative level. An example is the "Guidelines and Operational Strategies for Spatial Development in Germany" (Federal Ministry of Transport, Building and Urban Development 2006) issued by the Conference of Ministers for Spatial Planning. The bias of this document in favor of metropolitan regions led the Conference of Agricultural Ministers to criticize it as giving inadequate representation to rural areas—basically a quarrel about symbolic resources.

## 4.3.4 The Concept of Landscape in Schoolbooks and 'Fact' Books for Children and Young People

As the socially described, accepted, and experienced notion of landscape, the social landscape is the expression of a dominant symbolism assimilated in the process of socialization, of which education is an essential part. Here—especially in school—aesthetic, cognitive, and normative interpretations of landscape are systematically communicated and received "as a given result, a secure corpus of knowledge, a bundle of true statements" (Tillmann 2007, p. 179; see also Buttimer and Fahy 1999; Paasi 1999). In any social-constructivist analysis of the genesis of the concept of landscape in and through the process of subjectivation, critical scrutiny of one of the core media of this system—the school (here geography) textbook—is virtually indispensable. The process of subjectivation prepares school students to take their place in a "society of professors, doctors, teachers, and social workers", as well as landscape experts, "all of whom, in their own field, fulfill the role of judges upholding the realm of normativity" (Foucault 1977, pp. 392–393; Althusser 2011 [1970]; Sofsky 2007)—a function that inevitably entails a certain reduction in individually developed life opportunities.

### 4.3.4.1 The Concept of Landscape in School Textbooks

Based on an analysis—undertaken by the present author in 2007 (Kühne 2008a)—of 27 'fact' books for children and young people and school geography textbooks, the following sub-section presents a sample of the approaches offered to these readerships by German-speaking publishers. Three structural concepts can be distinguished:

(a) Landscape is treated in a limited and defined context. Thus Engelmann and Latz's school text Landschaftsgürtel—Ökologie und Nutzung ('Landscape belts—ecology and usage' 1997) deals with the ecology and usage of different landscape zones, whereas Wiese and Zils' 'fact' book Deutsche Kulturgeographie ('German cultural geography' 1987) is explicitly concerned with the genesis, change and preservation of German cultural landscapes. Degn et al. (1965) is an example of the many books that take landscape zones as a springboard for a discussion of the concept of appropriated physical landscape.
(b) Landscape is one among several focal topics, as in Landschaftszonen und Stadtökologie ('Landscape zones and urban ecology' Bender et al. 2000), a textbook for senior high school classes.
(c) Landscape is not itself a focal topic, but is treated in connection with other contexts, as in Heimat und Welt ('Home and world') edited by Kowalke (2001), or Richter et al.'s (2001) Geografie 5—both of them textbooks for senior high school classes. The children's 'fact' books Planet Erde ('Planet Earth': Parker 1996) and 'Wie ist das? Land, Meer und Luft' ('Land, sea and air —what are they?' Dixon 1991) take a similar line.

## 4.3 Landscape and Power

School textbooks based on landscape zones dispense altogether with an introduction to landscape or any discussion or definition of the term: they take it either as immediately understood or as implicitly defined by the treatment they provide. Thus, Engelmann and Latz (1997) begin their chapter on landscape zones with a map of the world divided into 'geo-ecological zones' and a table listing these. Bender et al. (2000) have neither a map nor a table but launch straight into a description of the tropics. Bständig et al. (2002, p. 146) take a naturalistic approach, describing appropriated physical landscapes as the result of the interaction of climate, soil, and vegetation. Accordingly, the opening paragraph of their section headed 'Climate and vegetation: a natural fit' defines landscape zones as zones "in which climate, soil, vegetation, and possible agricultural usage are similar." A table of these zones—representative of many such compilations—along with their specific types of vegetation is then presented (ibid. 148–149). As well as photographs of the typical landscapes of the different zones, the table details seasons, average annual temperatures, precipitation, possible plant growing periods, factors like cold or dry periods impacting growth, possible types of cultivation and its products, and livestock husbandry. The following climate and vegetation zones (other 'fact' books and school texts speak in this context of 'landscape zones') are listed: polar (frozen desert and tundra), subpolar (boreal coniferous forest), temperate (deciduous and mixed woodland, steppe), subtropical (Mediterranean vegetation, semi-desert and desert), and tropical (semi-desert and desert, savanna—subdivided into thorn-bush, dry, and wet savanna). The table is accompanied by two maps of the world, one of climate and the other of vegetation (or landscape) zones; in both maps the zones are strictly separate. Human impacts are as a rule considered as disturbing the ecological balance—see especially the subtitles in Bender et al. (2000): 'Destruction of the tropical rainforest for timber' (62), 'Interdependencies and interactions of natural factors in complex ecosystems: the example of woodland death in Central Europe' (100). Engelmann and Latz (1997) are less value-laden, with sections headed 'On the way to the ecological crisis?' (51), which introduces a treatment of rainforest protection and a possible transition to the shifting (migrant) cultivation of field and forest traditional in some parts of the tropics; and 'The origins of tree damage' (146), which covers such areas as the construction of filtration plants, raising fuel prices, and enacting stricter emission norms. Nowhere, however, is there any (even suggested) discussion of the reciprocal impact of such measures on politics and the economy.

A similar approach to landscape via zones can be found in some informative children's books. Thus Beautier and Derrien (1989) dispense with an introduction to landscape and describe the "typical landscapes" of the various zones from the perspective of two aliens called 'Buld' and 'Gorm'. These are illustrated with drawings (among other phenomena) of the pampas, Amazonian rainforest, south polar ice, Scandinavian tundra with coniferous forest (a highly improbable combination as depicted), and a European lowland plain.

In contrast to the foregoing, landscape is presented at least implicitly as a social construct in the senior school textbook *Diercke Erdkunde. Klasse 11* ('Diercke's geography for year 11': Claassen et al. 2005). Landscape is first presented in a classical geo-ecological perspective, with a chapter on 'Landscape zones—different usage, threats, and protective measures', before consideration is given to the agricultural consequences of these differences. The section on 'Tourism—Germany, Mallorca, and the world' (156 ff.) deals in a factual way with the social and ecological consequences of both short- and long-haul tourism, and with the artificial holiday worlds of, for example, Hawaii or the Alps, where "the tourist industry [...] has created roofed-in climate zones" (Claassen et al. 2005, p. 163). Without referring to these explicitly as sociophysical constructs, the book demonstrates at least implicitly that appropriated physical landscape is by no means only the result of geophysical factors.

Negative evaluation of human impacts on the landscape begins already in 'fact' books for children. Thus, after describing, in the chapter on 'Changing landscapes', the physical genesis of landscape elements like temperature, rainfall, frost, ice, wind, and water, Parker (1996) has a chapter on 'Destruction of the landscape' which (with an impressively superimposed transparent image) depicts the situation in a tropical landscape before and after human impacts. From a scientific viewpoint, however, Parker's presentation of tropical ecology is skimpy to the point of falsification. Kowalke's *Heimat und Welt* (2001), despite the absence of any explicit treatment of landscape (see above), has three sociologically interesting aspects: it is the only school textbook in this overview to provide—in its keyword list 'Geography at a glance' (adopted from Kissner et al. 1980)—an explicit definition of landscape as "a fundamental concept of the earth- and bio-sciences, as well as a colloquial term for 'terrestrial space'. In geography landscape is defined as an ecosystem in order to clarify the structural interaction of geosphere, biosphere, and anthroposphere which underlies that concept, and at the same time to allow for different ways of looking at it—for example as natural and/or cultural landscape (Kowalke 2001, p. 396). Kissner et al. (1980, p. 280) defines landscape as "a segment of the geosphere characterized by the structural holicity of its components"—an empirical postulate with exclusivist overtones. Both definitions derive from a naturalistic tradition that ignores or bypasses the radical post-1960s discussion of the concept of landscape. Thus, neither definition elaborates on the meaning of the term for the world in which we live, although Kowalke's reference to other possible perspectives indicates an opening in this direction. An implicit contradiction arises, however, when one compares two passages in his book referred to in the index under the term 'landscape'. On p. 290 the city is described as a man-made 'landscape', where the quotation marks suggest a conceptual disjunction between city and landscape—probably due to the author's ecosystemic definition of the city, which (see Luhmann 1984) entails a contrasting concept of the surrounding space as not-city (i.e. 'natural landscape'). An illustration on p. 383, on the other hand, explicitly shows settlements, common land and—as a sort of settlement subsystem—society as constituent components of landscape.

While all the books mentioned so far take an ecological or other natural-science-based approach to landscape, Wiese and Zils (1987) opt for a discussion of the German cultural landscape as a fundamentally human phenomenon: "Cultural landscape as a visible whole is an historically formed, continuously

## 4.3 Landscape and Power

changing structure that can be characterized in [spatial] terms ranging in extent from the regional to the national" (Wiese and Zils 1987, p. 9). Unlike the schoolbooks by Bender et al. (2000) and Engelmann and Latz (1997), Wiese and Zils do not focus evaluatively on the ecological side-effects of human (especially economic) activity, but simply on the genesis of a physical landscape as appropriated and changed by man. The care and preservation of the cultural landscape—and the term includes towns and cities—is regarded as part of the cultural heritage, which "as a sign of continuity amid dynamic transformation, is a task of European outreach" and central importance (Wiese and Zils 1987, p. 164).

Dixon (1991) also refrains in principle from any evaluation of the appropriated physical landscape. In his chapter on 'Changing landscapes' he briefly describes the motives for such change (population growth, quest for natural resources etc.) and their de facto ecological side-effects. Bständig et al. (2007, p. 108) deal with the use of space under the headings 'soil sealing' and 'urban sprawl' and with the ecological effects and side-effects of 'landscape usage' under 'landscape in stress'. This again implies a dichotomous construct of city and landscape, and this is further reflected in the illustration caption "Streets, highways, apartment blocks, and factories devour the landscape"—an emphatically moralizing, aesthetic perspective on human impacts. Brants et al. (2004, p. 165) use this same caption and approach the overall issue under the equally value-laden heading 'Steamrollered and urbanized' (ibid. 164). Frommelt-Beyer et al. (2003) deal in a rather less overtly moralistic tone with the 'consumption of landscape' by the automobile, comparing the space required for parking lots with that needed for public administrations. On the other hand, the cartographical comparison of a highway intersection with a small town (p. 19) seems irredeemably interest-laden.

An aesthetic rather than ecosystemic morality—this latter was a product of the somewhat later infiltration of geography by the natural sciences—is already evident in Degn et al., who see a need "to protect the most beautiful parts of a harmonious cultural landscape from ruthless disfigurement—to protect culture from devastation" (Degn et al. 1965, p. 161). That this attitudes is purveyed in books for grade (primary) school children can also be seen from Pommerening and Ritter's image of the effects of sand and gravel extraction: their description of flooded gravel pits speaks of "the destruction of landscape [...] and the desolate crater-filled landscape that now stands where fields and meadows once lay" (Pommerening and Ritter 1996, p. 12). The stereotype of an "intact cultural landscape" threatened by human activity is reinforced in a way that reflects the dichotomy of 'good' (fields and meadows) versus 'bad' (gravel pits and craters) without any indication of the criterion of differentiation. The same can be said of Auer et al., who describe Alpine hill farming in the following terms: "The farms are derelict and the pastures, no longer grazed, are overgrown. One begins to realize the importance of hill farmers for landscape conservation" (Auer et al. 2002, p. 45). Any discussion of the social reasons for these developments, and any reflection on the overall purpose of landscape conservation, is evidently considered obsolete.

Often implicit, sometimes even explicit in the schoolbooks examined in this subsection is a normative construct of landscape that reads like an expression of the archetypal Western values of 'the good, the true, and the beautiful'. A good landscape is one deriving from the immemorial structuring of regional nature and culture, which is necessarily authentic and hence also true. This good landscape, in contrast to one formed by the current needs of society, is ipso facto also beautiful. In this sense, the treatment accorded to landscape in most of these books oscillates between essentialism and positivism, reflecting a tradition that had its origins in German landscape research of the 19th century (see Sect. 3.1.4).

### 4.3.4.2 The Concept of Landscape in Schoolbooks and the Perpetuation of Social Power Structures

Against the background of Bourdieu's (e.g. 1973, 2001) and Althusser's (2011 [1970]) reflections on the significance of schooling for socialization, the foregoing comments on school and children's geography books shed light from several angles on the power relations implicit in concepts of landscape and their genesis (see Kühne 2008a):

(a) Almost all the illustrations in 'fact' books for children present stereotypically idealized landscapes that will condition the way children see and construct physical landscapes. This is reinforced by their later school textbooks and informative leisure reading. Landscape zones in particular are presented as inherent unities clearly distinct from one another (without any transitional zones). The implicit world-view conveyed by such books is dichotomous and exclusivist—as opposed to what Sloterdijk (1987) calls inclusive 'hybrid' thinking.
(b) All these books share an essentialist-positivist concept of landscape which as a rule bypasses the issue of social construction and, as such, socializes pre-school children (as well as school students in geography lessons) in existing structures of dominance. The subjective experience of real-world landscape is devalued in favor of assimilating current social concepts of objectivity.
(c) Also implicit in the positivist-essentialist attitude is the hierarchical distinction in rank and status between expert (systemic) and lay approaches.
(d) The emphasis on visual aspects of the appropriated physical landscape, especially in secondary socialization, tends to systematically exclude other, more transient sensory dimensions and hence to devalue as the product of multi-sensory perception not only the normal landscapes of daily life but also certain stereotypical landscape elements (see Fayet 2003).
(e) Especially the discussion of ecological modification of the appropriated physical landscape suffers from an implicit (and sometimes explicit) urban-rural dichotomy that stigmatizes the expansion of settlements as 'destruction of landscape'.
(f) This can be seen as the contribution of contemporary school geography to the fundamental dichotomies of Western thought (culture/nature, good/bad, masculine/feminine etc.).

## 4.3 Landscape and Power

(g) Discussion of ecological issues—also from the perspective of landscape experts—entails an implicit or explicit critique of prevalent political and/or economic systems. However, this is not accompanied by critical reflection on expertise in general—an omission that leads to the elevation of (landscape) experts to a status above that of elected politicians.

(h) The ecological perspective on landscape tends, even in scientific contexts, to introduce a moral slant into any discussion concerned with modifying the appropriated physical landscape. The shift in focus away from the substance of different views to their praise- or blameworthiness seriously impedes factual professional communication (see Luhmann 1993).

(i) With regard both to ecology and to the preservation of the—as a rule essentialistically conceived—cultural landscape, the solutions discussed and proposed in children's informative literature and schoolbooks are based on a linear concept of causality, to the exclusion of either feedback effects or causal networks. Presented as patent recipes, they perpetuate a cultural trust in the universal applicability of linear causal models, and as such serve the interests of those experts who both believe in and stand—in linear causal sequence—to gain from them.

(j) The texts in question show no sign of a socially systemic interpretation of landscape. Until well into the 1980s schoolbooks in both East and West Germany (the then GDR and FRG) treated landscape overwhelmingly as the result of natural geo-factors, along with occasional cultural influences. The ecosystemic perspective that entered West German geography in the 1970s filtered down into the schoolbooks of (reunified) Germany—usually with an admonishing slant—in the late 1980s, and with growing impact in the 1990s.

(k) Essentialist and positivist interpretations of landscape often (as in Degn et al. 1965) go hand in hand, but without any cross-referencing; nor do any of the books in the survey compare the two positions. The essentialist view of cultural landscape is, in fact, based on pure aesthetic exclusivism. Recent human changes to the physical substratum of landscape are rejected wholesale as a threat to the preservation—or, in the case of more massive intrusions, the full reinstatement—of the appropriated physical landscape (e.g. by renaturing flooded gravel pits or otherwise making human intrusions invisible—see Kühne 2013a, 2015a).

The perpetuation in young people's minds of stereotypical social target notions of landscape, whether ecological or historico-cultural (see Paasi 1999), can be understood as the individual and collective investment of symbolic capital by experts against the day when this capital might otherwise be devalued. Applying Habermas' theory of the systemic colonization of life-worlds (Habermas 1981), one could speak here of colonization: the colonization of the normal, stereotypical concept of landscape gained in primary socialization by the systemic scientific-ecological concept conveyed in the secondary socialization of the authors of the literature in question. Subjective notions of primary socialization are either dismissed as unscientific and inexact (see Fayet 2003), or replaced with essentialist ideas of cultural landscape as sacred and aesthetic. One way or the other, the individual experience of landscape is dismissed as culturally worthless (Kühne 2008a).

### 4.3.5 The New Governance Paradigm—Perpetuating or Overthrowing Power Structures?

The modern Western principles of order and tidiness, which long infused spatial and landscape planning in its methodology as well as execution, have experienced a crisis under the at least latent impact of inclusive postmodern thought. Purity sought to eradicate impurity: dirt "should be banished as unseemly, for fear it might disturb the basic order of things" (Bauman 2009 [1993], p. 241). Purity, however, has two other aspects, for "in the first place it produces waste, [...] and secondly it reduces the wealth and fruitfulness of accepted reality—it impoverishes the world" (Fayet 2003, p. 157; Engler 1997; Beck 2009). The new paradigm of governance can also be understood as a reaction to the criticism of an excessive bureaucracy in the modern welfare states of Europe (Jann and Wegrich 2004). On the practical level of interaction with the appropriated physical landscape, this implies a turning away from exclusive in favor of inclusive concepts that may well harbor contradictions— in contrast to the mostly normative, non-contradictory concepts of planners (Ipsen et al. 2003; see also Allmendinger and Haughton 2012; Schönwald 2015). An expression of these changes was "a new planning vocabulary [...]. Keywords such as networks, webs, corridors, hubs, flows, zones, and soft spaces soon began to characterize planning practice at various spatial scales and the representations of space embedded in such practice" (Paasi and Zimmerbauer 2016, p. 76).

Methodologically it implies a reduction in the role of experts in favor of the more intense integration of laypeople in planning processes. As an institutional framework, the so-called 'republican model' involves the consensual cooperation of citizens and corporate interests (Simmen and Walter 2007). In this context Gmeiner goes so far as to see lay participation as "both the goal and constitutive condition (conditio sine qua non)" of planning activities (Gmeiner 2005, p. 140; see also Ipsen et al. 2003). Two further models must also be mentioned: the expert model consisting of specialists from relevant disciplines, and the stakeholder model consisting of experts together with representatives of corporate interests. While these two latter models run a high (indeed virtually constitutive) risk of subjection to particular interests, the republican model allows control from above via selection of participants and determination of the agenda—but such control is comparatively indirect (see Gmeiner 2005).

Far from being an isolated phenomenon, the integration of laypeople in the planning process took place against the background of the Fordist crisis, with its reduction not only in the scope of fiscal action available to politics and the bureaucracy, but also in the fundamental significance of the national state. The

## 4.3 Landscape and Power

concomitant pluralization of lifestyles has seen a weakening both of the norms and values associated with great collective social ideas and of the "underlying concept of rationality, based on scientifically developed models of an 'optimal spatial order'" (Wood 2003, pp. 142–143; see also Reichert and Zierhofer 1993; Irrgang 2014). For all these reasons, postmodern planning has increasingly entailed lay involvement (see Brown 1989; Ryan 2011; Hartz and Kühne 2007; Stemmer 2016). In Sutter's words, "in view of the failing legitimacy of activist political and economic decision makers and technocrats, [participation] has recently become a course between Scylla and Charybdis" (Sutter 2005, p. 222), the consequence is a 'disorganized heterogeneity of situational projects' (Hannah 2009), instead of an apparently comprehensive planning (also: Allmendinger/Haughton 2012). The democratization of expertise is above all connected with the desire to enhance motivation, broaden the basis of knowledge and values, avoid—or more readily negotiate—potential conflicts, and heighten the legitimacy of political decisions, with processes that are visibly fair, inclusive, free from compulsion, and open to the aims and concepts of their participants (see Heinelt 1997; Kropp 2002; Abels and Bora 2004; Healey 2006; Fainstein 2010; Walk 2008).

Arthur Benz defines 'governance' as a general concept for "new forms of social, economic, and political regulation, coordination, and control in complex institutional structures commonly involving the cooperation of both state and private entities" (Benz 2004, pp. 12–13). In German the term is used in this way to refer to changes in the governmental practices of the modern state, concretized on the one hand in new forms of international politics, and on the other in changes to the organization and internal relations of public administrations, enterprises, markets, regions, and associations (see Diller 2005; Fürst 2007). As impacting the landscape, governance processes are also relevant to landscape research (see e.g. Healey 2006; Naranjo 2006; Ryan 2011; Gailing and Röhring 2008; Piniek et al. 2008; Säck-da Silva 2009; Leibenath and Otto 2011; Gailing 2012, 2015; Stemmer 2016), and in this context they have attained a status evident in the formation of international institutions: "Participation and cooperation have developed into constitutive principles of modern landscape management as understood by the European Landscape Convention" (Säck-da Silva 2009, p. 210). Huang (2010) even argues that participation is a precondition for sustainable landscape development as such (see also Wragg 2000; Jones 2007).

Walk (2008, 2012) distinguishes three sets of motives underlying governance processes:

(a) democratic—greater lay participation enhances legitimacy;
(b) economic—participation improves the efficiency of decision making processes and reduces the risk of planning errors; and
(c) emancipatory—participation increases the political leverage of ordinary citizens.

Broadening the basis of expertise is a governance measure that contrasts with the hierarchical, *dirigiste* procedures of central government: in a circular planning process the integration of civil forces entails a loss of power by elected politicians and public administrations. According to Habermas' theory of communicative action (Habermas 1981), it represents the attempt to integrate real-life perspectives into systemic structures, or at least to bring the two into greater proximity. In postmodern terms it can be understood as a de-differentiation of social subsystems (see Kühne 2006a, 2013).

In this context, Brown (1989) suggests postmodern symbolic realism as a theoretical framework for inclusive planning. Heuristically informed by the concept of the textual metaphor, and based methodologically on a combination of hermeneutic understanding of motives and structural analysis of factors unconsciously restricting action, Brown's proposal contrasts with the positivist model:

- The positivist model seeks to define a target state by means of a rational plan based on objective realities and allowing no alternative, whereas the symbolic realist model views reality as a construct, and any planning based on it as a process of step-by-step negotiation without any pre-defined target state.
- The positivist model distinguishes the roles of experts and laypeople: the former have specialist factual competency, the latter have value-competency. The symbolic realist model, on the other hand, is committed to learning from each other.
- The positivist model sees ends as independent of (value-free) means; symbolic realism calls for civil control of both ends and means.
- The quality of positivist planning is measured by the level of its target-attainment; that of symbolic realist planning by the degree of integration of those affected by the planning.
- Positivist planners base their decisions on criteria internalized in secondary socialization; symbolic rationalist planners base theirs on attainment of the negotiable framework.
- Positivist planning separates political and technical decisions, allocating the former to politicians, the latter to planners. Symbolic rationalist planning makes no such distinction, as citizens are involved in decisions on both levels.
- Where positivist planning seeks maximum rationality, symbolic rationalist planning combines rational with aesthetic elements (Brown 1989).

The circularity of a spatial planning process conducted according to these principles, with many different parties, ideas, and interests, virtually precludes the genesis of a carefully composed, comprehensive, and unified design. Instead of a univalent landscape with a consistent symbolic message (like the capital of an absolutist state), what comes into being is a polyvalent landscape with open symbolism. As such, the planning process makes high demands on all parties: not only the planners must be prepared to rethink and where necessary redefine their role; citizens, too, must be prepared to shed their role as *bourgeois* (in the sense of property-owning class) in favor of that of *citoyen*, and to make their decisions not so much in their own personal interests as in those of the community (Skorupinski and Ott 2002).

A neo-Marxist critique of participation in planning procedures has been advanced by Holm, who sees it as stabilizing existing structures of dominance: "Their disciplinary force lies in the procedures themselves, which—in the manner of a hidden curriculum—require and foster specific patterns of thought and behavior" (Holm 2006, p. 30). The mere avoidance of the use of restrictive legal instruments in negotiation processes between public and private interests does not render these instruments powerless: they remain "a behind-the-scenes threat indirectly serving the achievement of the desired goal" (ibid., p. 100). The internalization of the threat of sanctions means that controlling power is exercised by the controlled, not their masters (see Lemke 1997). Moreover, it is rarely possible for even suitably qualified laypeople "in their few after-work hours to embark on a sustained partnership on an equal footing in discussions and negotiations" (Holm 2006, p. 134). Less articulate and/or less powerful laypeople (e.g. the poor/uneducated/non-German-speaking etc.) will in any case be excluded from the process and their landscape requirements neglected.

New structures of governance do not automatically lessen the power of experts: in fact the body of experts is augmented by those whose professional skill lies in communicating the will of the people. Kühne (2013) points out that the involvement of citizens and NGOs in landscape planning may widen the outlook of planning departments, but it also tends to restrict the effectiveness of NGOs simply by overloading them with work. The answer may be to provide for appropriate remuneration of both NGOs and participating citizens in the planning budget.

Research into modern governance focuses powerfully on networking (Rhodes 1996), which is seen as an alternative structure to both the hierarchical state and the market (for further detail see Williamson 1991; Jansen and Wald 2007). While in the marketplace goods and services are exchanged for money and the social relation is confined to that transaction, and in a hierarchical structure money is exchanged for work, networks can develop other media of exchange like friendship, and their time span is correspondingly longer than that of market transactions (though not necessarily than that of hierarchies, which may well be permanent). Where markets operate with prices and hierarchies with instructions, networks are based on trust—which has its downside in terms of social closure, for trust in one's network partners is generally accompanied by mistrust in outsiders: "Networks that do not admit new members develop into closed circles" (Gottschick and Ette 2012, p. 27; see also Juarez and Brown 2008; Schnur and Drilling 2009; Meyer and Kühne 2012). And for potential members of the network, lack of social capital can lead to (at least temporary) exclusion. In the market, exclusion applies to those who lack financial means, while hierarchies admit on the merit principle only those who possess adequate cultural capital. In increasingly well educated societies (see e.g. Bell 1985 [1973]), the one-sidedness of actions with visible effects—construction of estates, infrastructural measures, patterns of agriculture etc.—will be increasingly questioned, and the right to be integrated in the decision making process will be demanded and (even if only partially) granted. But hierarchies rarely yield wholly to other modes of action; they are more often accompanied than replaced by network structures and non-official participants (see Diller 2005).

Nevertheless, current developments in governance can be seen on the one hand as both limiting the outreach of expertocratic articles of faith, and widening the scope and level of equality of opportunity. Yet much remains to be done to enable the inclusion of non-experts in landscape planning on equal terms (Nussbaum 2006)—from the use of a language open to all participants, through adequate remuneration for invested time, to an assurance of a (non-selective) hearing for non-expert voices. Conversely, lay participants must be prepared to critically question the stereotypical attitudes they bring from their primary socialization (itself often pre-formed by experts); it is, after all, these attitudes that to a great extent inform judgments about potential changes to the physical landscape.

## 4.4 Interim Summary with Further Reflections on the Interrelations of Socialization, Distinction, Power, and Landscape

The thesis presented here that landscape is a social—and socially derived individual—construct highlights the question of the development of that construct in the individual awareness and the manner of its social conditioning. Landscape experts stake a strong claim to interpretive hegemony over at least segments of the research field—an exclusivist claim (see Sloterdijk 1987) that entails a competitive stance not only toward other (geo-)disciplines but also toward notions instilled in primary socialization. If only for the sake of the sustained legitimacy of the landscape disciplines themselves, those strategies employed by experts in the contexts of planning and systematic socialization would appear most appropriate—in light of today's increasingly differentiated and pluralist society—that take seriously the need for acceptance by non-experts and/or seek a synthesis of expert and non-expert approaches. Landscape socialization should avoid the reflex of standardizing the requirements of the notional male whose family has been settled for generations in the same location—a reflex that de facto underlies both the paradigm of the historical cultural landscape and the imperative of its preservation. After all, such attitudes run counter to the formal principle, as well as the substance, of equality of opportunity (as propounded by Rawls 1971).

The increasing emphasis given to non-cognitive as well as non-expert dimensions of landscape awareness highlights the needs of those un- (or under-)represented in existing societal power structures. In the contemporary Western European context, three groups especially come to mind: recent migrants to the area and their descendants (with regard e.g. to the building of mosques and refugee accommodation), girls and women (who are still largely excluded from certain educational forms of landscape appropriation), and children in general (whose requirements are systematically under-represented in planning processes). A change in the prevailing construct of childhood could also do much to raise awareness among landscape experts that children, too, have a claim on their attention (see Nissen 1998).

A major step in this direction would be to see childhood as a phase of life in its own right—and one fundamental to the lifelong development of every individual biography—rather than merely as a lack of adulthood.

A problematic relation, not only in the context of landscape, is that between culture and power. Here Bourdieu does not mince his words: "In every respect culture is the result of a fight for power—a direct consequence of the fact that culture is intimately bound up with human dignity" (Bourdieu 1977, p. 26). Nassehi describes the principle of inclusion and exclusion as "a congenital defect of culture" arising from the disjunction between self and other implicit in any system of identity: "cultural differentiation, cultural not-belonging [...] is a virtual epistemological condition of any conceivable cultural perspective" (Nassehi 1999, p. 219). So when a landscape is described as 'cultural' or a new 'landscape culture' is called for (e.g. Haber 2000), power and dominance are always in play. Culture is "hierarchically organized and inclines to submission and domination" (Bourdieu 1977, p. 27): a condition evident with regard to landscape in levels and conditions of access to its cultural and communicative codes, as well as in rights to enjoy—and especially to change—the physical landscape.

Neither in its social nor in its physical dimension is landscape a value-free physical object that is simply given, simply 'there'. From a sociological point of view its—especially aesthetic—significance does not (*pace* Kant) lie in the simple enjoyment it evokes, without any further interest. On the contrary, landscape is the physical and symbolic expression of relations of social distinction and dominance, defined as distinctive by the legitimate and legitimating taste of the dominant class; for it is by implementing an aesthetic construct that 'landscape' is elevated above middle-class and popular taste. This process had its origins in (German and English) Romanticism, with the canonizing of both untamed and tamed landscapes as aesthetic; and a similar movement is afoot today with regard to the monuments of industrial archaeology (Kühne 2006c, 2007a, b). The socialization of the enjoyment and recreational use of landscape also serves to uphold social power structures: a country walk is so infused with sensory and cognitive relaxation that the social constraints of everyday are banished from the mind. In this sense the social landscape can also be seen as a disciplinary measure, for primary socialization not only teaches what one should feel in the face of the phenomena commonly synthesized as landscape; it also instills a healthy respect for the teacher-expert who, like a judge in court, lays down the standards of aesthetic interpretation and ecological concern.

The wealth of competing paradigms relating to landscape reflects the expansion of education and hence, too, the "debate about the 'right' education and the hegemony of the educated classes" (Resch 1999, p. 263). On both the physical level of landscape objects and the meta-level of their social and individual interpretation, different norms and values are in evidence. The postmodern leveling of the dichotomy between popular and high culture—also with regard to landscape—does not erase the potential of culture for creating social distinctions. As Liesmann puts it (2002, p. 17): "To maintain the opposition between kitsch and art is to stick fast in yesterday's—if not the day before yesterday's—mold" (albeit that too is a form of distinction). The predilection of today's intellectuals for distinctly profane landscapes (disused pitheads, suburban estates etc.) and 'artistic' stereotypes (South Sea sunsets, roaring stags against an Alpine background—see Sect. 4.2.1.1) indicates "that they are also familiar with popular culture, perhaps even more familiar—or at least additionally so" (Resch 1999, p. 282). Of equal importance as the choice of objects is the question how they are approached, and this is for the most part in the ironic postmodern mode, but sometimes through its reverse (see Liessmann 2002; Seidman 2012; Kühne and Schönwald 2015a); either way guarantees aesthetic (and hence also social) distinctiveness.

Moreover, either way the landscape perspective of postmodernism "takes sublime revenge on the presumptiveness of avant-garde Modernism" by indulging what that movement abhorred: "the fascination of things, garish pleasure, sensual religiosity, sentimental atmospherics, sunsets, C major chords, tears of joy, untrammeled delight in the exotic" (Liessmann 2002, pp. 73–74). Translated into landscape, this means fun pools, shopping malls, public parks, Alpine panoramas, subtropical holiday idylls, wall-to-wall TV landscapes, and model railways (see Sect. 5.5). Postmodernism has re-enabled experiential access to landscape via feeling (as proposed by Lipps 1891, 1902) and intuition (as proposed by Croce 1930); the cognitive approach (most commonly under expert positivist or essentialist guidance) is no longer *de rigeur* for the competent (and hence also distinctive) deployment of landscape semiotics.

The entire Western system of socialization and education is weighted toward an objectified concept of landscape. Socialized unsystematically in childhood and youth through the agency of the peer group and significant others, the objectivity of landscape is systematically reinforced by schooling and other educational measures —as demonstrated by Schultz (2008) with the example of wartime cartography in schoolbooks. Continued into adult life, the generally unsystematic acquisition of knowledge and development of aesthetic preferences and emotional ties informs the common lay understanding of the 'normal home landscape'. Alongside this, a stereotypical concept of landscape is conveyed both unsystematically by films, television programs, magazines, and travel brochures, and systematically by schoolbooks and teaching. The expert concepts mediated systematically in higher education are based on the norms, values, and standards of the specific (geo-) discipline whose internalization is a precondition for successful graduation and professional qualification. Finally, the mainstream research focus on a positivist-realist construct of landscape, together with the definition of the 'historical cultural landscape' as target state, implies an understanding of landscape as given that ipso facto discredits more recent vernacular intrusions as ahistorical.

Taken together, the standard landscapes of literature, painting, film, photography, computer games and cyberspace, as well as the landscape zones of school geography textbooks and factual books for children and young people, amalgamate to a stereotypical corpus creating a feedback relation to ever more films, computer games, paintings, and imaginative literature. Seemingly backed by exact scientific descriptions, these multiple images of landscapes and zones sediment into a diffuse backdrop of everyday knowledge that is continuously reflected and confirmed by visual impressions from photos and postcards, paintings, travel brochures, guidebooks (see Nohl 2004), novels, and films (both feature films and documentaries), as well as computer games, virtual landscape construction programs, and modern zoos which present animals as far as possible in their 'natural' habitats (Grotmann and Fuchs 2004; Wauschek 2004; Vogt 2007; Steinkrüger 2014). The recursive impact of these various factors is all the greater as it embodies the colonization of the everyday world by systemic landscape stereotypes in constantly recurring media and images, a process underpinned by the input of institutionalized education as described above (see Dreitzel 1962).

Kühne (2008a) distinguishes eight stereotypical landscapes:

(a) Central European: hills, woods, agricultural patchwork of fields and orchards, farms and villages 'set in the landscape', with meandering streams and rivers. Climatically characterized by four distinct seasons: spring blossoms, summer heat, autumn storms, and snowy winters. Its basic Arcadian charm is seen as aesthetically threatened (among other things) by industrial agriculture, urban sprawl (other than one's own suburb), traffic arteries, industrial plants and estates, acid rain, and the straightening of watercourses.
(b) Mediterranean: Arcadian archetype of the Italian/Provençal landscape in summer, conveyed in painting, literature and personal travel. Rolling hill country with Tuscan pines and acacias, beaches, small fishing/farming villages (again 'set in the landscape'); hot summers under blue skies and/or fields of lavender. Seen as aesthetically threatened by forest fires, karstification, and concrete hotel jungles (other than one's own favorite vacation habitat).
(c) Desert: for Central Europeans the dominant Saharan archetype is characterized by sand (esp. shifting dunes), extreme heat, and dryness, and aesthetically determined (if anything) by transcendent sublimity. Not generally thought of as itself threatened, it is conceived rather as a threat both on the individual/existential level—from which the only escape is an (equally stereotypical) oasis —and on the collective level, with expanding desertification due to climate change impacting esp. (from a European perspective) the Arcadian/Mediterranean landscape.
(d) Exotic/tropical: characterized by hot, humid climate and the contrast of impenetrable jungle and palm-lined beaches. Aesthetically sublime, yet harboring perceived individual threat from wild animals and indigenous peoples and collective threat of rainforest clearance and other exploitation.
(e) Steppe and savanna: European stereotypes are the Russian steppe and the Serengeti (popularized in Germany esp. by zoo-director and TV personality Bernhard Grzimek); both are characteristically flat, bare, dry, and either hot (Serengeti) or cold (Russian steppe). Aesthetic of sublime spatial openness threatened (if anything) by hunting/poaching of wildlife; individual threat from climatic exposure and/or wild animals.
(f) Subpolar: esp. (for the European) Siberia, Northern Canada, and Alaska. Basic landscape elements are boreal (coniferous) forest and open grassland (pampas), either flat or mountainous, and with lakes. Dominant aesthetic motif of

sublimity threatened by forest clearance (esp. Canada); individual threat from wild animals and cold.
(g) Arctic/Antarctic: stereotype of polar icecap, hostile to human life. Threat to global climate from polar ice melt is mitigated in common perception by enduring aesthetic of sublimity and environmental hostility.
(h) Mountain: for the European, archetypically the Alps, culturally transmitted in painting, literature, travelogues, and films (esp. the so-called German Heimat-movies); also the Rocky Mountains (esp. in Western movies). Characterized by steep peaks and rock faces, winter snow, lakes, glaciers, mountain villages and Alpine meadows; aesthetically either sublime or romantic. Threatened by global warming (vanishing glaciers) and intensive tourism; individual threat from avalanches, earth slippage, and sudden weather change.

Apart from these, another landscape with a distinctive appeal is that of industrial archaeology. Rejected by the common taste as an ugly memorial to social and ecological decadence, it is cultivated by the initiated as an aesthetically sublime symbol of desolation, the eschatologically charged expression of human economic transitoriness.

The social construct of landscape stereotypes outlined above arises from a process of abstraction which, driven by dominant interests, combines analytic description with a normative component dedicated to preservation of the construct and its physical (and economic) manifestation. Thus the descriptive model of (e.g.) an urban center becomes prescriptive when the actual state (the central district of Y is X) and target state (the central district of Y should really be Z) converge. All it needs for this to happen is the reinforcement of the central role of Z by urban planning measures. Moreover, the ascription of a distinctive aesthetic value to a landscape as a distraction from social constraints reveals two correlative dimensions in which power systems function in the constitution and use of landscape, both symbolically and really securing prevalent structures of social dominance. In short, appropriated landscape in its physical shape is the reification of social power. And here too there is a feedback process, for physical landscape symbolizes the (latent or manifest) drive for social distinction; and social distinction expresses itself in a differentiated scale of rights to ownership, access, and use of physical landscape. In fact, the socially constitutive structure of landscape is inherently recursive. What landscape is and what it should look like is defined by experts and communicated through the processes of primary and secondary socialization (see Sect. 4.3.4.2). The will of the governing class of experts becomes the will of the governed, inasmuch as they, too, seek to participate in the exclusive aesthetic of the dominant interpretation (see Foucault 1977). With the exception of intellectuals with critical landscape training, such participants are, however, conditioned by socialization to accept—in the direct encounter of a planning group—the interpretations of those who were at least indirectly responsible for their conditioning. The socially constitutive structure of landscape is in this sense inherently recursive. The apparent democratic legitimacy of landscape experts is, one may conclude, precisely that—apparent.

A central mechanism securing the power of experts, whether planners, geographers, or sociologists, is what Schimank calls "fictions of rationality" which, "however indispensable as practical guidelines, [...] harbor the chronic danger of infecting social activity with an unremarked loss of rationality" (Schimank 2012, p. 385). In the landscape context what this means is that paradigms established as fitting one context are applied unquestioningly to every context, with flagrant disregard for their contingency. In that sense, the very concept of cultural landscape is a euphemism (Kühne 2006c). The technical (structural) power (Popitz 1992) of landscape experts expresses itself in the manifold forms of a twofold mastery: over nature and over man—meaning, in the latter case, decisions affecting the life of local people (see Popitz 1992; DeMarrais et al. 1996). The sacralizing of this process as 'culture' is "one of the great achievements of the European imagination —the myth of a possible harmony between human beings and nature" (Hauser 2001, p. 240). Obscuring social realities, it underpins the hegemonic forces behind them.

Another historical manifestation of the same hegemonic tendency can be seen in the bureaucracy, whose growth to power is correlative with that of the experts that counsel and inform it. Expert social capital is a heady ingredient in the ethos of any public administration. Again, one can observe the diminished democratic legitimacy of the 'technical state' (Schelsky 1965). Already in 1911 Michels noted what he called the "iron law of oligarchy" as the universal fate of any political collective: "The organization is the mother of the government of the elected over the elector, of the agent over the principal, of the delegate over the delegator" (Michels 1911, p. 384). The same holds for the role of deputies in an organization, for the organization is more than "the sum of its parts—of its individual deputies" (Sofsky and Paris 1994, p. 178).

The measures taken to meet the transformational processes of society (whether growth, diminution, or stagnation—e.g. intensification or extension of agriculture, increase or shrinkage of population) suffer from a correlative imposition of standardized, institutionalized solution models, ranging from incentives to commands and prohibitions, which elevate the stereotypical to the norm and establish the ideology of an elite over the contingency of the particular situation. The only difference is that, unlike ideologies, the stereotypical norm knows no "complex reciprocal processes or multi-layered social structures, but follows the naïve law of first impressions that creates a world of likes and dislikes, of positive and negative [feelings], before launching an immediate search for a scapegoat" (Bergler 1976, p. 84)—a fairly typical bureaucratic reflex.

In the relationship between the worlds of politics and science, specialist knowledge serves as a principle of simplification as well as legitimacy, externalizing the responsibility of the political decision maker. Despite the skepticism with which experts are currently confronted, "scientific expertise remains the key resource of politicians in risk-fraught or controversial circumstances" (Bogner and Torgersen 2005, p. 7). And the key counterweight of a democratic society is on the one hand to know and critically analyze the vested interests, communication codes, and referential systems of the echelons of decision makers and experts ranged against it, and on the other to retain a clear focus on the particularities of the issue and its social contexts (see Saretzki 2005). What is needed is not a dichotomous moralistic reduction to good versus bad, but an awareness of the complex shades of knowledge, ranging from not knowing, through uncertainty, to reasoned conviction, that color the arguments and judgments of all parties—experts, politicians and citizens alike (see Paris 2005; Kühne 2014a).

As a last resort the less powerful in this hierarchical structure—the laity, the *demos*—retain what Hirschman (1970) calls the 'exit option': the withdrawal of loyalty. But this last resort will rarely be implemented for the simple reason that— for all their wooing of the public on the micro-level of case studies and individual expertise, as well as on the meso-level of institutional contact (open days and the like)—the elite can call on an overwhelming accumulation of symbolic capital (see Nowotny 2005). Only a massive transformation of the processes generating status and hierarchy could alter this. In the world in which we live, the non-expert is conditioned by socialization to trust the expert, and although expanding education has brought with it an expansion in the number and range of experts, and non-experts in matters of landscape may well be experts in another field, and as such—equipped with the self-confidence that goes with symbolic capital—may be prepared to stand up and speak, yet symbolic capital is in a sense undifferentiated, and it will be the rare expert who risks loss of face by embarking on a fundamental critique of institutionalized expertise.

## References

Abels, G., & Bora, A. (2004). Demokratische Technikbewertung. Bielefeld: Transcript.
Achleitner, F. (1997). *Region, ein Konstrukt? Regionalismus, eine Pleite?*. Basel, Boston, Berlin: Birkhäuser.
Adler, P., & Kwon, S. W. (2002). Social capital: Prospects for a new concept. *Academy of Management Review, 27*(1), 17–40.
Adorno, Th. W. (1977). Prolog zum Fernsehen. Gesammelte Schriften. Bd. 10.2. Frankfurt a. M.: Suhrkamp
Adorno, T. (1987). *Jargon der Eigentlichkeit*. Suhrkamp: Frankfurt a. M.
Ahrend, C. (1997). Lehren der Straße. Über Kinderöffentlichkeiten und Zwischenräume. In J. Ecadius & M. Löw (Eds.), Raumbildung Bildungsräume. Über die Verräumlichung sozialer Prozesse (pp. 197–212). Wiesbaden: Springer VS.
ARL – Akademie für Raumforschung und Landesplanung. (2011). Grundriss der Raumordnung und Raumentwicklung. Hannover: Verlag der ARL.
Allmendinger, P., & Haughton, G. (2012). Post-political spatial planning in England: A crisis of consensus? *Transactions of the Institute of British Geographers, 37*(1), 89–103.
Alonso, W. (1964). *Location and land use: Toward a general theory of land rent*. Cambridge: Harvard University Press.
Althusser, L. (2011 [1970]). Idéologie et appareils idéologiques d'État: (Notes pour une recherche). In L. Althusser (Ed.), Sur la reproduction (pp. 263–306). Paris: Presses Universitaires de France.
Arnim, H. V. (2007). Die deutsche Krankheit. Organisierte politische Unverantwortlichkeit? In G. Gornig, U. Kramer & U. Volkmann (Eds.), Staat – Wirtschaft – Gemeinde. Festschrift für Werner Frotscher zum 70. Geburtstag (pp. 267–284). Berlin: de Gruyter.

# References

Asmuth, C. (2005). Denklandschaften. In K. Röttgers & M. Schmitz-Emans (Eds.), Landschaft. Gesehen, beschrieben, erlebt (pp. 19–29). Essen: Die Blaue Eule.

Auer, H., Detsch, B., Hauck, A., Kirch, P., Lacler, R., Nebel, J., & Weigaert, C. (2002). Durchblick. Bayern, Hauptschule. Geschichte, Sozialkunde, Erdkunde. 6. Jahrgangsstufe. Braunschweig: Westermann.

Bacher, M., Walde, J., Pecher, C., Tasser, E., & Tappeiner, U. (2016). Are interest groups different in the factors determining landscape preferences? *Landscape Online, 47*, 1–18.

Bachtin, M. (1987). *Rabelais und seine Welt*. Suhrkamp: Volkskultur als Gegenkultur. Frankfurt a. M.

Bahrenberg, G. (1996). Die Länderkunde im Paradigmenstreit um 1970. *Berichte zur deutschen Landeskunde, 70*, 41–54.

Bathelt, H., & Glückler, J. (2003). Wirtschaftsgeographie. Ökonomische Beziehungen in räumlicher Perspektive. Stuttgart: utb.

Bätzing, W. (2000). Postmoderne Ästhetisierung von Natur versus ‚Schöne Landschaft' als Ganzheitserfahrung – von der Kompensation der ‚Einheit der Natur' zur Inszenierung von Natur als ‚Erlebnis'. In A. Arndt, K. Bal & H. Ottmann (Eds.), Hegels Ästhetik. Die Kunst der Politik – die Politik der Kunst. Zweiter Teil (pp. 196–201). Berlin: Akademie Verlag.

Baudrillard, J. (1978). *Agonie des Realen*. Berlin: Merve.

Bauman, Z. (2009 [1993]). Postmoderne Ethik. Hamburg: Hamburger Edition.

Beautier, F., & Derrien, L. (1989). *Atlas für Kinder. Eine Reise zur Erde*. Nürnberg, Hamburg: Neuer Tessloff Verlag.

Beck, U. (1986). *Risikogesellschaft. Auf dem Weg in eine andere Moderne*. Frankfurt a. M.: Suhrkamp

Beck, U. (1997). *Was ist Globalisierung? Irrtümer des Globalismus – Antworten auf die Globalisierung*. Frankfurt a. M.: Suhrkamp

Beck, J. (2009). *Dirty wars: Landscape, power, and waste in Western American literature*. Lincoln, London: Universitiy of Nebraska Press.

Belina, B. (2006). *Raum, Überwachung, Kontrolle*. Westfälisches Dampfboot: Vom staatlichen Zugriff auf städtische Bevölkerung. Münster.

Bell, D. (1973). *The coming of the post-industrial society*. New York: Basic Books.

Bender, G. (2004). Modus 2 – Wissenserzeugung in globalen Netzwerken? In U. Matthisen (Ed.), *Stadtregion und Wissen* (pp. 87–96). Wiesbaden: Springer VS.

Bender, H.-U., Korby, W., Kümmerle, U., Ruhen, N. v. d., Stein, Ch., & Viehof, W. (2000). Landschaftszonen und Stadtökologie. Gotha, Stuttgart: Ernst Klett Verlag.

Benz, A. (2004). Einleitung: Governance – Modebegriff oder nützliches sozialwissenschaftliches Konzept? In A. Benz (Ed.), Governance – Regieren in komplexen Regelsystemen. Eine Einführung (pp. 11–28). Wiesbaden: Springer VS.

Berger, P., & Luckmann, T. (1966 [1969]). *The Social Construction of Reality*. New York: Penguin Books.

Bergler, R. (1976). *Vorurteile – erkennen, verstehen, korrigieren*. Köln: Deutscher Instituts-Verlag.

Bernat, S., & Hernik, J. (2015). Polnische Klanglandschaft um die Jahrhundertwende. In O. Kühne, K. Gawroński, & J. Hernik (Eds.), *Transformation und Landschaft* (pp. 247–267). Wiesbaden: Springer VS.

Berr, K. (2013). Wahrheit und ‚Möglichkeitssinn'. Hegels Ästhetik im Kontext moderner Kultur. In H. Friesen & M. Wolf (Eds.), Kunst, Ästhetik, Philosophie. Im Spannungsfeld der Disziplinen (pp. 129–168). Münster: Mentis.

Berr, K. (2014). Zum ethischen Gehalt des Gebauten und Gestalteten. *Ausdruck und Gebrauch, 12*, 30–56.

Bertels, L. (1997). *Die dreiteilige Großstadt als Heimat. Ein Szenarium*. Opladen: Leske + Budrich.

Beyme, K. V. (2013). Von der Postdemokratie hin zur Neodemokratie. Wiesbaden: Springer VS.

Bezzenberger, A., Dutt, H., Hullmann, H., & Gimmler, J. (2003). Saul – Regionalpark Saar; Regionales Pilotprojekt Saarkohlenwald; Raumvision und Infrastruktur. Saarbrücken: Conte.

Bischoff, W. (2002). "Das ist ja wohl die Höhe". Höhe als Dimension der Geographie und Architektur. In J. Hasse (Ed.), Subjektivität in der Stadtforschung (pp. 41–60). Frankfurt a. M.: Suhrkamp.

Bischoff, W. (2003). Inspiration der Straße. Für eine Architektur "der Nase nach". In I. Flagge (Ed.), Architektur und Wahrnehmung. Jahrbuch Licht und Architektur 2003 (pp. 44–49). Darmstadt: WBG – Wissenschaftliche Buchgesellschaft.

Bischoff, W. (2005a). Wenn Landschaften ,näher' rücken – von den Merkwürdigkeiten des urbanen Geruchsraumes. In V. Denzer, J. Hasse, K. D. Kleefeld & U. Becker (Eds.), Kulturlandschaft. Wahrnehmen – Inventarisieren – Regionale Beispiele (pp. 89–99). Wiesbaden: Selbstverlag Landesamt für Denkmalpflege Hessen.

Bischoff, W. (2005b). Nicht-visuelle Dimensionen des Städtischen. Olfaktorische Wahrnehmung in Frankfurt, dargestellt an zwei Einzelstudien zum Frankfurter Westend und Ostend. Frankfurt a. M.: Suhrkamp.

Blackbourn, D. (2007). *Die Eroberung der Natur. Eine Geschichte der deutschen Landschaft.* München: Deutsche Verlags-Anstalt.

Bloor, D. (1982). Durkheim and Mauss revisited: Classification and the sociology of knowledge. *Studies in History and Philosophy of Science, 13,* 267–297.

Blotevogel, H. (1996). Theoretische Geographie Aufgaben und Probleme der Regionalen Geographie heute. Überlegungen zur Theorie der Landes- und Länderkunde anläßlich des Gründungskonzepts des Instituts für Länderkunde, Leipzig. *Berichte zur deutschen Landeskunde, 70*(1), 11–40.

Blotevogel, H. (2000). Geographische Erzählungen zwischen Moderne und Postmoderne. In H. H. Blotevogel, J. Ossenbrügge & G. Wood (Eds.), Lokal verankert, weltweit vernetzt. Tagungsbericht und wissenschaftliche Abhandlungen des 52. deutschen Geographentages (pp. 465–478). Stuttgart: Franz Steiner Verlag.

Blum, P., Kühnau, C., & Kühne, O. (2014). Energiewende braucht Bürgerpartizipation: Beteiligungsformen vor dem Hintergrund gesellschaftlicher Rahmenbedingungen. *Natur und Landschaft, 89*(6), 243–249.

Boczek, B. (2006). *Transformation urbaner Landschaft. Ansätze zur Gestaltung in der Rhein-Main-Region.* Wuppertal: Müller + Busmann.

Bodenschatz, H. (2001). Europäische Stadt, Zwischenstadt und New Urbanism. *Planerin, 3,* 24–26.

Bogner, A. (2005). Moralische Expertise? Zur Produktionsweise von Kommissionsethik. In A. Bogner & H. Torgersen (Eds.), *Wozu Experten? Ambivalenzen der Beziehung von Wissenschaft und Politik* (pp. 172–193). Wiesbaden: Springer VS.

Bogner, A. & Torgersen, H. (2005). Sozialwissenschaftliche Expertiseforschung. Zur Einleitung in ein expandierendes Forschungsfeld. In A. Bogner & H. Torgersen (Eds.), *Wozu Experten? Ambivalenzen der Beziehung von Wissenschaft und Politik* (pp. 7–32). Wiesbaden: Springer VS.

Böhme, G. (1995). *Atmosphäre.* Suhrkamp: Frankfurt a. M.

Böhnisch, L., Lenz, K., & Schröer, W. (2009). *Sozialisation und Bewältigung: eine Einführung in die Sozialisationstheorie der zweiten Moderne.* Weinheim: Beltz Juventa.

Born, K. (1995). Raumwirksames Handeln von Verwaltungen, Vereinen und Landschaftsarchitekten zur Erhaltung der Historischen Kulturlandschaft und ihrer Einzelelemente. Eine vergleichende Untersuchung in den nördlichen USA (New England) und der Bundesrepublik Deutschland. Göttingen: Universitätsverlag Göttingen.

Böse, H., Haas-Kirchner, U., Hülbusch, I.-M., & Hülbusch, K. (1981). *Untersuchungen zur Bundesgartenschau Frankfurt 1989.* Kassel: Selbstverlag.

Bourdieu, P. (1973). Kulturelle Reproduktion und soziale Reproduktion. In P. Bourdieu & J.-C. Passeron (Eds.), *Grundlagen einer Theorie der symbolischen Gewalt* (pp. 89–127). Frankfurt a. M.: Suhrkamp.

Bourdieu, P. (1976). *Entwurf einer Theorie der Praxis auf der ethnologischen Grundlage der kabylischen Gesellschaft.* Frankfurt a. M.: Suhrkamp.

Bourdieu, P. (1977). Politik, Bildung und Sprache. In M. Steinrücke (Ed.), *Die verborgenen Mechanismen der Macht* (pp. 13–30). Hamburg: VSA.

# References

Bourdieu, P. (1979). *La distinction: Critique sociale du jugement*. Paris: Les Editions de Minuit.
Bourdieu, P. (1982). Die verborgenen Mechanismen der Macht enthüllen. In M. Steinrücke (Ed.), *Die verborgenen Mechanismen der Macht* (pp. 81–87). Hamburg: VSA.
Bourdieu, P. (1983). Ökonomisches Kapital – Kulturelles Kapital – Soziales Kapital. In M. Steinrücke (Ed.), *Die verborgenen Mechanismen der Macht* (pp. 49–80). Hamburg: VSA.
Bourdieu, P. (1985). Sozialer Raum und „Klassen". Leçon sur la leçon. Zwei Vorlesungen. Frankfurt a. M.: Suhrkamp.
Bourdieu, P. (1990). *The logic of practice*. Stanford: University Press.
Bourdieu, P. (1991a). Physischer, sozialer und angeeigneter physischer Raum. In M. Wentz (Ed.), Stadt-Räume. Die Zukunft des Städtischen (pp. 25–34). Frankfurt a. M.: Suhrkamp.
Bourdieu, P. (1991b). Was anfangen mit Soziologie? In M. Steinrücke (Ed.), *Die verborgenen Mechanismen der Macht* (pp. 127–148). Hamburg: VSA.
Bourdieu, P. (1992 [1984]). Homo academicus. Paris: Les Editions de Minuit.
Bourdieu, P. (1992b). Keine wirkliche Demokratie ohne wahre kritische Gegenmacht. In M. Steinrücke (Ed.), *Die verborgenen Mechanismen der Macht* (pp. 149–160). Hamburg: VSA.
Bourdieu, P. (1997). Die männliche Herrschaft. In I. Dölling & B. Krais (Eds.), Ein alltägliches Spiel. Geschlechterkonstruktion in der sozialen Praxis (pp. 153–217). Frankfurt a. M.: Suhrkamp.
Bourdieu, P. (1998[1996]). *Über das Fernsehen*. Frankfurt a. M.: Suhrkamp.
Bourdieu, P. (2000). *Vom Gebrauch der Wissenschaft. Für den Gebauch einer klinischen Soziologie des wissenschaftlichen Felde*. Konstanz: UKV.
Bourdieu, P. (2001). *Meditationen*. Suhrkamp: Zur Kritik der scholastischen Vernunft. Frankfurt a. M.
Bourdieu, P. (2002). *Der einzige und sein Eigenheim*. Hamburg: VSA.
Bourdieu, P., & Eagleton, T. (1992). Doxa and common life. *New left Review, 191*(1), 111–121.
Brady, E. (2005). Sniffing and savoring. The aesthetics of smells and tastes. In A. Light & J. Smith (Eds.), *The aesthetics of everyday life* (pp. 177–193). New York, Chichester: Columbia University Press.
Brand, K.-W. (1999). Transformation der Ökologiebewegung. In A. Klein, H.-J. Legrand, & T. Leif (Eds.), *Neue soziale Bewegungen* (pp. 237–256). Opladen: Westdeutscher Verlag.
Brants, E., Gaffga, P., Geis, G., Kirch, P., Kreuzberger, N., & Nebel, J. (2004). Heimat und Welt. Erdkunde für Nordrhein-Westfalen 9/10. Braunschweig: Westermann.
Bray, B. (1995). Campagnes romanesques, aux charités bien ordonnées. In U. Dethloff (Ed.), Literarische Landschaft. Naturauffassung und Naturbeschreibung zwischen 1750 und 1830 (pp. 115–130). St. Ingbert: Röhrig Universitätsverlag.
Brown, R. (1989). Social Science as a Civic Discourse. Essays on the Invention, Legitimation and Uses of Social Theory. Chicago, London: University of Chicago Press.
Brunce, M. (1994). *The countryside ideal: Anglo-American images of landscape*. London: Psychology Press.
Bständig, V., Ernst, M., Priester, C., & Salzmann, W. (2002). *Diercke Erdkunde G8 für Gymnasien im Saarland 5/6*. Braunschwcig: Westermann.
Bständig, V., Ernst, M., Priester, C., & Salzmann, W. (2007). *Diercke Erdkunde G8 für Gymnasien im Saarland 7*. Braunschweig: Westermann.
Bucher, H., Losch, S., & Rach, D. (1982). Selektive Wanderungen, Wohnbautätigkeit und Bodenmarktprozesse als Determinanten der Suburbanisierung. *Informationen zur Raumentwicklung, 11*(12), 915–937.
BDLA – Bund Deutscher Landschaftsarchitekten. (2006a). bdla – der Verband. www.bdla.de/seite74.htm. Accessed April 30, 2007.
BDLA – Bund Deutscher Landschaftsarchitekten. (2006b). bdla – der Verband. www.bdla.de/seite60.htm. Accessed April 30, 2007.
Burckhardt, L. (1967). Bauen – ein Prozeß ohne Denkmalpflichten. In J. Fezer & M. Schmitz (Eds.), *Wer plant die Planung? Architektur, Politik, Mensch* (pp. 26–45). Kassel: Schmitz.
Burckhardt, L. (1970). Politische Entscheidungen der Bauplanung. In J. Fezer & M. Schmitz (Eds.), *Wer plant die Planung? Architektur, Politik, Mensch* (pp. 45–58). Kassel: Schmitz.

Burckhardt, L. (1974). Wer plant die Planung? In J. Fezer & M. Schmitz (Eds.), *Wer plant die Planung? Architektur, Politik, Mensch* (pp. 71–88). Kassel: Schmitz.

Burckhardt, L. (1977). Landschaftsentwicklung und Gesellschaftsstruktur. In M. Ritter & M. Schmitz (Eds.), *Warum ist Landschaft schön? Die Spaziergangswissenschaft* (pp. 19–33). Kassel: Schmitz.

Burckhardt, L. (1978a). Kommunikation und gebaute Umwelt. In M. Ritter & M. Schmitz (Eds.), *Wer plant die Planung? Architektur, Politik, Mensch* (pp. 88–99). Kassel: Schmitz.

Burckhardt, L. (1978b). Ästhetische Probleme des Bauens. In M. Ritter & M. Schmitz (Eds.), *Wer plant die Planung? Architektur, Politik, Mensch* (pp. 167–176). Kassel: Schmitz.

Burckhardt, L. (1978c). Von kleinen Schritten und großen Wirkungen. In M. Ritter & M. Schmitz (Eds.), *Wer plant die Planung? Architektur, Politik, Mensch* (pp. 176–187). Kassel: Schmitz.

Burckhardt, L. (1979). Warum ist Landschaft schön? In M. Ritter & M. Schmitz (Eds.), *Warum ist Landschaft schön? Die Spaziergangswissenschaft* (pp. 33–42). Kassel: Schmitz.

Burckhardt, L. (1980). Niemandsland. In M. Ritter & M. Schmitz (Eds.), *Warum ist Landschaft schön? Die Spaziergangswissenschaft* (pp. 140–141). Kassel: Schmitz.

Burckhardt, L. (1982). Zwischen Flickwerk und Gesamtkonzeption. In M. Ritter & M. Schmitz (Eds.), *Wer plant die Planung? Architektur, Politik, Mensch* (pp. 99–106). Kassel: Schmitz.

Burckhardt, L. (1991). Wertvoller Abfall, Grenzen der Pflege, Zerstörung durch Pflege. In J. Fezer & M. Schmitz (Eds.), *Wer plant die Planung? Architektur, Politik, Mensch* (pp. 221–335). Kassel: Schmitz.

Burckhardt, L. (1994). Landschaft ist transitorisch. In M. Ritter & M. Schmitz (Eds.), *Warum ist Landschaft schön? Die Spaziergangswissenschaft* (pp. 90–94). Kassel: Schmitz.

Burckhardt, L. (1995a). *Design ist unsichtbar*. Ostfildern: Hatje Cantz.

Burckhardt, L. (1995b). Spaziergangswissenschaft. In M. Ritter & M. Schmitz (Eds.), *Warum ist Landschaft schön? Die Spaziergangswissenschaft* (pp. 257–301). Kassel: Schmitz.

Butler, J. (2001 [1997]). Psyche der Macht. Das Subjekt der Unterwerfung. Frankfurt a. M.: Suhrkamp.

Buttimer, A., & Fahy, G. (1999). Imaging Ireland through geography texts. In A. Buttimer, St. D. Brunn & U. Wardenga (Eds.), *Text and image. Social construction of regional knowleges* (pp. 179–191). Leipzig: Institut für Länderkunde.

Chambart de Lowe, M.-J. (1977). Kinder-Welt und Umwelt Stadt. *ARCH+, 34*, 24–29.

Chilla, T., Kühne, O., & Neufeld, M. (2016). *Regionalentwicklung*. Stuttgart: Ulmer.

Claassen, K., Engelmann, D., Gaffga, P., Latz, W., & Weidner, W. (2005). Diercke Erdkunde. Klasse 11. Braunschweig: Westermann.

Clarke, S. (2008). Constructing the politics of landscape change. In J. Wescoat & D. Johnston (Eds.), *Political economies of landscape change: Places of integrative power* (Vol. 89, pp. 91–109). Dordrecht: Springer.

Colby, A., & Kohlberg, L. (1978). Das moralische Urteil. Der kognitionszentrierte entwicklungspsychologische Ansatz. In G. Steiner (Ed.), *Piaget und die Folgen* (pp. 348–365). Zürich: Kindler.

Cosgrove, D. E. (1984). *Social formation and symbolic landscape*. London, Sydney: University of Wisconsin Press.

Cosgrove, D. (1985). Prospect, perspective and the evolution of the landscape idea. *Transactions of the Institute of British Geographers, 10*(1), 45–62.

Cosgrove, D. (1993). *The Palladian landscape: geographical change an its representation in sixteenth century Italy*. Leicester: Leicester University Press.

Cosgrove, D., & Domosh, M. (1997). Author and authority. Writing the new cultural geography. In J. Duncan & D. Ley (Eds.), *Place, culture, representation* (pp. 25–38). London: Routledge.

Croce, B. (1930). *Aesthetik als Wissenschaft vom Ausdruck und allgemeine Sprachwissenschaft*. Tübingen: J. C. B. Mohr.

Crouch, C. (2008). *Postdemokratie*. Frankfurt a. M.: Suhrkamp

D'hulst, L. (2007). Analyse discursive et scénographie de l'espace dans deux romans d'Èdouard Glissant. In M. Schmeling & M. Schmitz-Emans (Eds.), *Das Paradigma der Landschaft in*

*Moderne und Postmoderne. (Post-)Modernist terrains: Landscapes—settings—spaces* (pp. 93–104). Würzburg: Augustinus.

Dangschat, J. (1997). Wohlstands- und Armutsentwicklung in deutschen Städten. In Th Krämer-Badoni & W. Petrowsky (Eds.), *Das Verschwinden der Städte* (pp. 168–206). Bremen: Temmen.

Dangschat, J. (2000). Segregation. In H. Häußermann (Ed.), *Großstadt. Soziologische Stichworte* (pp. 209–221). Opladen: Leske + Budrich.

Davis, M. (2004 [1998]). Ökologie der Angst. Das Leben mit der Katastrophe. München, Zürich: Antje Kunstmann.

De Visscher, S., & Bouverne-De Bie, M. (2008). Recognizing urban public space as a co-educator: Children's socialization in Ghent. *International Journal of Urban and Regional Research, 32* (3), 604–616.

Debes, C. (2005). Landschaft versus Raum. Der Widerstand gegen Windenergieanlagen als Akzeptanzproblem ästhetisch-symbolischern Ursprungs – eine raumplanerische Herausforderung. In I. Kazal, A. Voigt, A. Weil & A. Zutz (Eds.), *Kulturen der Landschaft. Ideen von Kulturlandschaft zwischen Tradition und Modernisierung* (pp. 111–128). Berlin: Universitätsverlag TU Berlin.

Degn, C., Eggert, E., Kolb, A., Bartels, H., & Baur, L. (Eds.). (1965). *Seydlitz. Sechster Teil. Das Weltbild der Gegenwart*. Kiel, Hannover: Ferdinand Hirt-Hermann Schroedel Verlag.

Deleuze, G. (1975). Écrivain non: un nouveau cartographe. *Critique, 343,* 1207–1227.

DeMarrais, E., Castillo, L., & Earle, T. (1996). Ideology, materialization, and power strategies. *Current Anthropology, 37*(1), 15–31.

Denecke, D. (2000). Geographische Kulturlandschaftsforschung für eine Kulturlandschaftspflege bezogen auf unterschiedliche Landschaftsräume. Ein Beitrag zur Diskussion. *Berichte zur deutschen Landeskunde, 74,* 197–219.

Dethloff, U. (1995a). Vorwort. In U. Dethloff (Ed.), *Literarische Landschaft. Naturauffassung und Naturbeschreibung zwischen 1750 und 1830* (pp. 7–10). St. Ingbert: Conte.

Dethloff, U. (1995b). Naturerlebnis und Landschaftsdarstellung im französischen Roman der Frühromantik – ein Beitrag zur Einführung. In U. Dethloff (Ed.), *Literarische Landschaft. Naturauffassung und Naturbeschreibung zwischen 1750 und 1830* (pp. 15–32). St. Ingbert: conte.

Diller, C. (2005). Regional governance by and with government: Die Rolle staatlicher Rahmensetzungen und Akteure in drei Prozessen der Regionsbildung. http://fss.plone.unigiessen.de/fss/fbz/fb07/fachgebiete/geographie/bereiche/lehrstuhl/planung/pdf-bilder/HabilitationsschriftDiller.pdf/file/HabilitationsschriftDiller.pdf. Accessed December 11, 2012.

Dingler, J. (1998). Die diskursive Konstruktion von Natur als Produkt von Macht-Wissen-Diskursen: Für eine postmoderne Wende in der ökologischen Theorienbildung. www.gradnet.der/pa-pers/pomo2.archives/pomo98.papers/jsdingle98.htm. Accessed December 5, 2006.

Dixon, D. (1991). *Wie ist das? Land, Meer und Luft*. Wien, Stuttgart: Österreichischer Bundesverlag.

Dosch, F., & Beckmann, G. (1999). Strategien künftiger Landnutzung – ist Landschaft planbar? *Informationen zur Raumentwicklung, 5–6,* 381–398.

Dreitzel, H. (1962). Selbstbild und Gesellschaftsbild. *Europäisches Archiv für Soziologie, 3,* 181–228.

Drepper, T. (2003). Der Raum der Organisation – Annäherung an ein Thema. In Th. Krämer-Badoni & K. Kuhm (Eds.), *Die Gesellschaft und ihr Raum. Raum als Gegenstand der Soziologie* (pp. 103–129). Opladen: ulb.

Duncan, J. (1973). Landscape taste as a symbol of group identity: A Westchester County village. *Geographical Review, 63,* 334–355.

Duncan, J. (1999). Elite landscapes as cultural (re)productions: The case of shaugunessy heights. In K. Anderson & F. Gale (Eds.), *Cultural geographies* (pp. 53–70). London: Longman.

Duncan, J., & Duncan, N. (2004). *Landscapes of Privilege. The Politics of the Aesthetic in an American Suburb*. New York, London: Routledge.

Durkheim, É. (1984[1912]). *Die elementaren Formen des religiösen Lebens*. Frankfurt a.M.: Verlag der Weltreligionen.

Durkheim, É. (2013 [1912]). *Les formes élémentaires de la vie religieuse*. Paris: PUF.

EASAC—European Academies Science Advisory Council. (2009). *Ecosystem services and biodiversity in Europe*. EASAC Policy Report 09. Cardiff: EASAC Secretariat, The Royal Society.

Eisel, U. (1980). *Die Entwicklung der Anthropogeographie von einer, Raumwissenschaft' zur Gesellschaftswissenschaft*. Kassel: Gesamthochschul-Bibliothek.

Eisenberg, N., Cumberland, A., & Spinrad, T. (1998). Parental socialization of emotion. *Psychological Inquiry, 9*(4), 241–273.

Eissing, H., & Franke, N. (2015). Orte in der Landschaft. Anmerkungen über die Macht von Institutionen. In S. Kost & A. Schönwald (Eds.), *Landschaftswandel – Wandel von Machtstrukturen* (pp. 55–63). Wiesbaden: Springer VS.

Elder, G. (2000). Das Lebenslaufs-Paradigma: Sozialer Wandel und individuelle Entwicklung. In M. Grundmann & K. Lüscher (Eds.), *Sozialökologische Sozialisationsforschung. Ein anwendungsorientiertes Lehr- und Studienbuch* (pp. 167–199). Konstanz: Universitäts-Verlag.

Elias, N. (2002). Kitschstil und Kitschzeitalter. In N. Elias (Ed.), *Gesammelte Schriften* (Vol. 1, pp. 148–163). Frankfurt a. M.: Suhrkamp.

Engelmann, D., & Latz, W. (1997). *Landschaftsgürtel. Ökologie und Nutzung*. Braunschweig: Westermann.

Engler, M. (1997). Repulsive matter. Landscapes of waste in the American middle-class reidential domain. *Landscape Journal, 16*(2), 60–79.

Ewald, K. (1996). Traditionelle Kulturlandschaften. Elemente und Bedeutung. In W. Konold (Ed.), *Naturlandschaft. Kulturlandschaft. Die Veränderung der Landschaften nach der Nutzbarmachung durch den Menschen* (pp. 99–120). Landsberg: ecomed.

Fainstein, S. (2010). *The just city*. Ithacta: Cornell University Press.

Fayet, R. (2003). *Reinigungen. Vom Abfall der Moderne zum Kompost der Nachmoderne*. Wien: Passagen-Verlag.

Fend, H. (1981). *Theorie der Schule*. München et al.: Deutsches Jugendinstitut e.V.

Fezer, J. (2006). Politik – Umwelt – Mensch. In J. Fezer & M. Schmitz (Eds.), *Wer plant die Planung? Architektur, Politik, Mensch* (pp. 5–11). Kassel: Schmitz.

Fischer, R. (2005). Regulierter Rinderwahnsinn. Die Reform der wissenschaftlichen Politikberatung innerhalb der Europäischen Union. In A. Bogner & H. Torgersen (Eds.), Wozu Experten? Ambivalenzen der Beziehung von Wissenschaft und Politik (pp. 109–130). Wiesbaden: Springer VS.

Forkel, J., & Grimm, M. (2014). Die Emotionalisierung durch Landschaft oder das Glück in der Natur. *Sozialwissenschaften & Berufspraxis, 37*(2), 251–266.

Forsyth, T. (2004). *Critical political ecology: The politics of environmental science*. London, New York: Routledge.

Foucault, M. (1977 [1975]). Surveiller et punir. Naissance de la prison. Paris: Éditions Gallimard.

Foucault, M. (1978). *Dispositive der Macht*. Berlin: Merve.

Foucault, M. (2006 [1976]). La Volonté de savoir: Droit de mort et pouvoir sur la vie. Paris: Folio.

Freidson, E. (1986). *Professional powers. A study of the institutionalization of formal knowledge*. Chicago: University of Chicago Press.

Friedrichs, J. (1995). *Stadtsoziologie*. Opladen: Leske + Budrich.

Fröhlich, G. (2002). *Anonyme Kritik: Peer Review auf dem Prüfstand der Wissenschaftsforschung*. Wien: Phoibos.

Frommelt-Beyer, R., Hofmeister, U., Sajak, D., & Tiefke, R. (2003). Durchblick. GSW Erdkunde 9/10. Realschule Niedersachsen. Braunschweig: Westermann.

Fuchs, T. (2000). *Leib, Raum, Person. Entwurf einer phänomenologischen Anthropologie*. Klett-Cotta: Stuttgart.

Fuhrer, U. (1997). *Von den sozialen Grundlagen des Umweltbewusstseins zum verantwortlichen Umwelthandeln*. Bern, Göttingen, Toronto, Seattle: Huber.

# References

Füller, H., & Marquardt, N. (2010). *Die Sicherstellung von Urbanität. Innerstädtische Restrukturierung und soziale Kontrolle in Los Angeles*. Münster: Verlag Westfälisches Dampfboot.

Funtowicz, S., & Ravetz, J. (1990). *Uncertainty and quality in science for policy*. Dordrecht: Kluwer Academic Publishers.

Fürst, D. (2007). Regional governance. In A. Benz, S. Lütz, U. Schimank, & G. Simonis (Eds.), *Handbuch governance – Theoretische Grundlagen und empirische Anwendungsfelder* (pp. 353–366). Wiesbaden: Springer VS.

Gailing, L. (2012). Sektorale Institutionensysteme und die Governance kulturlandschaftlicher Handlungsräume. Eine institutionen- und steuerungstheoretische Perspektive auf die Konstruktion von Kulturlandschaft. *Raumforschung und Raumordnung, 70*(2), 147–160.

Gailing, L. (2014). *Kulturlandschaftspolitik. Die gesellschaftliche Konstituierung von Kulturlandschaft durch Institutionen und Governance*. Detmold: Rohn-Verlag.

Gailing, L. (2015). Landschaft und productive Macht. Auf dem Weg zur Analyse landschaftlicher Gouvernementalität. In S. Kost & A. Schönwald (Eds.), Landschaftswandel – Wandel von Machtstrukturen (pp. 37–41). Wiesbaden: Springer VS.

Gailing, L., & Röhring, A. (2008). Kulturlandschaften als Handlungsräume der Regionalentwicklung. Implikationen des neuen Leitbildes zur Kulturlandschaftsgestaltung. *RaumPlanung, 136*, 5–10.

Gallagher, L. (2013). *The end of the suburbs. Where the American dream is moving*. New York: Penguin.

Gehlen, A. (1956). *Urmensch und Spätkultur. Philosophische Ergebnisse und Aussagen*. Frankfurt a. M., Bonn: Athenäum.

Geiger, T. (1947). *Vorstudien zu einer Soziologie des Rechts*. Kopenhagen: Nyt Nordisk Forlag.

Gelfert, H.-G. (2000). *Was ist Kitsch?*. Göttingen: Vandenhoeck & Ruprecht.

Geulen, D. (1991). Die historische Entwicklung sozialisationstheoretischer Ansätze. In K. Hurrelmann & D. Ulrich (Eds.), *Neues Handbuch der Sozialisationsforschung* (pp. 21–54). Weinheim: Beltz Verlag.

Geulen, D. (2005). *Subjektorientierte Sozialisationstheorie. Sozialisation als Epigenese des Subjekts in Interaktion mit der gesellschaftlichen Umwelt*. Weinheim: Juventa.

Geulen, D., & Hurrelmann, K. (1980). Zur Programmatik einer umfassenden Sozialisationstheorie. K. Hurrelmann, D. Ulrich (Eds.), *Handbuch der Sozialisationsforschung* (pp. 51–67). Weinheim: Beltz Verlag.

Gibbons, M., Limoges, C., Nowotny, H., Schwartzman, S., Scott, P., & Trow, M. (1994). *the new production of knowledge. The dynamics of science and research in contemporary societies*. London: SAGE.

Giddens, A. (1990). *The consequences of modernity*. Palo Alto: Stanford University Press.

Glaser, R., & Chi, M. (1988). Overview. In M. Chi, R. Glaser, & M. Farr (Eds.), *The nature of expertise*. Hillsdale, New York: Lawrence Erlbaum Associates.

Gmeiner, R. (2005). Nationale Ethikkommissionen: Aufgaben, Formen, Funktionen. In A. Bogner & H. Torgersen (Eds.), *Wozu Experten? Ambivalenzen der Beziehung von Wissenschaft und Politik* (pp. 133–148). Wiesbaden: Springer VS.

Goodman, R. (1971). *After the planners*. New York: Simon and Schuster.

Goss, J. (1993). The „magic of the mall": an analysis of form, function, and meaning in the contemporary retail built environment. *Annals of American Geographers, 83*(1), 18–47.

Gottschick, M., & Ette, J. (2012). Etablierte Partizipationslandschaften. Hemmnis für Innovationen zur nachhaltigen regionalen Entwicklung und Anpassung an den Klimawandel? In A. Knierim, St. Baasch & M. Gottschick (Eds.), *Partizipationsforschung und Partizipationsverfahren in der sozialwissenschaftlichen Klimaforschung* (pp. 26–39). Müncheberg: Plenum Press.

Graham, B. (1998). The past in Europe's present: diversity, identity and the construction of place. In B. Graham (Ed.), *Modern Europe. Place. Culture. Identity* (pp. 19–49). London: Routledge.

Graham, B., Ashworth, G. J., & Tunbridge, J. E. (2000). *A geography of heritage: Power, culture, and economy*. London et al.: Arnold.

Greenberg, C. (2007 [1939]). Avantgarde und Kitsch. In U. Dettmar & Th. Küpper (Eds.), Kitsch. Texte und Theorien (pp. 203–212). Stuttgart: Reclam.
Gregory, D. (1994). *Geographical imaginations*. Oxford: Blackwell.
Greider, T., & Garkovich, L. (1994). Landscapes: The social construction of nature and the environment. *Rural Sociology, 59*(1), 1–24.
Groß, M. (2006). Natur. Bielefeld: transcript.
Groth, P., & Wilson, C. (2003). Die Polyphonie der Cultural Landscape Studies. In B. Franzen & St. Krebs (Eds.), *Landschaftstheorie. Texte der Cultural Landscape Studies* (pp. 58–90). Köln: Krebs.
Grotmann, N., & Fuchs, S. (2004), Auf Safari in Afrika... Die neue ‚Kiwara-Savanne' im Zoo Leipzig. Stadt + Grün, *53*, 45–45.
Grundmann, M. (2006). Sozialisation. Skizze einer allgemeinen Theorie. Konstanz: utb.
Grunewald, K., & Bastian, O. (2013). Ökosystemdienstleistungen (ÖSD) – mehr als ein Modewort? In K. Grunewald & O. Bastian (Eds.), *Ökosystemdienstleistungen. Konzept, Methoden und Fallbeispiele* (pp. 1–11). Berlin, Heidelberg: Spektrum.
Güth, J. (2004). Kultur kaputt – vom Umgang mit unserer Umgebung. In A. Berger & M. v. Hohnhorst (Eds.), *Heimat – die Wiederentdeckung einer Utopie* (pp. 156–176). Blieskastel: Gollenstein.
Haber, W. (1992). Über die Entwicklung der Naturschutzgesetzgebung. In Bayrische Akademie der Wissenschaften (Ed.), *Probleme der Umweltforschung in historischer Sicht. Rundgespräche der Kommission für Ökologie*, Volume 7 (pp. 221–231). München: Verlag Dr. Friedrich Pfeil.
Haber, W. (2000). Die Kultur der Landschaft. Von der Ästhetik zur Nachhaltigkeit. In St. Appel, E. gr. Duman, F. Kohorst & F. Schafranski (Eds.), *Wege zu einer neuen Planungs- und Landschaftskultur. Festschrift für Hanns Stephan Wüst* (pp. 1–19). Kaiserslautern: Universität, ZBT.
Habermas, J. (1970). *Zur Logik der Sozialwissenschaften*. Suhrkamp: Frankfurt a. M.
Habermas, J. (1981). *Theorie des kommunikativen Handelns* (2 Vols). Frankfurt a. M.: Suhrkamp.
Habermas, J. (1982). *Zur Logik der Sozialwissenschaften*. Suhrkamp: Frankfurt a. M.
Hagemann-White, C. (1993). Wie (un)gesund ist Weiblichkeit? *Zeitschrift für Frauenforschung, 12*, 20–27.
Hahn, A. (2001). Lebenswelten im Wandel. Interpretationen zum kulturellen Ausdruck von Wohnsuburbanisierung. In K. Brake, J. Dangschat & G. Herfert (Eds.), *Suburbanisierung in Deutschland – aktuelle Tendenzen* (pp. 223–233). Opladen: Leske + Budrich.
Hahn, A. (2014). Entwerfen, Planen und Entscheiden. *Ausdruck und Gebrauch, 12,* 71–95.
Hammerschmidt, V., & Wilke, J. (1990). *Die Entdeckung der Landschaft. Englische Gärten des 18. Jahrhunderts*. Stuttgart: Deutsche Verlags-Anstalt.
Hanlon, B. (2012). *Once the American dream. Inner-ring suburbs of the Metropolitan United States*. Philadelphia: Temple University Press.
Hannah, M. (2009). Calculable territory and the West German census boycott movements of the 1980s. *Political Geography, 28*(1), 66–75.
Hannigan, J. (2014). *Environmental sociology*. London, New York: Routledge, Taylor & Francis Group.
Hansjürgens, B., & Schröter-Schlaack, C. (2012). Die ökonomische Bedeutung der Natur. In B. Hansjürgens, C. Nesshöver & I. Schniewind (Eds.), *Der Nutzen von Ökonomie und Ökosystemleistungen für die Naturschutzpraxis. Workshop I: Einführung und Grundlagen. BfN-Skripten 318* (2nd ed., pp. 16–21). Bonn-Bad Godesberg: Bundesamt für Naturschutz.
Hard, G. (1965). Arkadien in Deutschland. Bemerkungen zu einem landschaftlichen Reiz. In G. Hard (Ed.) (2002), *Landschaft und Raum. Aufsätze zur Theorie der Geographie* (pp. 11–34). Osnabrück: University Press Rasch.
Hard, G. (1969). Das Wort Landschaft und sein semantischer Hof. Zur Methode und Ergebnis eines linguistischen Tests. *Wirkendes Wort, 9,* 3–14.

# References

Hard, G. (1977). Zu den Landschaftsbegriffen der Geographie. In A. V. Wallthor & H. Quirin (Eds.), ‚Landschaft' als interdisziplinäres Forschungsproblem (pp. 13–24). Münster: Aschendorff.

Hard, G. (1985). Städtische Rasen – hermeneutisch betrachtet. Ein Kapitel aus der Geschichte der Verleugnung der Stadt durch die Städter. Klagenfurter Geographische Schriften (Vol. 6). Klagefurt: Herbert Wichmann.

Hard, G. (1991). Landschaft als professionelles Idol. Garten und Landschaft, 3, 13–18.

Hard, G. (2002a [1969a]). „Dunstige Klarheit". Zu Goethes Beschreibung der italienischen Landschaft. In G. Hard (Ed.), Landschaft und Raum. Aufsätze zur Theorie der Geographie (pp. 49–68). Osnabrück: University Press Rasch.

Hard, G. (2002b [1969b]). Die Diffusion der Idee der Landschaft. Präliminarien zu einer Geschichte der Landschaftsgeographie. In G. Hard (Ed.), Landschaft und Raum. Aufsätze zur Theorie der Geographie (pp. 103–132). Osnabrück: University Press Rasch.

Hard, G. (2002c [1970]). „Was ist eine Landschaft?" Über Etymologie als Denkform in der geographischen Literatur. In G. Hard (Ed.), Landschaft und Raum. Aufsätze zur Theorie der Geographie (pp. 133–154). Osnabrück: University Press Rasch.

Hard, G. (2002d [1971]). Über die Gleichzeitigkeit des Ungleichzeitigen. Anmerkungen zur jüngsten methodologischen Literatur in der deutschen Geographie. In G. Hard (Ed.), Landschaft und Raum. Aufsätze zur Theorie der Geographie (pp. 155–170). Osnabrück: University Press Rasch.

Härle, J. (2004). Landschaftspflege. Gegen die Verarmung der Kulturlandschaft. Praxis Geographie, 34, 4–11.

Hartmann, G. (1981). Die Ruine im Landschaftsgarten. Worms: Werner'sche Verlagsgesellschaft.

Hartz, A. (2003). Neue Perspektiven für die Stadtlandschaft. Garten und Landschaft, 9, 16–17.

Hartz, A., & Kühne, O. (2006). Der Regionalpark Saar – eine Betrachtung aus postmoderner Perspektive. Raumforschung und Raumordnung, 65(1), 30–43.

Hartz, A., & Kühne, O. (2007). Regionalpark Saarkohlewald – eine Bestandsaufnahme aus postmoderner Perspektive. Raumforschung und Raumordnung, 65, 30–43.

Harvey, D. (1990). Between space and time: Reflections on the geographical imagination. Annals of the Association of American Geographers, 80, 418–434.

Harvey, D. (1991). Geld, Zeit, Raum und die Stadt. In M. Wentz (Ed.), Stadt-Räume (pp. 149–168). Frankfurt a. M.: WW Norton, New York.

Harvey, D. (1996). Justice, nature, and the geography of difference. Oxford: Wiley-Blackwell.

Hasse, J. (1993). Heimat und Landschaft. Über Gartenzwerge, Center Parcs und andere Ästhetisierungen. Wien: Passagen Publishing Company.

Hasse, J. (2000). Die Wunden der Stadt. Für eine neue Ästhetik unserer Städte. Wien: Passagen Publishing Company.

Haug, W. (1986). Die Faschisierung des bürgerlichen Subjektes. Die Ideologisierung der gesunden Normalität und die Ausrottungspolitik im deutschen Faschismus: Materialanalysen. Berlin: Argument.

Haupt, H. (1998). Der Bürger. In F. Furet (Ed.), Der Mensch der Romantik (pp. 23–67). Frankfurt a. M.: Campus.

Hauser, S. (2001). Metamorphosen des Abfalls. Campus: Konzepte für alte Industrieareale. Frankfurt a. M.

Hauser, S. (2004). Industrieareale als urbane Räume. In W. Siebel (Ed.), Die europäische Stadt (pp. 146–157). Frankfurt a. M.: Suhrkamp.

Hauser, S., & Kamleithner, C. (2006). Ästhetik der Agglomeration. Wuppertal: Müller und Busmann.

Häußermann, H., & Kapphan, A. (2000). Berlin. Von der geteilten zur gespaltenen Stadt? Sozialräumlicher Wandel seit 1990. Opladen: Springer.

Häußermann, H., & Siebel, W. (2004). Stadtsoziologie. Eine Einführung. Frankfurt, New York: Campus.

Healey, P. (2006). Urban complexity and spatial strategies: Towards a relational planning for our times. London: Routledge.

Hegel, G. (1970). Vorlesungen über die Ästhetik I-III. In E. Moldenhauer & K. Michel (Eds.), *Werke in 20 Bänden* (pp. 13–15). Frankfurt a. M.: Suhrkamp.

Heiland, S. (1999). *Voraussetzungen erfolgreichen Naturschutzes. Individuelle und Gesellschaftliche Bedingungen umweltgerechten Verhaltens, ihre Bedeutung für den Naturschutz und die Durchsetzbarkeit seiner Ziele.* Landsberg: Ecomed.

Heiland, S. (2006). Zwischen Wandel und Bewahrung, zwischen Sein und Sollen: Kulturlandschaft als Thema und Schutzgut in Naturschutz und Landschaftsplanung. In U. Matthiesen, R. Danielzyk, S. Heiland & S. Tzschaschel (Eds.), *Kulturlandschaften als Herausforderung für die Raumplanung. Verständnisse – Erfahrungen – Perspektiven* (pp. 43–70). Hannover: Verlag der ARL.

Heineberg, H. (1989). *Stadtgeographie.* Paderborn, München, Wien: utb.

Heinelt, H. (1997). *Die Transformation der Demokratie und die Bedeutung des zivilgesellschaftlichen Sektors im politischen System moderner Gesellschaften.* In H. Heinelt & K. Schals (Eds.), *Zivile Gesellschaft. Entwicklung, Defizite und Potentiale* (pp. 323–339). Opladen: Leske + Budrich.

Henkel, G. (1997). Kann die überlieferte Kulturlandschaft ein Leitbild für die Planung sein? *Berichte zur deutschen Landeskunde, 71,* 27–39.

Heringer, J. (2005). Inszenierung von Kulturlandschaft. In Deutscher Rat für Landespflege (Ed.), *Landschaft und Heimat. Schriftenreihe des Deutschen Rates für Landespflege* (Vol. 77, pp. 77–85). Meckenheim: Deutscher Rat für Landespflege.

Hesse, M. (2008). Reurbanisierung? Urbane Diskurse, Deutungskonkurrenzen, konzeptuelle Konfusion. *Raumforschung und Raumordnung, 5,* 415–428.

Hesse, M. (2010). Suburbs: The next slum? Explorations into the contested terrain of social construction and political discourse. *Articulo - revue de sciences humaines.* http://articulo.revues.org/1552. Accessed September 27, 2013.

Higley, S. (1995). *Privilege, power, and place: The geography of the American upper class.* Lanham: Rowman & Littlefield Publishers.

Hilbig, H. (2014). Warum es keine Architekturethik braucht – Und warum vielleicht doch. *Ausdruck und Gebrauch, 12,* 96–106.

Hirschauer, S. (2004). Peer Review Verfahren auf dem Prüfstand. Zum Soziologiedefizit der Wissenschaftsevaluation. *Zeitschrift für Soziologie, 33*(1), 62–83.

Hirschman, A. (1970). *Exit, voice, and loyalty: Responses to decline in firms, organizations, and states.* Cambridge: Harvard University Press.

Hoeres, W. (2004). *Der Weg der Anschauung. Landschaft zwischen Ästhetik und Metaphysik.* Kusterdingen: Graue Edition.

Höfer, W. (2001). *Natur als Gestaltungsfrage. Zum Einfluss aktueller gesellschaftlicher Veränderungen auf die Idee von Natur und Landschaft als Gegenstand der Landschaftsarchitektur.* München: Herbert Utz.

Höfer, W. (2004). Landschaft in Bewegung – Die Konfusion mit der Konversion. *Stadt + Grün, 53,* 28–33.

Hofmeister, S.,& Kühne, O. (2016). StadtLandschaften: Die neue Hybridität von Stadt und Land. In S. Hofmeister & O. Kühne (Eds.), *StadtLandschaften. Die neue Hybridität von Stadt und Land* (pp. 1–10). Wiesbaden: Springer VS.

Hoisl, R., Nohl, W., Zekorn, S., & Zöllner, G. (1987). *Landschaftsästhetik in der Flurbereinigung. Empirische Grundlagen zum Erleben der Agrarlandschaft.* München: Bayerisches Staatsministerium für Ernährung, Landwirtschaft und Forsten.

Hoisl, R., Nohl, W., Zekorn, S., & Zöllner, G. (1989). *Verfahren zur landschaftsästhetischen Vorbilanz.* München: Bayerisches Staatsministerium für Ernährung, Landwirtschaft und Forsten.

Hokema, D. (2009). Die Landschaft der Regionalentwicklung: Wie flexibel ist der Landschaftsbegriff? *Raumforschung und Raumordnung, 67*(3), 239–249.

Hokema, D. (2015). Landscape is everywhere. The construction of landscape by US-American laypersons. *Geographische Zeitschrift, 103*(3), 151–170.

Holm, A. (2006). Die Restrukturierung des Raumes. Stadterneuerung der 90er Jahre in Ostberlin. Interessen und Machtverhältnisse. Bielefeld: Transkript.

Holzinger, M. (2004). *Natur als sozialer Akteur. Realismus und Konstruktivismus in der Wissenschafts- und Gesellschaftstheorie.* Opladen: Leske + Budrich.

Holzner, L. (1996). *Stadtland USA: Die Kulturlandschaft des American Way of Life.* Gotha: Klett/SVK.

Hoppmann, H. (2000). *pro:Vision. Postmoderne Taktiken in einer strategischen Gegenwartsgesellschaft. Eine soziologische Analyse.* Berlin: wvb.

Horkheimer, M. (1977 [1937]). *Traditionelle und kritische Theorie.* Frankfurt a. M: Fischer.

Huang, S.-C. (2010). The impact of public participation on the effectiveness of, and users' attachment to, urban neighbourhood parks. *Landscape Research, 35*(5), 551–562.

Hugill, P. (1986). English landscape tastes in the United States. *Geographical Review, 76,* 408–423.

Hugill, P. (1995). Upstate Arcadia. Landscape, aesthetics, and the triumph of social differentiation in America. Lanham, London: Rowman & Littlefield Publishers.

Hunziker, M. (1995). The spontaneous reafforestation in abandoned agricultural lands: Perception and aesthetic assessment by locals and tourists. *Landscape and Urban Planning, 31,* 399–410.

Hunziker, M. (2010). Die Bedeutungen der Landschaft für den Menschen: objektive Eigenschaft der Landschaft oder individuelle Wahrnehmung des Menschen? *Forum für Wissen, 2010,* 33–41.

Hunziker, M., Felber, P., Gehring, K., Buchecker, M., Bauer, N., & Kienast, F. (2008). How do different societal groups evaluate past and future landscape changes? Results of two empirical studies in Switzerland. *Mountain Research and Development, 28*(2), 140–147.

Hunziker, M., & Kienast, F. (1999). Impacts of changing agricultural activities on scenic beauty— A prototype of an automated rapid assessment technique. *Landscape Ecology, 14,* 161–176.

Hüppauf, B. (2007). Heimat – die Wiederkehr eines verpönten Wortes. Ein Populärmythos im Zeitalter der Globalisierung. In G. Gebhard, O. Geisler & S. Schröter (Eds.), *Heimat. Konturen und Konjunkturen eines umstrittenen Konzepts* (pp. 109–140). Bielefeld: transkript.

Hurrelmann, K. (2006). *Einführung in die Sozialisationstheorie.* Basel, Weinheim: Beltz.

Hussong, H. (2006). *Auf dem Weg zur Biosphäre Bliesgau. Der Einfluss regionaler Akteure auf den Prozess der Implementierung eines Biospährenreservates im Bliesgau. Saarbrücker Landeskundliche Arbeiten.* http://www.iflis.de/resources/Bd1_Bliesgau.pdf. Accessed May 14, 2007.

Illich, I. (1979). *Entmündigung durch Experten. Zur Kritik der Dienstleistungsberufe.* Reinbeck bei Hamburg: Rowohlt.

Illing, F. (2006). *Kitsch, Kommerz und Kult. Soziologie des schlechten Geschmacks.* UVK: Konstanz.

Ingarden, R. (1992). Prinzipien einer erkenntnistheoretischen Betrachtung der ästhetischen Erfahrung. In D. Henrich & W. Iser (Eds.), *Theorien der Kunst* (pp. 70–80). Stuttgart: Reclam.

Ipsen, D. (2002). Raum als Landschaft. In D. Ipsen & D. Läpple (Eds.), *Soziologie des Raumes – Soziologische Perspektiven* (pp. 86–111). Hagen: Centaurus.

Ipsen, D. (2006). *Ort und Landschaft.* Wiesbaden: Springer VS.

Ipsen, D., Reichhardt, U., Schuster, St, Wehrle, A., & Weichler, H. (2003). *Zukunft Landschaft.* Bürgerszenarien zur Landschaftsentwicklung. Kassel: Kassel University Press.

Irrgang, B. (2014). Architekturethik oder Gestaltung von Wohnen. Über das Bauen und die Einbettung von Architektur in Natur und Kultur. *Ausdruck und Gebrauch, 12,* 10–29.

Jackson, J. (1984). Concluding with Landscapes. In J. Jackson (Ed.), *Discovering the vernacular landscape* (pp. 145–157). New Haven: Yale University Press.

Jackson, J. (1990). Die Zukunft des Vernakulären. In B. Franzen & S. Krebs (Eds.), *Landschaftstheorie. Texte der Cultural Landscape Studies* (pp. 45–56). Köln: König, Walther.

Jain, A. (2000). *Reflexiv-deflexive Modernisierung und die Diffusion des Politischen.* München: edition fatal.

Jann, W., & Wegrich, K. (2004). Governance und verwaltungspolitik. In A. Benz (Ed.), *Governance—Regieren in komplexen Regelsystemen* (pp. 193–214). Wiesbaden: VS Verlag für Sozialwissenschaften.

Jansen, D., & Wald, A. (2007). Netzwerktheorien. In A. Benz, S. Lütz, U. Schimank, & G. Simonis (Eds.), *Handbuch Governance – Theoretische Grundlagen und empirische Anwendungsfelder* (pp. 188–199). Wiesbaden: Springer VS.

Jessel, B. (2000). ‚Landschaft' – zum Gebrauch mit einem als selbstverständlich gebrauchten Begriff. In S. Appel, E. Duman, F. Große-Kohorst & F. Schafranski (Eds.), *Wege zu einer neuen Planungs- und Landschaftskultur. Festschrift für Hanns Stephan Wüst* (pp. 143–160). Kaiserslautern: Universität Kaiserslautern.

Jessel, B. (2004). Von der Kulturlandschaft zur Landschafts-Kultur in Europa. Für die Zukunft: Handlungsmaximen statt fester Leitbilder. *Stadt + Grün, 53,* 20–27.

Jessel, B., Tschimpke, O., & Waiser, M. (2009). *Produktivkraft Natur*. Hamburg: Hoffmann und Campe.

Jones, M. (2007). The European landscape convention and the question of public participation. *Landscape Research, 32*(5), 613–633.

Jordan, D. (1996). *Die Neuerschaffung von Paris*. Frankfurt a. M.: S. Fischer.

Jörke, D. (2010). Die Versprechen der Demokratie und die Grenzen der Deliberation. *Zeitschrift für Politikwissenschaft, 20*(3–4), 269–290.

Juarez, J., & Brown, K. (2008). Extracting or empowering? A critique of participatory methods for marginalized populations. *Landscape Journal, 27*(2), 190–203.

Kaplan, R., & Kaplan, S. (1989). *The Experience of Nature. A Psychological Perspective*. New York: Cambridge University Press.

Kaufmann, S. (2004). Umstrittenes Grün. Naturpolitik am Golfplatz. In W. Eßbach, St. Kaufmann, D. Verdicchio, W. Lutterer, S. Bellanger & G. Uerz (Eds.), *Landschaft, Geschlecht, Artefakte. Zur Soziologie naturaler und artifizieller Alteritäten* (pp. 79–98). Würzburg: Ergon.

Kaufmann, S. (2005). *Soziologie der Landschaft*. Wiesbaden: Springer VS.

Kazig, R. (2007). Atmosphären – Konzept für einen nicht repräsentationellen Zugang zum Raum. In R. Pütz & C. Berndt (Eds.), *Kulturelle Geographien. Zur Beschäftigung mit Raum und Ort nach dem Cultural Turn* (pp. 167–187). Bielefeld: transcript.

Kazig, R. (2008). Typische Atmosphären städtischer Plätze. Auf dem Weg zu einer anwendungsorientierten Atmosphärenforschung. *Die alte Stadt, 2,* 148–160.

Keupp, H. (1992). Verunsicherungen. Risiken und Chancen des Subjekts in der Postmoderne. In Th. Rauschenbach & H. Gängler (Eds.), *Soziale Arbeit und Erziehung in der Risikogesellschaft* (pp. 165–183). Neuwied: Hermann Luchterhand.

Kissner, K.-H., Dischereit, M., & Eigenfeld, F. (1980). *Geographie in Übersichten. Wissensspeicher für den Unterricht*. Berlin (Ost): Volk und Wissen Volkseigener Verlag.

Kneer, G., & Nassehi, A. (1997). *Niklas Luhmanns Theorie sozialer Systeme*. München: Fink.

Knorr Cetina, K. (2002a). *Wissenskulturen. Ein Vergleich naturwissenschaftlicher Wissensformen*. Suhrkamp: Frankfurt a. M.

Knorr Cetina, K. (2002b). *Die Fabrikation von Erkenntnis. Zur Anthropologie der Naturwissenschaft*. Suhrkamp: Frankfurt a. M.

Knorr Cetina, K. (2006). Sozialität mit Objekten. Soziale Beziehungen in post-traditionalen Wissensgesellschaften. In D. Tänzler, H. Knoblauch & H.-G. Soeffner (Eds.), *Zur Kritik der Wissensgesellschaft* (pp. 101–138). Konstanz: UVK.

Knorring, E. v. (1995). *Quantifizierung des Umweltproblems durch Monetarisierung?* (p. 128). Volkswirtschaftliche Diskussionsreihe: Institut für Volkswirtschaftslehre der Universität Augsburg, Article Nr.

Knox, P., & Pinch, S. (2010). *Urban social geography. An introduction*. Harlow et al.: Routledge.

Kohlberg, L. (1974). *Zur kognitiven Entwicklung des Kindes*. Suhrkamp: Frankfurt a. M.

Konold, W. (1996). Von der Dynamik einer Kulturlandschaft. Das Allgäu als Beispiel. In W. Konold (Ed.), *Naturlandschaft – Kulturlandschaft. Die Veränderung der Landschaft nach Nutzbarmachung des Menschen* (pp. 121–228). Landsberg: Hüthig Jehle Rehm.

# References

Kook, K. (2008). Zum Landschaftsverständnis von Kindern: Aussichten – Ansichten – Einsichten. In R. Schindler, J. Stadelbauer, & W. Konold (Eds.), *Points of view – Landschaft verstehen – Geographie und Ästhetik, Energie und Technik* (pp. 107–124). Freiburg im Breisgau: Modo.

Korff, C. (2008). Kulturlandschaft im Blick der Urlaubsgäste: Aussichten – Ansichten – Einsichten. In R. Schindler, J. Stadelbauer, & W. Konold (Eds.), *Points of view – Landschaft verstehen – Geographie und Ästhetik, Energie und Technik* (pp. 99–106). Freiburg im Breisgau: Modo.

Körner, S. (2005). Landschaft und Raum im Heimat- und Naturschutz. In M. Weingarten (Ed.), *Strukturierung von Raum und Landschaft. Konzepte in Ökologie und der Theorie gesellschaftlicher Naturverhältnisse* (pp. 107–117). Münster: Westfälisches Dampfboot.

Körner, S. (2006). Eine neue Landschaftstheorie? Eine Kritik am Begriff „Landschaft Drei". *Stadt und Grün, 55,* 18–25.

Kortländer, B. (1977). Die Landschaft der Literatur des ausgehenden 18. und beginnenden 19. Jahrhunderts. In A. v. Wallthor & H. Quirin (Eds.), *Landschaft als interdisziplinäres Forschungsproblem. Vorträge und Diskussion des Kolloquiums am 7./8. November 1975 in Münster* (pp. 26–44). Münster: Aschendorff Verlag.

Kost, S., & Schönwald, A. (Eds.). (2015). *Landschaftswandel – Wandel von Machtstrukturen*. Wiesbaden: Springer VS.

Kowalke, H. (Ed.). (2001). *Heimat und Welt. Oberstufe*. Braunschweig: Westermann.

Krackhardt, D. (1990). Assessing the political landscape: Structure, cognition, and power in organizations. *Administrative Science Quarterly, 35*(2), 342–369.

Krämer-Badoni, T. (2003). Die Gesellschaft und ihr Raum – kleines verwundertes Nachwort zu einem großen Thema. In T. Krämer-Badoni & K. Kuhm (Eds.), *Die Gesellschaft und ihr Raum. Raum als Gegenstand der Soziologie* (pp. 275–286). Opladen: Leske + Budrich.

Krätke, S. (1995). *Stadt – Raum – Ökonomie*. Basel, Boston, Berlin: Birkhäuser Verlag.

Kropp, C. (2002). ‚Natur' – Soziologische Konzepte, politische Konsequenzen. *Soziologie und Ökologie 9*. Opladen: Leske und Budrich.

Kropp, C. (2015). Regionale StadtLandschaften – Muster der lebensweltlichen Erfahrung postindustrieller Raumproduktion zwischen Homogenisierung und Fragmentierung. *Raumforschung und Raumordnung, 73*(2), 91–106.

Kruse-Graumann, L. (1996). Umweltschutz aus psychologischer Perspektive: Bewusstsein und Verhalten. In K.-H. Erdmann & J. Nauber (Eds.), *Beiträge zur Ökosystemforschung und Umwelterziehung* (pp. 171–179). Bonn: MAB.

Kuhn, T. (1973 [1962]). *Die Struktur wissenschaftlicher Revolutionen*. Frankfurt a. M.: Suhrkamp.

Kühn, M. (2002). Landschaft in der Regionalstadt – zwischen Grüngürteln und grünen Herzen. In D. Kornhardt, G. Pütz & T. Schröder (Eds.), *Mögliche Räume. Stadt schafft Landschaft* (pp. 93–99). Hamburg: Junius.

Kühne, O. (2003). *Transformation und Umwelt in Polen. Eine kybernetisch-systemtheoretische Analyse. Mainzer Geographische Studien 51*. Mainz: Geographisches Institut Johannes-Gutenberg-Universität.

Kühne, O. (2004). *Monetarisierung der Umwelt - Chancen und Probleme aus raumwissenschaftlich-systemtheorietischer Perspektive. Beiträge zur kritischen Geographie 3*. Wien: Verein kritische Geographie.

Kühne, O. (2005). *Landschaft als Konstrukt und die Fragwürdigkeit der Grundlagen der konservierenden Landschaftserhaltung – eine konstruktivistisch-systemtheoretische Betrachtung. Beiträge zur Kritischen Geographie 4*. Wien: Verein kritische Geographie.

Kühne, O. (2006a). *Landschaft in der Postmoderne. Das Beispiel des Saarlandes*. Wiesbaden: Deutscher Universitätsverlag.

Kühne, O. (2006b). Landschaft und ihre Konstruktion – theoretische Überlegungen und empirische Befunde. *Naturschutz und Landschaftsplanung, 38,* 146–152.

Kühne, O. (2006c). *Landschaft, Geschmack, soziale Distinktion und Macht – von der romantischen Landschaft zur Industriekultur. Eine Betrachtung auf Grundlage der Soziologie Pierre Bourdieus. Beiträge zur Kritischen Geographie 6*. Wien: Verein kritische Geographie.

Kühne, O. (2006d). Soziale Distinktion und Landschaft. Eine landschaftssoziologische Betrachtung. Stadt + Grün, 56(12), 40–43.
Kühne, O. (2007a). Soziale Akzeptanz und Perspektiven der Altindustrielandschaft. Das Beispiel des Saarlandes. RaumPlanung, 132(133), 156–160.
Kühne, O. (2007b). Das Ende der europäischen Stadt? Von der Suburbanisierung zur Stadtlandschaft. Studienbrief der FernUniversität in Hagen. Hagen: FernUniversität.
Kühne, O. (2008a). Distinktion, Macht, Landschaft. Zur sozialen Definition von Landschaft. Wiesbaden: Springer VS.
Kühne, O. (2008b). Die Sozialisation von Landschaft – sozialkonstruktivistische Überlegungen, empirische Befunde und Konsequenzen für den Umgang mit dem Thema Landschaft in Geographie und räumlicher Planung. Geographische Zeitschrift, 96(4), 189–206.
Kühne, O. (2008c). Landschaft und Kitsch – Anmerkungen zu impliziten und expliziten Landschaftsvorstellungen. Naturschutz und Landschaftsplanung, 44(12), 403–408.
Kühne, O. (2010). UNESCO-Biosphärenreservat Bliesgau – Entwicklungen, Beteiligungen und Verfahren in einer Modellregion. Standort – Zeitschrift für angewandte Geographie, 34(1), 27–33.
Kühne, O. (2012). Stadt – Landschaft – Hybridität. Ästhetische Bezüge im postmodernen Los Angeles mit seinen modernen Persistenzen. Wiesbaden: Springer VS.
Kühne, O. (2013). Landschaftstheorie und Landschaftspraxis. Eine Einführung aus sozialkonstruktivistischer Perspektive. Wiesbaden: Springer VS.
Kühne, O. (2014a). Das Konzept der Ökosystemdienstleistungen als Ausdruck ökologischer Kommunikation. Betrachtungen aus der Perspektive Luhmannscher Systemtheorie. Naturschutz und Landschaftsplanung, 46(1), 17–22.
Kühne, O. (2014b). Die intergenerationell differenzierte Konstruktion von Landschaft. Naturschutz und Landschaftsplanung, 46(10), 297–302.
Kühne, O. (2014c). Landschaft und Macht: von Eigenlogiken und Ästhetiken in der Raumentwicklung. Ausdruck und Gebrauch, 12, 151–172.
Kühne, O. (2015a). The streets of Los Angeles—about the integration of infrastructure and power. Landscape Research, 40(2), 139–153.
Kühne, O. (2015b). Wasser in der Landschaft – Deutungen und symbolische Aufladungen. In C. Kressin (Ed.), Oberwasser – Kulturlandschaft mit Kieswirtschaft (pp. 35–48). Rees: PIUS.
Kühne, O. (2015c). Landschaftsforschung und Landschaftspraxis aus konstruktivistischer Sicht. Am Beispiel der landschaftlichen Folgen der Energiewende. Geographie aktuell und Schule, 37(213), 9–15.
Kühne, O. (2015d). Weltanschauungen in regionalentwickelndem Handeln – die Beispiele liberaler und konservativer Ideensysteme. In O. Kühne & F. Weber (Eds.), Bausteine der Regionalentwicklung (pp. 55–69). Wiesbaden: Springer VS.
Kühne, O. (2016a). Los Angeles – machtspezifische Implikationen einer Verkehrsinfrastruktur. In S. Hofmeister & O. Kühne (Eds.), StadtLandschaften. Die neue Hybridität von Stadt und Land (pp. 253–281). Wiesbaden: Springer.
Kühne, O. (2016b). Räume, Grenzen und Ränder – Aspekte gesellschaftlicher Raumorganisation. In A. Schaffer, E. Lang, & S. Hartard (Eds.), An und in Grenzen – Entfaltungsräume für eine nachhaltige Entwicklung (pp. 303–326). Marburg: Metropolis.
Kühne, O., & Schönwald, A. (2014). Landschaften und Identitäten in Zeiten offener Grenzen und fortgeschrittener Hybridisierung – das Beispiel der Großregion. Europa Regional, 20(1), 3–14.
Kühne, O., & Schönwald, A. (2015a). San Diego – Eigenlogiken, Widersprüche und Entwicklungen in und von, America's finest city'. Wiesbaden: Springer VS.
Kühne, O., & Schönwald, A. (2015b). Biographische Konstruktionen von Mobilität und Landschaft in der Grenzregion San Diego-Tijuana. In J. Schreiner & Ch. Holz-Rau (Eds.), Räumliche Mobilität und Lebenslauf. Studien zu Mobilitätsbiographien und Mobilitätssozialisation (pp. 221–238). Wiesbaden: Springer VS.
Kühne, O., & Schönwald, A. (2015c). Identität, Heimat sowie In- und Exklusion: Aspekte der sozialen Konstruktion von Eigenem und Fremdem als Herausforderung des Migrationszeitalters. In B. Nienaber & U. Roos (Eds.), Internationalisierung der

# References

*Gesellschaft und die Auswirkungen auf die Raumentwicklung Beispiele aus Hessen, Rheinland-Pfalz und dem Saarland. Arbeitsberichte der ARL 13* (pp. 100–110). Hannover: Verlag der ARL.

Kukla, A. (2000). *Social constructivism and the philosophy of science*. New York: Psychology Press, Routledge.

Kunzmann, K. (2001). Welche Zukünfte für Suburbia? Acht Inseln im Archipel der Stadtregion. In K. Brake, J. Dangschat, & G. Herfert (Eds.), *Suburbanisierung in Deutschland – aktuelle Tendenzen* (pp. 213–221). Opladen: Leske & Budrich.

Kuphal, A. (2006). Jeder Verein ist ein Unikat – alle Vereine sind gleich! In Ministerium für Umwelt (Ed.), *Dorf und Verein – Tradition und Zukunft* (pp. 34–40). Saarbrücken: Ministerium für Umwelt.

Lacoste, Y. (1990). *Geographie und politisches Handeln*. Berlin: Wagenbach.

Lahnstein, M. (2000). *Die Feuerwehr als Brandstifter. Die unheimliche Macht der Experten*. München: Droemer.

Larson, M. (1977). *The rise of professionalism*. Berkeley: University of California Press.

Lash, S., & Urry, J. (1994). *Economies of signs and space*. London: Sage.

Latour, B. (1999). *Pandora's hope. Essays on the reality of science studies*. Cambridge, Massachusetts: Harvard University Press.

Latour, B. (2002 [1999]). *Die Hoffnung der Pandora*. Frankfurt a. M.: Suhrkamp.

Lefèbvre, H. (1972). *Die Revolution der Städte*. München: Hanser Verlag.

Lefèbvre, H. (1974). *La production de l'espace*. Paris: Anthropos.

Leggewie, C. (1998). *Internet & Politik: von der Zuschauer- zur Beteiligungsdemokratie?*. Köln: Bollmann.

Leibenath, M., & Otto, A. (2011). *Diskursive Konstituierung von Kulturlandschaft am Beispiel politischer Windenergiediskurse*. Lecture at the Kulakon-final conference on 13th May 2011 in Hannover.

Lemke, T. (1997). *Eine Kritik der politischen Vernunft. Foucaults Analyse der modernen Gouvernementalität*. Hamburg, Berlin: Argument.

Lenski, G. (2013). *Power and privilege: A theory of social stratification*. Chapel Hill: UNC Press Books.

Lerf, A. (2016). Grenzen des Wissens. Anmerkungen eines Naturwissenschaftlers. In A. Schaffer, E. Lang & S. Hartard (Eds.), *An und in Grenzen – Entfaltungsräume für eine nachhaltige Entwicklung* (pp. 237–254). Marburg: Metropolis.

Leser, H. (1984). Der ökologische Natur- und Landschaftsbegriff. Überlegungen zu seiner Bedeutung für Nutzung, Planung und Entwicklung des Lebensraums. In J. Zimmermann (Ed.), *Das Naturbild des Menschen* (pp. 74–117). München: Fink.

Levidow, L. (2005). Expert-based policy or policy-based expertise? Regulating GM grops in Europe. In A. Bogner & H. Torgersen (Eds.), *Wozu Experten? Ambivalenzen der Beziehung von Wissenschaft und Politik* (pp. 86–108). Wiesbaden: Springer VS.

Liddiard, R. (2005). *Castles in context: Power, symbolism and landscape, 1066 to 1500*. Barnsley: David Brown Book Company.

Liessmann, K. (2002). *Kitsch! Oder warum der schlechte Geschmack der gute ist*. Wien: Christian Brandstätter Verlag.

Lippard, L. (1999). Park-Plätze. In B. Franzen & S. Krebs (Eds.), *Landschaftstheorie. Texte der Cultural Landscape Studies* (pp. 110–138). Köln: König.

Lipps, T. (1891). *Ästhetische Faktoren der Raumanschauung*. Hamburg: L. Voss.

Lipps, T. (1902). *Vom Fühlen, Wollen und Denken*. Hamburg: JA Barth.

Luhmann, N. (1984). *Soziale Systeme. Grundriß einer allgemeinen Theorie*. Suhrkamp: Frankfurt a. M.

Luhmann, N. (1986). *Ökologische Kommunikation*. Opladen: Westdeutscher Verlag.

Luhmann, N. (1989). *Gesellschaftsstruktur und Semantik. Studien zur Wissenssoziologie der modernen Gesellschaft* (Vol. 3). Frankfurt a. M.: Suhrkamp.

Luhmann, N. (1990a). *Die Wissenschaft der Gesellschaft*. Suhrkamp: Frankfurt a. M.

Luhmann, N. (1990b [1986]). *Ökologische Kommunikation. Kann die moderne Gesellschaft sich auf ökologische Gefährdungen einstellen?* Opladen: Westdeutscher Verlag.

Luhmann, N. (1993). Die Moral des Risikos und das Risiko der Moral. In G. Bechmann (Ed.), *Risiko und Gesellschaft* (pp. 327–338). Opladen: Westdeutscher Verlag.

Luhmann, N. (1997). *Die Gesellschaft der Gesellschaft.* Suhrkamp: Frankfurt a. M.

Lynch, M. (2016). Social constructivism in science and technology studies. *Human Studies, 39,* 101–112.

Lyotard, F. (1991). *Leçons sur l'Analytique du sublime.* Paris: Galilée.

Maasen, S., & Weingart, P. (2013). *Metaphor and the dynamics of knowledge.* London, New York: Routledge.

Maier-Solgk, F., & Greuter, A. (1997). *Landschaftsgärten in Deutschland.* Stuttgart: Deutsche Verlagsanstalt.

Mansel, J., & Hurrelmann, K. (2003). Jugendforschung und Sozialisationstheorie. Über Möglichkeiten und Grenzen der Lebensgestaltung im Jugendalter. In J. Mansel, H. Giese & A. Scherr (Eds.), *Theoriedefizite der Jungendforschung. Standortbestimmung und Perspektiven.* München, Weinheim: Juventa.

Manwaring, E. (1965 [1925]). *Italian Landscape in Eighteenth Century England. A Study chiefly on the Influence of Claude Lorrain and Salvator Rosa on English Taste 1700–1800.* London: Russell & Russell.

Markowitz, I. (1995). Ausblicke in die Landschaft. In H. Wunderlich (Ed.), *Landschaft' und Landschaften im achtzehnten Jahrhundert* (pp. 121–156). Heidelberg: Winter.

Massey, D. (2006). *For space.* London: Sage.

Matheis, A. (2016). Vernunft, Moral, Handeln – Grenzverläufe. Anmerkungen zu einem abendländischen kulturellen Selbstverständnis. In A. Schaffer, E. Lang & S. Hartard (Eds.), *An und in Grenzen – Entfaltungsräume für eine nachhaltige Entwicklung* (pp. 107–125). Marburg: Metropolis.

Matless, D. (2005). *Landscape and englishness.* London: Reaktion Books.

MEA – Millennium Ecosystem Assessment. (2005). *Ecosystem and human well-being: Scenarios* (Vol. 2). Washington: Island Press.

Mead, G., & Morris, C. (1967). *Mind, self & society from the standpoint of a social behaviorist.* Chicago: University of Chicago Press.

Megerle, H. (2015). Landschaftswandel in den Savoyer Alpen als Resultat geo- und wirtschaftspolitischer Machtstrukturen. In S. Kost & A. Schönwald (Eds.), *Landschaftswandel – Wandel von Machtstrukturen* (pp. 141–163). Wiesbaden: Springer.

Mekdjian, S. (2008). Les quartiers urbains des diasporas: perennité et reinvention des identifications. Le cas des Arméniens à Los Angeles. In F. Dufaux (Ed.), *Pérennité urbaine ou la ville par-delà ses métamorphoses* (pp. 179–191). Paris: L'Harmattan.

Menzl, M. (2006). Alltag in Suburbia–Betrachtungen zu einer Schlusselkategorie in der Konkurrenz um junge Familien. *Berichte zur deutschen Landeskunde, 80*(4), 433.

Meyer, W., & Kühne, O. (2012). Nachhaltige Entwicklung durch gerechte Beteiligung im grenzenlosen Raum: Herausforderungen des Klimawandels und Perspektiven für neue institutionelle Lösungen. In A. Knierim, S. Baasch & M. Gottschick (Eds.), *Partizipationsforschung und Partizipationsverfahren in der sozialwissenschaftlichen Klimafolgenforschung. Diskussionspapiere zum Workshop* (pp. 78–84). Müncheberg: Leibniz-Zentrum.

Micheel, M. (2012). Alltagsweltliche Konstruktionen von Kulturlandschaft. *Raumforschung und Raumordnung, 70*(2), 107–117.

Michels, R. (1911). *Zur Soziologie des Parteiwesens in der modernen Demokratie. Untersuchungen über die oligarchischen Tendenzen des Gruppenlebens.* Leipzig: Verlag von Dr. Werner Klinkhardt.

Michelsen, D., & Walter, F. (2013). *Unpolitische Demokratie. Zur Krise der Repräsentation.* Suhrkamp: Frankfurt a. M.

Michert, J. (2000). Erinnerung und Zukunftsfähigkeit. Landschaftsarchitektur im Rahmen der IBA Ruhrgebiet. *Stadt + Grün, 48,* 13–21.

Mills, C. (1997a). Myths and meanings of gentrification. In J. S. Duncan & D. Ley (Eds.), *Place, culture, representation* (pp. 149–170). London: Routledge.
Mills, S. (1997b). *The American landscape*. Edinburgh: Keele.
Mitchell, D. (2001). *Cultural geography. A critical introduction*. Oxford: Blackwell.
Mitchell, D. (2005). Landscape. In D. Atkinson, P. Jackson, D. Sibley & N. Washbourne (Eds.), *Cultural geography. A critical dictionary of key concepts* (pp. 49–56). London: Tauris.
Mitchell, D., & Staeheli, L. (2009). Turning social relations into space: Property, law and the Plaza of Santa Fe, New Mexico. In K. Olwig & D. Mitchell (Eds.), *Justice, power and the political landscape* (pp. 73–90). London, New York: Routledge.
Mitscherlich, A. (1980). *Die Unwirtlichkeit unserer Städte*. Frankfurt a. M.: suhrkamp.
Molina, J. (2001). The informal organizational chart in organizations: An approach from the social network analysis. *Connections, 24*(1), 78–91.
Monk, J. (1992). Gender in the landscape: expressions of power and meaning. In K. Anderson & F. Gale (Eds.), *Inventing Places. Studies in Cultural Geography* (pp. 123–138). Kings Gardens: Longman Cheshire.
Mukarovsky, J. (1970 [1942]). *Kapitel aus der Ästhetik*. Frankfurt a. M.: Suhrkamp.
Myers, G. (1996). Naming and placing the other: Power and the urban landscape in Zanzibar. *Tijdschrift voor economische en sociale geografie, 87*(3), 237–246.
Naranjo, F. (2006). Landscape and spatial planning politics. In Council of Europe (Ed.), *Landscape and sustainable development. Challenges of the European Landscape Convention* (pp. 55–82). Strasbourg: Council of Europe Publishing.
Nassauer, J., & Wascher, D. (2008). The globalized landscape: Rural landcape change and politcy in the united states and European Union. In J. Wescoat & D. Johnston (Eds.), *Political economies of landscape change: Places of integrative power* (Vol. 89, pp. 169–194). Dordrecht: Springer Science + Business Media.
Nassehi, A. (1999). *Differenzierungsfolgen. Beiträge zur Soziologie der Moderne*. Opladen, Wiesbaden: Westdeutscher Verlag.
Nennen, H.-U., & Garbe, D. (Eds.). (1996). *Das Expertendilemma. Zur Rolle wissenschaftlicher Gutachter in der öffentlichen Meinungsbildung*. Berlin: Springer.
Nienaber, B., & Lübke, S. (2012). Die Akzeptanz der Bevölkerung ländlicher Gemeinden zur Ausweisung eines UNESCO-Biosphärenreservates am Beispiel der saarländischen Biosphäre Bliesgau. *Europa Regional, 18,* 2–3.
Nissen, U. (1998). *Kindheit, Geschlecht und Raum. Sozialisationstheoretische Zusammenhänge geschlechtsspezifischer Raumaneignung*. Weinheim, München: Beltz Juventa.
Nohl, W. (2004). Landschaft und Erinnerung. *Stadt + Grün, 53,* 37–44.
Nowotny, H. (1995). The need for socially robust knowledge. *TA-Datenbanknachrichten, 8,* 12–16.
Nowotny, H. (2005). Experten, Expertisen und imaginierte Laien. In A. Bogner & H. Torgersen (Eds.), *Wozu Experten? Ambivalenzen der Beziehung von Wissenschaft und Politik* (pp. 33–44). Wiesbaden: Springer VS.
Nowotny, H., Scott, P., & Gibbons, M. (2001). *Re-thinking science: Knowledge and the public in an age of uncertainty*. Cambridge: Wiley.
Nussbaum, M. (2006). *Frontiers of justice: Disability, nationality, species membership*. Cambridge, London: Harvard University Press.
Ó Tuathail, G. (1996). *Critical geopolitics*. London: Routledge.
Ode, Å., Hagerhall, C., & Sang, N. (2010). Analysing visual landscape complexity. Theory and application. *Landscape Research, 35*(1), 111–131.
Ohmae, K. (1999). *The borderless world: Power and strategy in the interlinked economy, management lessons in the new logic of the global marketplace*. New York: McKensey & Company.
Paasi, A. (1999). The changing pedagogies of space: Representation of the other in finnish school geography textbooks. In A. Buttimer, S. Brunn & U. Wardenga (Eds.), *Text and Image. Social Construction of Regional Knowleges* (pp. 226–237). Leipzig: Selbstverlag Institut fur Landerkunde.

Paasi, A. (2003). Region and place: Regional identity in question. *Progress in Human Geography, 27*(4), 475–485.
Paasi, A. (2009). Bounded spaces in a 'borderless world': Border studies, power and the anatomy of territory. *Journal of Power, 2*(2), 213–234.
Paasi, A. (2011). The region, identity, and power. *Procedia-Social and Behavioral Sciences, 14*, 9–16.
Paasi, A., & Zimmerbauer, K. (2016). Penumbral borders and planning paradoxes: Relational thinking and the question of borders in spatial planning. *Environment and Planning A, 48*(1), 75–93.
Palang, H., & Sooväli-Sepping, H. (2011). Are there counter-landscapes? On milk trestles and invisible powe-lines. *Landscape Research, 37*, 467–482.
Palen, J. (1995). *The suburbs*. New York et al.: Mc-Graw-Hill Humanities.
Papadimitriou, F. (2010). Conceptual modelling of landscape complexity. *Landscape Research, 35* (5), 563–570.
Paris, R. (2005). *Normale Macht. Soziologische Essays*. Konstanz: UVK Verlagsgesellschaft mbH.
Parker, S. (1996). *Planet Erde*. Erlangen: Heidecker.
Peatross, F., & Peponis, J. (1995). Space, education, and socialization. *Journal of Architectural and Planning Research, 12*, 366–385.
Piaget, J. (1937). *La construction du réel chez l'enfant*. Paris: Neuchâtel.
Piaget, J., & Inhelder, B. (1948). *La représentation de l'espace chez l'enfant*. Paris: Presses universitaires de France.
Piepmeier, R. (1980). Das Ende der ästhetischen Kategorie, Landschaft'. Zu einem Aspekt neuzeitlichen Naturverhältnisses. *Westfälische Forschungen, 30*, 8–46.
Pietila, A. (2010). *Not in my neighborhood. How bigotry shaped a Great American City*. Chicago: Rowman & Littlefield.
Piniek, S., Prey, G., & Güles, O. (2008). Zukunftsperspektiven urbaner Brachflächen. Wahrnehmung, Bewertung und Aneignung durch türkische Migranten im nördlichen Ruhrgebiet. *Berichte zur deutschen Landeskunde, 82*(3), 267–284.
Plessner, H. (1924). *Grenzen der Gemeinschaft. Eine Kritik des sozialen Radikalismus*. Bonn: Cohen.
Pollard, S. (1998). Der Arbeiter. In F. Furet (Ed.), *Der Mensch der Romantik* (pp. 68–110). Frankfurt a. M.: Suhrkamp.
Pommerening, R., & Ritter, J. (Eds.). (1996). *Pusteblume. Das Sachbuch. 4. Schuljahr*. Hannover: Schroedel.
Popitz, H. (1992). *Phänomene der Macht*. Tübingen: Mohr Siebeck.
Popitz, H. (1995). *Der Aufbruch zur Artifiziellen Gesellschaft. Zur Anthropologie der Technik*. Tübingen: Mohr Siebeck.
Pred, A. (1990). *Lost Words and lost worlds: Modernity and the language of everyday life in late nineteenth-century Stockholm*. Cambridge et al.: Cambridge University Press.
Pred, A., & Watts, M. (1992). *Reworking modernity. Campitalism and symbolic discontent*. New Brunswick, New Jersey: Rutgers University Press.
Pregernig, M. (2005). Wissenschaftliche Politikberatung als kulturgebundene Grenzarbeit. Vergleich der Interaktionsmuster in den USA und Österreich. In A. Bogner & H. Torgersen (Eds.), *Wozu Experten? Ambivalenzen der Beziehung von Wissenschaft und Politik* (pp. 267–290). Wiesbaden: Springer VS.
Prengel, A. (1994). Perspektiven der feministischen Pädagogik in der Erziehung von Mädchen und Jungen. In E. Glücks & F. Ottemeier-Glücks (Eds.), *Geschlechtsbezogene Pädagogik* (pp. 62–76). Münster: Waxmann.
Prigge, W. (1991). Die Revolution der Städte lesen. Raum und Präsentation. M. Wentz (Ed.), *Stadt-Räume* (pp. 99–112). Frankfurt a. M., New York: WW Norton.
Prominski, M. (2004). *Landschaft entwerfen. Zur Theorie aktueller Landschaftsarchitektur*. Berlin: Reimer-Verlag.

# References

Prominski, M. (2006a). Landschaft – warum weiter denken? Eine Antwort auf Stefan Körners Kritik am Begriff ‚Landschaft Drei'. *Stadt + Grün, 55*, 34–39.

Prominski, M. (2006b). Landschaft drei. In Perspektive Landschaft (Ed.), *Institut für Landschaftsarchitektur und Umweltplanung – Technische Universität Berlin* (pp. 241–251). Berlin: Sprinter-Verlag.

Proshansky, H., Fabian, A., & Kaminoff, R. (1983). Place-identity: Physical world socialization of the self. *Journal of Environmental Psychology, 3*(1), 57–83.

Quasten, H. (1997). Grundsätze und Methoden der Erfassung und Bewertung kulturhistorischer Phänomene der Kulturlandschaft. In W. Schenk, K. Fehn & D. Denecke (Eds.), *Kulturlandschaftspflege. Beiträge der Geographie zur räumlichen Planung* (pp. 19–34). Berlin, Stuttgart: Gebrüder Borntraeger.

Raab, J. (1998). *Die soziale Konstruktion olfaktorischer Wahrnehmung. Eine Soziologie des Geruchs*. Konstanz: UVK Verlagsgesesellschaft mbH.

Radding, C. (2005). *Landscapes of power and identity: Comparative histories in the Sonoran desert and the forests of Amazonia from colony to republic*. Durham, London: Duke University Press.

Ratzel, F. (1904). *Über Naturschilderung*. München, Berlin: Reimer-Verlag.

Rawls, J. (1971). *A theory of justice*. Cambridge, Massachusetts: Harvard University Press.

Reichert, D., & Zierhofer, W. (1993). *Umwelt zur Sprache bringen: über umweltverantwortliches Handeln, die Wahrnehmung der Waldsterbensdiskussion und den Umgang mit Unsicherheit*. Opladen: Westdeutscher Verlag.

Resch, C. (1999). *Die Schönen Guten Waren: Die Kunstwelt und ihre Selbstdarsteller*. Münster: Westfälisches Dampfboot.

Rhodes, R. (1996). The new governance: Governing without government. *Political Studies, 44*, 652–667.

Richter, D., Freey, K., & Funken, W. (2001). *Geografie 5. Sachsen*. Ernst Klett: Berlin.

Riedel, W. (1989). *„Der Spaziergang". Ästhetik der Landschaft und Geschichtsphilosophie der Natur bei Schiller*. Würzburg: Königshausen & Neumann.

Rodenstein, M. (2000). Von der ‚Hochhausseuche' zur ‚Skyline als Markenzeichen' – die steile Karriere der Hochhäuser in Frankfurt am Main. In *Hochhäuser in Deutschland* (pp. 15–70). Wiesbaden: Vieweg + Teubner Verlag.

Rosa, H. (2013). *Beschleunigung und Entfremdung*. Suhrkamp: Frankfurt a. M.

Rose, G. (1995). Place and identity: A sense of place. In D. Massey & P. Jess (Eds.), *A place in the world? Place, cultures, and globalization* (pp. 87–132). Oxford: Oxford University Press.

Rotenberg, R. (1995). *Landscape and power in Vienna*. Baltimore, London: Rutgers University Press.

Ryan, R. (2011). The social landscape of planning: Integrating social and perceptual research with spatial planning information. *Landscape and Urban Planning, 100*(4), 361–363.

Saar, M. (2010). Power and critique. *Journal of Power, 3*, 7–20.

Sachs, H. (2007 [1932]). Kitsch. In U. Dettmar & T. Küpper (Eds.), *Kitsch. Texte und Theorien* (pp. 184–192). Stuttgart: Reclam.

Säck-da Silva, S. (2009). *MitWirkung Zukunft gestalten. Prozessmanagement in der räumlichen Planung*. Kassel: Kassel University.

Safranski, R. (2007). *Romantik. Eine deutsche Affäre*. München: Carl Hanser.

Said, E. (2000). Invention, memory, and place. *Critical Inquiry, 26*(2), 175–192.

Saretzki, T. (2005). Welches Wissen – wessen Entscheidung? Kontroverse Expertise im Spannungsfeld von Wissenschaft, Öffentlichkeit und Politik. In A. Bogner & H. Torgersen (Eds.), *Wozu Experten? Ambivalenzen der Beziehung von Wissenschaft und Politik* (pp. 345–369). Wiesbaden: Springer VS.

Saugeres, L. (2002). The cultural representation of the farming landscape: masculinity, power and nature. *Journal of rural studies, 18*(4), 373–384.

Schäfer, H. (1981). Soziale Determinanten der Wohnungsnutzung. In J. Brech (Ed.), *Wohnen zur Miete* (pp. 248–262). Weinheim, Basel: Beltz.

Schein, R. (1997). The place of landscape: a conceptual framework for interpreting an American scene. *Annals of the Association of American Geographers, 87*(4), 660–680.
Schelling, T. (1960). *The strategy of conflict.* Cambridge: Harvard University Press.
Schelsky, H. (1965). *Auf der Suche nach Wirklichkeit: Gesammelte Aufsätze.* Düsseldorf, Köln: E. Diederichs.
Schenk, W. (1997). Unsere historischen Kulturlandschaften sind ernsthaft bedroht! Beispiel Kloster Bronnbach und Umgebung. *Schwäbische Heimat, 48,* 355–359.
Schenk, W. (2006). Der Terminus ‚gewachsene Kulturlandschaft' im Kontext öffentlicher und raumwissenschaftlicher Diskurse zu ‚Landschaft' und ‚Kulturlandschaft'. In U. Matthiesen, R. Danielzyk, St. Heiland & S. Tzschaschel (Eds.), *Kulturlandschaften als Herausforderung für die Raumplanung. Verständnisse – Erfahrungen – Perspektiven* (pp. 9–21). Hannover: Verlag der ARL.
Schiller, F. (1970 [1802]). Über das Erhabene. In K. Berghahn (Ed.), *Vom Pathetischen und Erhabenen. Schriften zur Dramentheorie* (pp. 83–100). Stuttgart: Reclam.
Schimank, U. (2012). *Die Entscheidungsgesellschaft: Komplexität und Rationalität der Moderne.* Wiesbaden: Springer VS.
Schluchter, W. (2009). Was heißt politische Führung? Max Weber über Politik als Beruf. *Zeitschrift für Politikberatung, 2,* 230–250.
Schneider, G. (1987). Frauen an der Hochschule – Geschichten aus dem Patriachat. In Frauengruppen in Freising und Wien (Ed.), *Planungsfrauen – Frauenplanung. Symposiumsbeträge von Landschaftsplanerinnen.* Wien: Eigenverlag.
Schneider, G. (1989). *Die Liebe zur Macht. Über die Reproduktion der Enteignung der Landespflege.* Kassel: Eigenverlag.
Schneider, M. (2016). Der Raum – ein Gemeingut? Die Grenzen einer artorientierten Raumverteilung. In F. Weber & O. Kühne (Eds.), *Fraktale Metropolen. Stadtentwicklung zwischen Devianz, Polarisierung und Hybridisierung* (pp. 179–214). Wiesbaden: Springer VS.
Schnittger, A., & Schubert, H. (2005). Kriterien für Kriminalprävention im Städtebau und in der Wohnbewirtschaftung. In H. Schubert (Ed.), *Sicherheit durch Stadtgestaltung. Städtebauliche und wohnungswirtschaftliche Kriminalprävention. Konzepte und Verfahren, Grundlagen und Anwendungen* (pp. 33–108). Köln: Technische Hochschule Köln.
Schnur, O., & Drilling, M. (2009). Governance – ein neues Zauberwort auch in der Quartiersentwicklung? In O. Schnur & M. Drilling (Eds.), *Governance der Quartiersentwicklung. Theoretische und praktische Zugänge zu neuen Steuerungsformen* (pp. 11–26). Wiesbaden: Springer CA.
Schönenborn, M. (2005). La bellezar insular. Blicke auf Kuba bei Carpentier, Hemingway, Arenas. In M. Schmeling & M. Schmitz-Emans (Eds.), *Das Paradigma der Landschaft in Moderne und Postmoderne. (Post-)Modernist terrains: Landscapes—settings—spaces* (pp. 122–132). Würzburg: Königshausen & Neumann.
Schönwald, A. (2015). Bedeutungsveränderungen der Symboliken von Landschaften als Zeichen eines veränderten Verständnisses von Macht über Natur. In S. Kost & A. Schönwald (Eds.), *Landschaftswandel – Wandel von Machtstrukturen* (pp. 127–138). Wiesbaden: Springer VS.
Schroeder, U. (1994). Siedlung in der Landschaft. In Deutscher Heimatbund (Ed.), *Plädoyer für Umwelt und Kulturlandschaft* (pp. 77–82). Hennef: GFA-Verlag.
Schroer, M. (2006). *Räume, Orte, Grenzen. Auf dem Weg zu einer Soziologie des Raumes.* Frankfurt a. M.: Suhrkamp.
Schröter-Schlaack, C. (2012). Das Konzept der Ökosystemleistungen. In B. Hansjürgens, C. Nesshöver & I. Schniewind (Eds.), *Der Nutzen von Ökonomie und Ökosystemleistungen für die Naturschutzpraxis. Workshop I: Einführung und Grundlagen.* BfN-Skripten 318 (pp. 8–15). Bonn-Bad Godesberg: Bundesamt für Naturschutz.
Schultheiß, G. (2007). Alles Landschaft? Zur Konjunktur eines Begriff sin der Urbanistik. In U. Eisel & S. Körner (Eds.), *Landschaft in einer Kultur der Nachhaltigkeit. Landschaftsgestaltung im Spannungsfeld zwischen Ästhetik und Nutzen* (Vol. 2, pp. 86–104). Kassel: Universität Kassel.

# References

Schultz, H.-D. (2006). Geographie für Mädchen: Eine andere Geographie? Die Geschlechterfrage in der älteren Geographiedidaktik. In U. Horst, D. Kanwischer & D. Stratenwerth (Eds.), *Die Kunst sich einzumischen. Vom vielfältigen und kreativen Wirken des Geographen Tilman Rhode-Jüchtern* (pp. 111–125). Berlin: Mensch & Buch-Verlag.

Schultz, H.-D. (2008). Das Kartenbild als Waffe im Geographieunterricht der Zwischenkriegszeit. *Kartographische Nachrichten, 58*, 19–27.

Schütz, A., & Luckmann, T. (1973). *The structures of the life-world* (Vol. 1). Evanston, Illinois: Northwestern University Press.

Scott, A. J. (1986). High technology industry and territorial development. The rise of the orange county country complex, 1955–1984. *Urban Geography, 7*, 3–45.

Seel, M. (1996). *Eine Ästhetik der Natur*. Suhrkamp: Frankfurt a. M.

Seidman, S. (2012). *Contested knowledge: Social theory today*. Malden, Oxford: Wiley-Blackwell.

Selle, K. (2002). Öffentliche Räume. Drei Annäherungen. In K. Selle (Ed.), *Was ist los mit den öffentlichen Räumen? Analysen, Positionen, Konzepte. AGB-Bericht, Nr. 49*. Aachen, Dortmund: Dortmunder Vertrieb für Bau- und Planungsliteratur.

Selle, K. (2004). Öffentliche Räume in der europäischen Stadt – Verfall und Ende oder Wandel und Belebung? Reden und Gegenreden. In W. Siebel (Ed.), *Die europäische Stadt* (pp. 131–145). Frankfurt a. M.: Suhrkamp.

Siebel, W. (2004). Einleitung: Die europäische Stadt. In W. Siebel (Ed.), *Die europäische Stadt* (pp. 11–50). Frankfurt a. M.: Suhrkamp.

Sieferle, R. (1984). *Fortschrittsfeinde? Opposition gegen Technik und Industrie von der Romantik bis zu Gegenwart*. München: C. H. Beck.

Siekmann, R. (2004). *Eigenartige Senne. Zur Kulturgeschichte der Wahrnehmung einer peripheren Landschaft*. Lemgo: Landesverband Lippe, Institut für Lippische Landeskunde.

Sieverts, T. (2001). *Zwischenstadt. Zwischen Ort und Welt, Raum und Zeit, Stadt und Land*. Braunschweig, Wiesbaden: Vieweg + Teubner Verlag.

Sieverts, T. (2004). Die Kultivierung von Suburbia. In W. Siebel (Ed.), *Die europäische Stadt* (pp. 85–91). Frankfurt a. M.: Suhrkamp.

Simmel, G. (1908). *Soziologie. Untersuchungen über die Formen der Vergesellschaftung*. Berlin: Duncker & Humblot.

Simmel, G. (1999 [1908]). Exkurs über die Soziologie der Sinne. In G. Simmel (Ed.), *Soziologie. Untersuchungen über die Formen der Vergesellschaftung* (pp. 734–735). Frankfurt a. M.: Suhrkamp.

Simmen, H., & Walter, F. (2007). *Landschaft gemeinsam gestalten. Möglichkeiten und Grenzen der Partizipation*. Zürich: vdf Hochschulenverlag AG.

Skorupinski, B., & Ott, K. (2002). *Partizipative Technikfolgenabschätzung als ethische Erfordernis*. www.ta-swiss.ch/www-remain/reports_archive/publications/2002/DT31_Bericht_kompl.pdf. Accessed March 23, 2005.

Sloterdijk, P. (1987). *Kopernikanische Mobilmachung und ptolemäische Abrüstung. Ästhetischer Versuch*. Suhrkamp: Frankfurt a. M.

Smith, N. (1984). *Uneven development*. Oxford: Blackwell.

Smith, N. (2010). Classical peer review: an empty gun. *Breast Cancer Research, 12*(4), 1–4. https://breast-cancer-research.biomedcentral.com/articles/10.1186/bcr2742. Accessed June 26, 2017.

Sofsky, W. (2007). *Verteidigung des Privaten. Eine Streitschrift*. München: C. H. Beck.

Sofsky, W., & Paris, R. (1994). Figurationen sozialer Macht. Autorität, Stellvertretung, Koalition. Frankfurt a. M.: Suhrkamp.

Soja, E. (1993). Los Angeles, eine nach außen gekehrte Stadt: Die Entwicklung der postmodernen Metropole in den USA. In V. Kreibich, B. Krella, U. v. Petz & P. Potz (Eds.), *Rom – Madrid – Athen. Die neue Rolle der städtischen Peripherie. Dortmunder Beiträge zur Raumplanung 62* (pp. 213–228.). Dortmund: Institut für Raumplanung Dortmund.

Soja, E. (1995). Postmodern urbanization: The six restructurings of Los Angeles. In S. Watson & K. Gibson (Eds.), *Postmodern cities and spaces* (pp. 125–137). Oxford: Blackwell.

Somerville, M., Power, K., & de Carteret, P. (2009). *Landscapes and learning: Place studies for a global world.* Rotterdam: Sense.

Spanier, H. (2006). Pathos der Nachhaltigkeit. Von der Schwierigkeit, „Nachhaltigkeit" zu kommunizieren. *Stadt + Grün, 55,* 26–33.

Spitzer, H. (1996). *Einführung in die räumliche Planung.* Stuttgart: Ulmer.

Stehr, N. (2006). Wissenspolitik. In D. Tänzler, H. Knoblauch, & H.-G. Soeffner (Eds.), *Zur Kritik der Wissensgesellschaft* (pp. 31–56). Konstanz: UVK Verlagsgesellschaft mbH.

Steinkrüger, J. (2014). *Thematisierte Welten: Über Darstellungspraxen in Zoologischen Gärten und Vergnügungsparks.* Bielefeld: transcript.

Stemmer, B. (2016). *Kooperative Landschaftsbewertung in der räumlichen Planung. Sozialkonstruktivistische Analyse der Landschaftswahrnehmung der Öffentlichkeit.* Wiesbaden: Springer VS.

Stevens, G. (2002). *The favored circle: The social foundations of architectural distinction.* Cambridge, Massachusetts: MIT Press.

Stichweh, R. (1997). Professions in modern society. *International Review of Sociology, 7,* 95–102.

Stremlow, M. (2001). Landschaft und nachhaltige Entwicklung – Ein Diskussionsbeitrag. *GAIA-Ecological Perspectives for Science and Society, 10*(2), 85–88.

Sutter, B. (2005). Von Laien und guten Bürgern. Partizipation als politische Technologie. In A. Bogner & H. Torgersen (Eds.), *Wozu Experten? Ambivalenzen der Beziehung von Wissenschaft und Politik* (pp. 220–240). Wiesbaden: Springer VS.

Swyngedouw, E. (2013). The non-political politics of climate change. ACME: An International E-Journal for Critical Geographies, 12, 8–11.

Tamayo, M. (1998). Die Entdeckung der Eliten – Gaetano Mosca und Vilfredo Pareto über Macht und Herrschaft. In P. Imbusch (Ed.), *Macht und Herrschaft – sozialwissenschaftliche Konzeptionen und Theorien* (pp. 61–75). Opladen: Leske + Budrich.

Tänzler, D. (2007). Politisches Charisma in der entzauberten Welt. In P. Gostmann & P.-U. Merz-Benz (Eds.), *Macht und Herrschaft. Zur Revision zweier soziologischer Grundbegriffe* (pp. 107–138). Wiesbaden: Springer VS.

TEEB—The Economics of Ecosystems and Biodiversity. (2009). *An interim report.* European Comission. www.teebweb.org. Accessed 23.07.2013.

Tessin, W. (2008). *Ästhetik des Angenehmen. Städtische Freiräume zwischen professioneller Ästhetik und Laiengeschmack.* Wiesbaden: Springer VS.

Thabe, S. (2002). *Raum(de)konstruktionen. Reflexionen zu einer Philosophie des Raumes.* Opladen: Leske + Budrich.

Thacker, C. (1995). The Role of the Antique in the Landscape Garden. In H. Wunderlich (Ed.), *Landschaft' und Landschaften im achtzehnten Jahrhundert* (pp. 67–78). Heidelberg: Winter-Verlag.

Thibaud, J.-P. (2003). Die sinnliche Umwelt von Städten. Zum Verständnis urbaner Atmosphären. In M. Hauskeller (Ed.), *Die sinnliche Umwelt der Wahrnehmung. Beiträge zu einer sinnlichen Erkenntnis.* (pp. 280–297). Kusterdingen: Die Graue Edition.

Thileking, K. (2006). Erhaltung der Kulturlandschaft braucht regionale Identität. Aus der Praxis der Regionalentwicklung in Niedersachsen. In D. Hornung & I. Grotzmann (Eds.), *Erhaltung der Natur- und Kulturlandschaft und regionale Identität* (pp. 51–55). Bonn: Bund Heimat und Umwelt.

Thompson, M. (1979). *Rubbish Theory.* Oxford: Oxford University Press.

Thrift, N. (1983). On the determination of social action in space and time. *Environment and Planning D: Society and Space, 1*(1), 23–57.

Tillmann, K.-J. (2007). *Sozialisationstheorien.* Reinbeck bei Hamburg: Rowohlt.

Trepl, L. (2012). *Die Idee der Landschaft. Eine Kulturgeschichte von der Aufklärung bis zurÖkologiebewegung.* Bielefeld: transcript.

Tuan, Y.-F. (1979). Sight and pictures. *Geographical Revue, 69*(4), 413–422.

Ulrich, R. (1977). Visual landscape preference: A model and application. *Man-Environment Systems, 7,* 279–293.

# References

Ulrich, R. (1979). Visual landscapes and psychological well-being. *Landscape Research, 4*(1), 17–23.
Van Assche, K., & Verschraegen, G. (2008). The limits of planning: Niklas Luhmann's systems theory and the analysis of planning and planning ambitions. *Planning Theory, 7*(3), 263–283.
Veblen, T. (1899). *The theory of the leisure class. An economic study of the evolution of institutions*. New York: B.W. Huebsch.
Veith, H. (2008). *Sozialisation*. München, Basel: Ernst Reinhardt Verlag.
Vervloet, A. J. (1999). Kulturlandschaftspflege im europäischen Vergleich. In Natur- und Umweltschutzakademie des Landes Nordrhein-Westfalen (Ed.), *Kulturlandschaftspflege Sukzession – contra Erhalten* (pp. 18–23). Recklinghausen: NUA-Seminarbericht 3.
Vicenzotti, V. (2005). Kulturlandschaft und Stadt-Wildnis. In I. Kazal, A. Voigt, A. Weil & A. Zutz (Eds.), *Kulturen der Landschaft. Ideen von Kulturlandschaft zwischen Tradition und Modernisierung* (pp. 221–236). Berlin: Universitätsverlag der TU Berlin.
Vicenzotti, V. (2008). ‚Stadt-Wildnis'. Bedeutungen, Phänomene und gestalterische Strategien. In Bayerische Akademie für Naturschutz und Landschaftspflege (Ed.), *Die Zukunft der Kulturlandschaft – Entwicklungsräume und Handlungsfelder* (pp. 29–37). Laufen: vwf.
Vicenzotti, V. (2011). *Der ‚Zwischenstadt'-Diskurs. Eine Analyse zwischen Wildnis, Kulturlandschaft und Stadt*. Bielefeld: transcript.
Vicenzotti, V. (2012). Gestalterische Zugänge zum suburbanen Raum. Eine Typisierung. In M. Schenk, M. Kühn, M. Leibenath & S. Tzschaschel (Eds.), *Suburbane Räume als Kulturlandschaften* (pp. 252–275). Hannover: Verlag der ARL.
Viehöver, W. (2005). Der Experte als Platzhalter und Interpret moderner Mythen. Das Beispiel der Stammzellendebatte. In A. Bogner & H. Torgersen (Eds.), *Wozu Experten? Ambivalenzen der Beziehung von Wissenschaft und Politik* (pp. 149–171). Wiesbaden: Springer VS.
Vöckler, K. (1998). Psychoscape. In W. Prigge (Ed.), *Peripherie ist überall* (pp. 276–288). Frankfurt a. M., New York: Campus.
Vogt, G. (2007). Die Landschaft im Zoo – eine touristische Erkundung des Vertrauten. In A. Künzel (Ed.), *Vom Ort zur Landschaft* (pp. 14–16). Zürich: Artemis.
Voigt, A. (2015). Die Macht des Ökonomischen im Blick auf Natur und Landschaft. In S. Kost & A. Schönwald (Eds.), *Landschaftswandel – Wandel von Machtstrukturen* (pp. 202–219). Wiesbaden: Springer VS.
Wagner, J.-M. (1997). Zur emotionalen Wirksamkeit von Kulturlandschaft. In W. Schenk, K. Fehn & D. Denecke (Eds.), *Kulturlandschaftspflege. Beiträge der Geographie zur räumlichen Planung* (pp. 59–66). Stuttgart: Gebrüder Bohrntraeger Berlin.
Wagner, J.-M. (1999). *Schutz der Kulturlandschaft – Erfassung, Bewertung und Sicherung schutzwürdiger Gebiete und Objekte im Rahmen des Aufgabenbereichs von Naturschutz und Landschaftspflege*. Saarbrücken: Selbstverlag der Fachrichtung Geographie der Universität des Saarlandes.
Walgenbach, P. (1999). Institutionalistische Ansätze in der Organisationstheorie. In A. Kieser (Ed.), *Organisationstheorien* (pp. 319–353). Stuttgart: Kohlhammer.
Walk, H. (2008). *Partizipative Governance. Beteiligungsrechte und Beteiligungsformen im Mehrebenensystem der Klimapolitik*. Wiesbaden: Springer VS.
Walk, H. (2012). Partizipationsforschung und Partizipationsverfahren – Herausforderungen für eine integrative Perspektive. In A. Knierim, St. Baasch & M. Gottschick (Eds.), *Partizipationsforschung und Partizipationsverfahren in der sozialwissenschaftlichen Klimaforschung* (pp. 5–6). München: Oekom-Verlag.
Warnke, M. (1992). *Politische Landschaft. Zur Kunstgeschichte der Natur*. München, Wien: Carl Hanser Verlag.
Wauschek, D. (2004). Pflanzenauswahl für Zoologische Gärten. *Stadt + Grün, 53*, 54–57.
WBGU—Wissenschaftlicher Beirat der Bundesregierung Globale Umweltveränderungen (1999). *Welt im Wandel. Erhaltung und nachhaltige Nutzung der Biosphäre. Jahresgutachten 1999*. Berlin et al.: Springer.
Weber, M. (1976 [1922]). *Wirtschaft und Gesellschaft. Grundriß der verstehenden Soziologie*. Tübingen: Mohr Siebeck.

Weber, I. (2007). *Die Natur des Naturschutzes. Wie Naturkonzepte und Geschlechtskodierungen das Schützenswerte bestimmen.* München: Oekom-Verlag.

Weber, F. (2015). Diskurs – Macht – Landschaft. Potenziale der Diskurs- und Hegemonietheorie von Ernesto Laclau und Chantal Mouffe für die Landschaftsforschung. In S. Kost & A. Schönwald (Eds.), *Landschaftswandel – Wandel von Machtstrukturen* (pp. 97– 112). Wiesbaden: Springer VS.

Weilacher, U. (2008). *Syntax der Landschaft. Die Landschaftsarchitektur von Latz und Partner.* Basel, Boston, Berlin: Birkhäuser Verlag.

Weingart, P. (2001). *Die Stunde der Wahrheit? Zum Verhältnis der Wissenschaft zu Politik, Wirtschaft und Medien in der Wissensgesellschaft.* Weilerswist: Velbrück Wissenschaft.

Weingart, P. (2003). *Wissenschaftssoziologie.* Bielefeld: Transcript.

Weingart, P. (2012). The lure of the mass media and its repercussions on science. In S. Rödder, M. Franzen & P. Weingarten (Eds.), *The sciences' media connection—public communication and its repercussions. Sociology of the Sciences Yearbook* (Vol. 28, pp. 17–32). Dordrecht et al.: Springer VS.

Weingart, P., Engels, A., & Pansegrau, P. (2008). Von der Hypothese zur Katastrophe. Der anthropogene Klimawandel im Diskurs zwischen Wissenschaft, Politik und Massenmedien. Opladen: Leske + Budrich.

Weis, M. (2008). *Methode zur Entwicklung von Landschaftsleitbildern mithilfe einer dynamischen Landschaftsmodellierung – erarbeitet am Fallbeispiel Hinterzarten im Hochschwarzwald.* Freiburg i. Br: Institut für Physische Geographie.

Werlen, B. (1995). Landschafts- und Länderkunde in der Spät-Moderne. In U. Wardenga & I. Hönsch (Eds.), *Kontinuität und Diskontinuität der deutschen Geographie in Umbruchphasen. Studien zur Geschichte der Geographie* (pp. 161–176). Münster: Institut für Geographie der Westfälischen Wilhelms-Universität.

Werlen, B. (1998). Landschaft, Raum und Gesellschaft. Entstehungs- und Entwicklungsgeschichte wissenschaftlicher Sozialgeographie. *Jenaer Geographische Manuskripte, 18,* 7–34.

Wescoat, J. (2008). Introduction. Three Faces of Power in Landscape Change. In J. L. Wescoat & D. M. Johnston (Eds.), *Political economies of landscape change: Places of integrative power. GeoJournal library* (Vol. 89, pp. 1–27). Dordrecht: Springer Science & Business Media.

Wiese, B., & Zils, N. (1987). *Deutsche Kulturgeographie. Werden, Wandel und Bewahrung deutscher Kulturlandschaften.* Herford: Busse-Seewald.

Williamson, O. (1991). Comparative Economic Organization. The Analysis of Discrete Structural Alternatives. *Administrative Science Quarterly, 36*(2), 269–296.

Wöbse, H. (1991). Landschaftsästhetik und ihre Operationalisierungsmöglichkeiten bei der Anwendung von § 8 Bundesnaturschutzgesetz. In Bundesforschungsanstalt für Naturschutz und Landschaftsökologie (Ed.), *Landschaftsbild – Eingriff – Ausgleich* (pp. 31–35). Bonn-Bad Godesberg: BFANL – Bundesforschungsanstalt für Naturschutz und Landschaftsökologie.

Wöbse, H. (1994). Die Erhaltung historischer Kulturlandschaften und ihrer Elemente. In Deutscher Heimatbund (Ed.), *Plädoyer für Umwelt und Kulturlandschaft* (pp. 37–44). Bonn: Deutscher Heimatbund.

Wöbse, H. (1999). Kulturlandschaft' und, historische Kulturlandschaft'. *Informationen zur Raumentwicklung, 5,* 269–278.

Wöbse, H. (2002). *Landschaftsästhetik. Über das Wesen, die Bedeutung und den Umgang mit landschaftlicher Schönheit.* Stuttgart: Ulmer Verlag.

Wöbse, H. (2006). Die sinnliche Erkenntnis von Eigen-Art und Schönheit historischer Kulturlandschaftselemente. In D. Hornung & I. Grotzmann (Eds.), *Erhaltung der Natur- und Kulturlandschaft und regionale Identität* (pp. 41–50). Bonn: Bund Heimat und Umwelt.

Wood, G. (2003). Die postmoderne Stadt: Neue Formen der Urbanität im Übergang vom zweiten ins dritte Jahrtausend. In H. Gebhard, P. Reuber & G. Wolkersdorfer (Eds.), *Kulturgeographie. Aktuelle Ansätze und Entwicklungen* (pp. 131–148). Heidelberg, Berlin: Spektrum Akademischer Verlag.

Wragg, A. (2000). Towards sustainable landscape planning. Experiences from the Wye Valley area of outstanding natural beauty. *Landscape Research, 25*(2), 183–200.

Wyckoff, W. (1990). Landscapes of private power and wealth. In D. Conzen (Ed.), *The making of the American landscape* (pp. 335–354). London: Unwin-Hyman.

Zeller, T. (2002). *Straße, Bahn, Panorama. Verkehrswege und Landschaftsveränderung in Deutschland von 1930 bis 1990*. Frankfurt a. M., New York: Campus.

Zierhofer, W. (2002). *Gesellschaft. Transformation eines Problems*. Oldenburg: Bibliotheks- und Informationssystem der Universität.

Zierhofer, W. (2003). Natur – das Andere der Kultur? Konturen einer nicht-essentialistischen Geographie. In H. Gebhard, P. Reuber & G. Wolkersdorfer (Eds.), *Kulturgeographie. Aktuelle Ansätze und Entwicklungen* (pp. 193–212). Heidelberg, Berlin: Spektrum.

Zillich, S. (2004). Heimische Vielfalt bewahren: Naturschutz im Garten. *BUND Magazin in Bayern, 86*(3), B15.

Ziman, J. (2000). *Real science. What it is and what it means*. Cambridge: Cambridge University Press.

Zimmermann, J. (1982). *Landschaft: verwandelt, misshandelt: Mensch und Umwelt in nordrheinischen Ballungszentren*. Neuss: Neusser Druckerei und Verlag GmbH.

Zukin, S. (1993). *Landscapes of Power. From Detroit to Disney World*. Berkeley, Los Angeles, London: University of California Press.

Zukin, S. (2009). Changing landscapes of power: Opulence and the urge for authenticity. *International Journal of Urban and Regional Research, 33*(2), 543–553.

Zukin, S., Lash, S., & Friedman, J. (1992). Postmodern urban landscapes: mapping culture and power. In S. Lash & J. Friedman (Eds.), *Modernity and identity* (pp. 221–247). Oxford et al.: Blackwell Publishers.

Zutz, A. (2005). „Heimatliche Landschaftsgestaltung". Die Herausbildung des Prinzips der landschaftlichen Eingliederung, dargestellt am Beispiel der Flugschriften der Fürst Pückler-Gesellschaft 1931–1934. In I. Kazal, A. Voigt, A. Weil & A. Zutz (Eds.), *Kulturen der Landschaft. Ideen von Kulturlandschaft zwischen Tradition und Modernisierung* (pp. 39–58). Berlin: Technische Universität Berlin.

Zutz, A. (2015). Von der Ohnmacht über die Macht zur demokratischen Neuaushandlung. Die geschichtliche Herausbildung der Position des Planers zur Gewährleistung ‚Landschaftlicher Daseinsvorsorge'. In S. Kost & A. Schönwald (Eds.), *Landschaftswandel – Wandel von Machtstrukturen* (pp. 65–94). Wiesbaden: Springer VS.

# Chapter 5
# Case Studies

The following case studies are intended both to give a more concrete form to the argument of the preceding chapters and to further explore the multidimensional correlations between landscape and power. The concentration on German-speaking Central Europe will in this chapter yield to a wider perspective embracing the United States to the west and Poland to the east.

## 5.1 Landscape Between Modernization and Mystification: The American Grid and the Frontier

The spatial organization of the USA is for historical reasons quite different from anything in Europe, and hence of especial interest for any international comparison of the relation between power and landscape. There is a close connection between the appropriation of the physical landscape of the USA, with its gradually advancing boundary between cultivated territory and 'the wild', and the role of landscape in the 'frontier' myth—a role that bears closely on the significance of the social landscape for the process of national self-definition (Slotkin 1973; Knox et al. 1988; Clarke 1993; Jarvis 1998; Pregill and Volkman 1999; Hardinghaus 2004; for the UK: Matless 2005). The concept of the American frontier was formulated by Frederick Jackson Turner in the closing decade of the 19th century as a constitutive myth of the American construct of identity (see Slotkin 1973; Kocks 2000). It referred to an idealized line—or more likely a relatively wide transitional zone—between the territory controlled by European Americans and the areas they did not yet control. Turner derived "the uniqueness of American democracy from the material and physical experience of life on the frontier, in which the individual could rely only on himself to secure the basic necessities of life" (Schneider-Sliwa 2005, p. 11; see also Knox et al. 1988; Kocks 2000). Life on the frontier was construed in diametrical opposition to the decadence of Europe as "the secular counterpart to the religious foundation of America. According to Turner, American democracy began not with the Pilgrim Fathers but in the forests [of the mid-West], where it was strengthened with every mile the frontier advanced" (Schneider-Sliwa 2005, p. 11).

Confrontation with the alien, the 'wild', was stylized into a catharsis in which the European immigrant was deprived of his old identity and reborn as the 'American Adam' (Hardinghaus 2004, p. 14). Consistently with this interpretation, Williams (2010) sees not the building of cities but the clearance of the native forests that covered almost half the country as the seminal factor in the appropriation of America's physical landscape. A second, somewhat different frontier was opened up with the cultural assimilation of the Western deserts and semi-deserts (Wescoat 2010). So powerful was the frontier myth that, together with the Calvinist concept of religious election (or predestination), it formed the basis for "American exceptionalism" (Schneider-Sliwa 2005, p. 11; Madsen 1998)—a construct whose roots reached back to the biblical narrative of the exodus of the People of Israel from their Egyptian slavery and its analogue in the 17th century Puritan flight to America (Hardinghaus 2004)—in both cases destination and destiny being a 'Promised Land'. For in their 'manifest destiny' as settlers of the American continent, the (European) Americans, in pursuit of the divine plan of salvation, had become the new "*chosen people*" (Hardinghaus 2004, p. 46; original emphasis). In accordance with this construct, 'the West' was a great deal more than a point on the compass or a segment of physical space: it was the symbol of a dynamic transformation, a "space of possibilities" (Campbell 2000, p. 2; Madsen 1998).

The cultivation of the wilderness as the result of the westward movement of the frontier resulted in the glorification of the West as a topos of challenge and potentiality (Culver 2010). Thus the myth of the frontier became a recurrent ingredient of 19th century American literature and painting, giving birth to the lonesome hero "stalwartly facing the tribulations of the wild" (Schneider-Sliwa 2005, p. 7), the lone rider of innumerable Hollywood films through to the present day. On a broader canvas, it was the myth of the simple hard life of individual labor and cooperation in a community of equals, a life of frugal contentment whose ideals, both religious and secular, "were projected onto the physical landscape" (Kocks 2000, p. 44; and see Corner 1992, 2002a, b; Pregill and Volkman 1999). Just as Adam and Eve had had to leave their earthly Paradise for the wilderness where evil reigned, so the new Adam and Eve were tasked with the transformation of that wilderness into a second Paradise. This meant not only the discovery and extraction of economic resources and the appropriation of land for an ever-growing population; the "experience of the frontier, constructed and reconstructed as struggle and conquest" (Egner 2006, p. 60) and mythologized in the era of the Western movie also entailed an ongoing process of destruction of flora, fauna, and preexisting cultures (Pregill and Volkman 1999; and see Knox et al. 1988; Kocks 2000; Körner 2010; on frontiers in general Newman and Paasi 1998). Moreover, the 19th century was in pioneering America a time of deep-seated fear (Moïsi 2009)—fear of wild animals, of native Americans, of other pioneers, of witches, and of slaves (Tuan 1979a, b; Schäfer 1998)—a fear that found expression in the "largely unrestricted right to bear firearms" (Moïsi 2009, p. 161). This right, which is "still today typical of the United States, reflects not only a cultural individualism that embraces the right to self-defense. More than that, it is heir to a wild, violent and dangerous past, when 'man was wolf to man' and fear an inalienable aspect of everyday life" (ibid.).

When the frontier reached the Pacific, there was no more room for "alternative land usage, like that of the indigenous peoples" (Pregill and Volkman 1999, p. 435). So far as the physical substrate of landscape, as well as its social appropriation and interpretation was concerned, this entailed the revocation of all contingency in favor of an exclusive rational modernity. Against the anti-aesthetic of the wild as a space inhabited solely by fear was set the aesthetic of a landscape formed by human hands. And this, too, had a religious dimension: the construct of a "Christian garden landscape" (Schneider-Sliwa 2005, p. 47), a space free from fear (see Tuan 1979a, b), was propagated against that of the untamed wilderness with religious fervor, even "at the cost of demeaning and destroying existing cultures" (Schneider-Sliwa 2005, p. 47; and see Knox et al. 1988; Weinstein 1996). A further feature of this Christian garden was an aesthetic of individuality in its physical manifestations "as the [particular] incarnation of God's will—against the abstract generalities of the natural sciences which imply the equality of all phenomena before the law" (Eisel 2009, p. 233; see also Cosgrove 1984).

Schneider-Sliwa (2005, p. 51) sees the suppression of the physical spaces staked out by indigenous cultures as the result of a specifically White Anglo-Saxon Protestant (so-called WASP) dominance: "Belief in their own [divinely] chosen status, combined with the rigid economic interests of settlers, merchants, and land-developing enterprises [...] underpinned the monocultural structuring of society and legitimated the repression of native societies" (Schneider-Sliwa 2005, p. 51; and see Madsen 1998). In this sense the frontier is for Varnelis the site of a dramatic encounter between infrastructure and theology, inasmuch as the inscription of the settled landscape as a new Garden of Eden amounted to the sacralization of the per se profane infrastructure that had made it so (on this process see Durkheim 2013[1912]; Campbell 2000). Consistently, therefore, modernity dawned on the American people in the form of roads and dams before it did so in the form of buildings (Varnelis 2009a).

Key factors in the planned transformation of the wilderness were on the one hand—under the influence of Romanticism—the conscious appropriation of the cultivated physical landscape as picturesque (partly by settlement of the best available parcel of land), and on the other hand the 'American grid' (Pregill and Volkman 1999; Kaufmann 2005). Imposed systematically and comprehensively—with the exception of the 13 New England states, as well as Texas and some mountainous territory to the west (Johnson 2010 [1990])—the American grid, with its uniform squares of owned and developed land, accounts for the checkerboard pattern of much of the appropriated physical landscape of the USA (Fig. 5.1). Implicit in the strict rationality of this pattern, which Waldie (2005, p. 4) has described as an "encompassing of [concrete] possibilities," is also, as Cronon (1996b) has observed, a gender-specific stereotype: the 'male' cultivation of a 'virgin' wilderness typified as wild and emotional (see also Merchant 1996; Cronon 1996b; Campbell 2000; Hardinghaus 2004; Radding 2005).

From the end of the 18th century onward, the grid was superimposed on the entire land, including the towns (Fehl 2004), which were arranged (with some

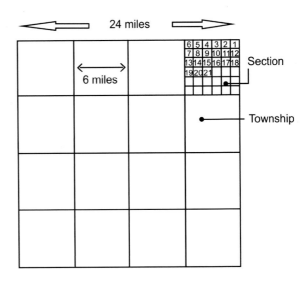

**Fig. 5.1** Abstract structure of the American Grid (after Pregill and Volkman 1999, p. 448)

exceptions) not to take best advantage of the local topography, but in accordance with the dictates of the grid—and hence, too, with the symbolic dictates of an egalitarian, democratic political system (Jackson 1970). Town and country, urban and rural space were—in contrast to European models—"subjected broadly speaking to one and the same principle of order in the apportioning of land" (Fehl 2004, p. 43). The American grid stood for egalitarian access, free from privilege or position, to urban as well as rural plots and sections (Fehl 2004). Kaufmann (2005) sees further political aspects of this system as the relative cheapness of land and the ease with which every member of society could know who owned what.

The village no longer played any role in this order: the regular patchwork of farms with roads along the edges of each parcel of land was the physical realization of a liberal individualism based on private property (Kaufmann 2005), a geometrical "objectification of the Enlightenment on American soil" (Kaufmann 2005, p. 163; and see Mills 1997; Muller 2010). Looking forward in history, the grid pattern can be seen as a foreshadowing of Fordist functionalism applied to physical space: a structured ordering of territorial limits "within whose expansive framework the future nation could freely determine its own development" (Kaufmann 2005, p. 331; and see Kühne 2008a; Fig. 5.2).

The aesthetic dimension of the twin concepts of frontier and American grid can be seen in the modernist principle of beauty as incorporating order and cultivation—even at the cost of a certain "monotony" (Muller 2010, p. 303). The counter-principle was that of the wild—a realm of uncultivated disorder, which, as the 'urban jungle', was later to find new relevance in the social landscape of the American city. The westward expansion of the American grid imposed on physical space the aesthetic metaphors of wilderness and ocean, the latter above all in respect of "a landscape that was free from distinctive features and hence open to [the] purely formal, rational inscription" of the grid (Kaufmann 2005, p. 164; see also Knox et al. 1988; Zukin

5.1 Landscape Between Modernization and Mystification ...

**Fig. 5.2** Actual structure of the grid: even in rectangular parcels the underlying squares remain recognizable (*Photo* Olaf Kühne)

et al. 1992; Zukin 1993; Campbell 2000; Massey 2006; Conzen 2010). Kaufmann sees "the imagination of the American landscape as a sea" especially in terms of the experience of land surveyors, whose search for a triangulation point in the endless expanses "found only emptiness" (Kaufmann 2005, p. 211), and who were forced to resort, instead of the theodolite, to the nautical compass.

In the course of the 19th century the ocean metaphor, whether applied to hill country, plains or mountains, became a cliché which, according to Kaufmann (2005, p. 213) "dissolved all natural diversity and every cultural boundary in a limitless, uniform perspective that effectively neutralized the external world" and enabled Americans in general, and new settlers in particular, to minimize the complexity and diversity of their environment (Sennett 1991). Sennett (1991) interprets the drive for simplification of the external landscape as a transference of the Protestant ethic—an inherently inward-looking cultural product, which bore within it, however, the seminal command to "fill the earth and subdue it" (Sennett 1991, p. 28). The fact that, faced concretely with the Green Mountains in western Vermont, the counter-pole of the wild "harbored fears that in the final analysis could not be tamed" (Kaufmann 2005, p. 227)—at least not with the available technologies of the 19th century—allowed the WASPs to revert on the one hand to "puritan fear of the godless wilderness [which] forbade any transformation into a Garden of Eden" (Kaufmann 2005, p. 227; and see Knox et al. 1988; Merchant 1996; Hardinghaus 2004; Kotkin 2006), and on the other to awe-filled contemplation in the face of the dynamic-sublime. In this myth-charged narrative the American grid functioned as an idyllic "garden fence, enclosing a space in which democracy rose in an integrated hierarchy from the family on the farm, through the wards and counties, up to the national administration" (Kaufmann 2005, p. 315; and see Knox et al. 1988; for Europe see Johler 2001).

In its ineluctable progress, the American grid "imposed an aesthetic that drew the sting of fear from the endless expanses" of the land (Kaufmann 2005, p. 228) and, along with the Industrial Revolution, gradually absorbed the remaining wilderness into a converse aesthetic. For in the national parks the wilderness was no longer the uncouth, overwhelming 'other'—it was subsumed into human action (in the form of conscious non-action rather than inaction) and hence, too, into human culture. Purposely left aside by the grid, even when the technology was available that could have subjected it to that ordering principle, the remaining wilderness, left untamed especially in the national parks, signified human superiority, inasmuch as what had once threatened it now fulfilled the aesthetic desire for transcendence. Charged with central cultural values, the wild was sacralized as the symbol of freedom and independence, of personal development and (transcendent) identity (Sennett 1991; Cronon 1996b). In this sense John Muir—and after him Arnold Schwarzenegger—cited Yosemite National Park (Fig. 5.3), and in the same breath the condor, as the symbol of a "threatened but at the same time resilient U.S. environment" (Starr 2007, p. 153). That access to this symbol was as a rule by automobile is an expression of the spatial logic that will be analyzed in the coming section.

**Fig. 5.3** One of the best-known views of Yosemite National Park—an element of myth and mystification in the Californian (and to a lesser extent U.S. American) identity (*Photo* Olaf Kühne)

## 5.2 Motorized Space—The Development of Los Angeles as an Urban-Rural Hybrid

Despite its crucial importance for both modern and postmodern societies, the basic dependence of infrastructure on power has received little attention from scholars, exceptions being, in the German-language context, e.g. Bruns and Gee (2009), Engels (2010), Kühne (2015, 2016a); and in English (which has seen considerably more research and over a longer period of time) e.g. Swyngedouw (1997) and Graham and Marvin (2001). The particular history of Los Angeles makes it an ideal subject for an examination of the relations between infrastructure and landscape, and many studies have been devoted in recent decades to a city that is seen "on the one hand as a metropolitan hub of economic and social modernization, and on the other as a monster of poverty, criminality, fragmentation, and chaos" (Schwentker 2006, p. 15). As an urban-rural hybrid (Kühne 2012), Los Angeles has developed into a "prototype and […] symbol of the automobile society" (Bratzel 1995, p. 11), uniquely accommodated to the logic of individually motorized mobility. Significantly, the Los Angeles freeway system, which bears the brunt of the city's physical traffic, has developed a symbolic force similar to that of the Eiffel Tower for Paris (Wachs 1998, p. 106). The process of postmodernization in the Los

Angeles agglomeration has also been intensively investigated (see e.g. Soja 1995; Dear 1998; Jarvis 1998; Fröhlich 2003; Davis 2004 [1998]; Culver 2010; Kühne 2012, 2015), as has the specific transportation infrastructure that evolved alongside it (see e.g. Wachs 1984, 1998; Bottles 1987; Halle 2003a, b; Banham 2009 [1971]; Varnelis 2009c; in more general terms Swift 2012; Kühne 2015, 2016a).

### 5.2.1 Historical Aspects of the Co-evolution of Settlement and Transportation Infrastructure in LA

The characteristically polycentric and at the same time decentralized agglomeration of Los Angeles that has contributed much to "its image as a featureless, contourless city" was not "brought into being by the automobile, although the automobile certainly fostered its growth" (Thieme and Laux 1996, p. 82). The subjection of the entire region to the automobile occurred more intensely, and its impact on the infrastructure was more crucial than in other parts of the USA. Already in 1920 Los Angeles had more licensed vehicles than any other major U.S. city (Soja and Scott 2006), and by 1925 every third citizen used an automobile (Bottles 1987; Bierling 2006). The consequences for municipal administration, as well as citizens, were horrendous: on an average working day some 260,000 vehicles were transfixed in a downtown gridlock (Buntin 2009). This was not simply because of the number of vehicles, but also because in the early 1920s the streets of the city's Central Business District (CBD) were more confined than in other cities. Less than a fifth of the surface area of the CBD was dedicated to traffic; San Diego, for example, had almost double that space (Bottles 1987). Moreover, the decentralized structure of the agglomeration—with its quality (in Popitz's terms) of both authoritative (directive) and technical (structural) power—positively promoted the use of the individual automobile. By the 1920s, the city had also been served for some thirty years by an extensive electric streetcar system: a busy 1600 km network connecting fifty centers from San Bernadino in the east through to the coast (Bierling 2006; Wachs 1984, 1998). Already in 1915 this was reckoned the best public transportation system in the country, with around 600 trolley cars a day passing the central station (Bierling 2006); in 1924 the system is reputed to have carried a total of 110 million passengers (Buntin 2009). It was run by two companies: Los Angeles Railway served the city of LA with its 'Big Yellow Cars', while the Pacific Electric Railway covered a wide area of the Los Angeles Basin with its 'Big Red Cars' (see Fogelson 1993 [1967]; Karrasch 2000; Starr 2007, 2009).

During and after the 1930s the network was gradually run down. The blocked streets (especially of downtown LA) increasingly prevented overfilled trolley cars from getting through. The packed rush-hour streetcars of the 1910s and early 1920s, and the numerous accidents caused when they did manage to travel fast (Bottles 1987), had given rise to heated public discussion, and in any case falling population densities made extension of the rail network unprofitable. The city was expanding outward, and by the later 1920s, the automobile was taking over. Its advantages were obvious: "No fixed timetable or route, no crush: the automobile seemed to restore the lost sovereignty of the horse-drawn carriage" (Sachs 1989, p. 106). In spite of these developments—which were compounded by the deteriorating condition of the rolling stock (Bottles 1987; Wachs 1998)—LA's streetcars survived another thirty years, until "on April 8, 1961 the last line of the Big Red Cars was closed" (Bratzel 1995, p. 37; Light 1988; Fogelson 1993 [1967]). Regional public transportation was now based exclusively on buses and (a few) trains. The automobile's ascent as the prime expression of the norms and values of WASP culture (as authoritative power) was now unstoppable. Freed from compliance with the restrictive timetables of a public system (the embodiment of technical power), people could now travel when and where they wanted and by whatever route they pleased. The automobile symbolized the autonomy of the individual—a value sacred to American culture.

Southern California had, in addition, three natural advantages for the automobile: its climate, its oil, and its topographical simplicity. Early automobiles offered little protection from the weather: snow and rain made driving uncomfortable and, with the primitive tires of the 1920s dangerous (Wachs 1998); so, unsurprisingly, Californians took to the private car more readily than did East Coasters (ibid.). California's oil production—1.5 billion tonnes between 1920 and 1926, which amounted at the time to about a fifth of U.S. consumption—meant abundant gasoline, and local refineries, combined with intense competition between the oil companies, brought prices down at times to less than a dollar a tankful (Olessak 1981; Ruchala 2009). All of this promoted the automobile to the prime symbol of urban and industrial progress (Bottles 1987), and it gradually became the normal means of daily travel, supplanting the streetcar system as structural artifact (technical power) (Kühne 2016a). The third factor was LA's relatively simple topography, which allowed easy access to the national highway system (as both technical and authoritative power)—far easier, for example, than in San Diego, the city's nearest competitor in Southern California for political and economic dominance (Kühne and Schönwald 2015). The physical situation of San Diego delayed its link-up to the U.S. highway system, feeding traffic instead toward Los Angeles. In fact, in the mid-1920s, twice as much LA-bound traffic came from Arizona as from San Diego (Baker 2007; Starr 1986; Kühne and Schönwald 2015).

The phase of mass automobility coincided in LA with a phase of intense urban expansion (Varnelis 2009a, b)—in contrast, for example, with New York, which had been planned in the age of pedestrian and horse-drawn traffic. Pre-existent settlement structures were small and decentralized, so they could easily be incorporated into modern transportation structures (Kühne 2015), and without incurring strong resistance an infrastructure could be installed which (as technical power) normalized everyday travel and laid down a development trajectory for the automobile that has remained valid until the present day.

The development of LA's traffic network was not, however, only the result (or side effect) of the physical terrain, nor solely of economic interests, or of the sum of its population's habits. It was also the product of a conscious political decision (as instrumental power), backed by a 1910 referendum, to conceive Los Angeles not as a traditional compact city—described in an accompanying document to the referendum as 'a pool for the ills of Europe and the vices of America'—but as a looser, decentralized structure, a synthesis of urban and rural elements promoted in the same quasi-religious terms that defined the settlement program of the United States as a whole. The Los Angeles Basin was to be 'a new Garden of Eden', a place where life should prosper, and in particular a perfect environment for bringing up children (Wachs 1998; Fogelson 1993 [1967]; Molotch 1998). Advertisement of this sort clearly intended to set the direction in which the city should develop (again exemplifying authoritative power).

The referendum determining the city's fundamental settlement structure was followed in 1924 by a 'Major Traffic Street Plan for Los Angeles' (Olmsted et al. 1924) laying out the infrastructural requirements in terms of multi-lane traffic arteries free from intersections (Wachs 1984; for the USA as a whole Swift 2012). These were seen in three particular respects to be a direct consequence of prior shortcomings: 'unscientific' street width, consequent inadequate traffic carrying capacity, and the mixing of different types of transportation: private automobiles, trucks and trolley cars. All three aspects demonstrate the penetration into urban planning of modernist-rationalist ideals: the unquestioned primacy of science, the separation of functions, and the economies of scale to be gained from increased traffic volumes.

In accordance with this governing concept, the freeways of the Los Angeles Basin followed the standards not of densely populated urban agglomerations but of sparsely settled areas where speeds of 60–70 mph were normal (Wachs 1998, p. 140). This meant wide traffic lanes, gentle bends, hard shoulders, and extended entrance and exit lanes. Taken together, these significantly increased the spatial requirements of the system (Wachs 1998, p. 140), whose comprehensive outreach boosted the development of further decentralized settlements in a recursive pattern that exemplified modernist thinking and practices. In terms of the functions of power (Popitz 1992), Los Angeles' freeway system was the technical-structural realization of authoritative (directive) power, implemented with the aid of instrumental (persuasive) and active (coercive) power.

In keeping with automotive logic, the infrastructural development was promoted by the federal decision to support highway construction from public funds. Highways were a public good, and as such not to be left solely to private interests, as had been proposed in early discussions of interregional traffic requirements (Swift 2012). Set up for this purpose in 1916, the federal highway fund still operates today. The 1921 Federal Aid Highway Act, which saw a need to connect all cities with a population of more than 50,000, initiated a highway construction program covering some 200,000 miles (U.S. Department of Transportation 1921). The Federal Aid Highway Act of 1956—known as the Interstate and Defense Highways Act (U.S. Department of Transportation 1956)—accelerated the construction program still further. Meanwhile regional public transportation measures were left to local authorities (Zelinsky 2010; see Brodsly 1981; Wachs 1984, 1998; Bottles 1987; Bratzel 1995; Kühne 2015).

## 5.2 Motorized Space—The Development of Los Angeles ...

The recursive pattern of freeway extension and decentralized settlement (see Brodsly 1981; Wachs 1984; Kühne 2012, 2015) meant that Los Angeles itself became a network with no clearly preferred center (see Fig. 5.4 and Holzner 1996; Kühne 2012, 2015). The choice of the private automobile as preferred means of transportation brought with it a concentration of services—gas stations, fast-food outlets, mini-malls and motels, as well as office buildings—along major traffic routes, especially near intersections (see Sloane 2003). Shopping became a function of driving, a normal and normative by-product of participation in the traffic flow. The concentration of functions that arose in 'edgeless cities' (Lang 2003) of this sort exhibits a type of centralization, and with it of technical-structural power, that cannot be reduced to that of the traditional monocentric city.

In accordance with the logic of automobile and freeway, physical space in the Los Angeles agglomeration was structured so that "drivers should be prevented from falling asleep at the wheel, which meant that dead-straight highways were out" (Hanich 2007, p. 35). In addition, this in turn meant that the freeways of the urban-rural hybrid broke the otherwise rigid formal pattern of the American grid (see Sect. 5.1; Kaufmann 2005; Kühne 2012). The construction of the freeways involved "cutting through established community areas and [...] destroying residual public space" (Hanich 2007, p. 35); 60% of the land needed for the Hollywood Freeway, for example, was already occupied and had to be cleared through dispossession orders and evictions (Keil 1993). After its completion, the elevated freeway cut across existing road systems and severed local contexts—quite apart from the environmental load it imposed in terms of noise and emissions, and the deterioration of property values that came in its wake (Jakle 2010). Residents with

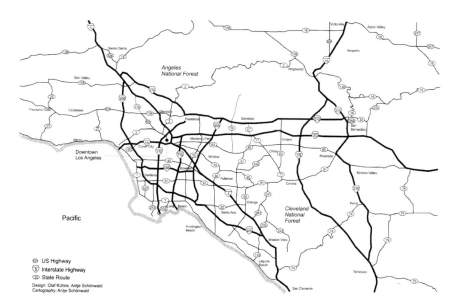

**Fig. 5.4** Los Angeles highway system (Kühne 2012, p. 279)

adequate symbolic capital quickly took the opportunity to move elsewhere, with predictable consequences for social segregation.

Olessak illustrates the symbolic as well as economic priority of LA's automobile-centered infrastructure in the singular spatial definition of downtown Los Angeles, not in functional terms but purely physiognomically as the area between four freeways: the Hollywood Freeway to the north, Harbor Freeway to the west, Santa Ana Freeway to the east, and Santa Monica Freeway to the south. All of these "seem to exercise a centrifugal rather than centripetal force" (Olessak 1981, p. 210; see Fig. 5.4). Excessive traffic volumes and consequent air pollution have led to the current decision of LA municipal authorities to strengthen and extend the public transportation system by both road and rail. The Big Yellow and Big Red Cars have recently been revived as a historical myth, with municipal PR attributing their demise to the interests of the automobile lobby. Elevated into a glorified symbol of the golden days of downtown LA during the Second World War, they take their place alongside movie theaters, clubs, and elegant restaurants jostling on the broadways with newspaper publishing houses and adobes from the Spanish and Mexican eras against a backdrop of American Neo-Renaissance architecture and one of the country's biggest assemblages of public buildings, crowned by a City Hall radiating urban self-confidence and populated by fashionably dressed people enthusiastically demonstrating the advantages of public transportation (Starr 2006). The mythologizing of LA's downtown was reflected in public referendum votes in the 1980s and 1990s for tax increases to finance a rail-based system for the agglomeration: the 30-year plan presented in 1990 by the Los Angeles County Transportation Commission (LACTC) envisaged costs of 184 billion dollars (Starr 2006). The actual effect of these measures, however—especially when compared with other U.S. cities of similar size, was relatively modest: in July 1990 the 'Blue Line' (streetcars and subway combined) between downtown LA and Long Beach carried 30,000 people daily, which was about the same as the local bus route it superseded (Bratzel 1995).

### 5.2.2 Impact of the Private Automobile on the Life and Environment of Los Angeles

Varnelis expresses the primacy of (and trust in) technical infrastructure in a quasi-religious metaphor: "If the West is dominated by the theology of infrastructure, Los Angeles is its Rome" (Varnelis 2009a, p. 6)—a reading that demonstrates the powerful co-construction (Engels 2010) of physical entities and symbolic meaning characteristic of technical power. Without a private automobile —in Bourdieu's terms a deficiency in economic, social and cultural capital (Bourdieu 1979)—one is largely deprived of mobility, including its social dimension, which in turn indicates the active (coercive) power of LA's automobile ideology.

## 5.2 Motorized Space—The Development of Los Angeles ...

A further result of the automobile culture is that some fifty percent of Los Angeles County is taken up by roads and parking lots (see Keil 1993), and public space is largely reduced to serving private mobility. Nooteboom describes Los Angeles as "a landscape of roads" (Nooteboom 2002 [1973, 1987], p. 13), and this is also the image it conveys in the media: an analysis of 216 YouTube videos showed that LA's freeways, next to downtown, were the most frequently pictured feature of the city (Kühne 2012, pp. 408–410; see Fig. 5.5; on the analysis of Internet videos see also Sect. 5.6).

The relatively weak differentiation of public from private space in Los Angeles (Jacobs 1961) is further evident in the functional transformation (or de-functionalization) of public space (Sennett 1977). Hence Nooteboom's remark that LA is meaningless without a private car (2002 [1973, 1987], p. 15), and Olessak's ironic definition of a pedestrian in that city as "someone who is just going to, or coming from, his automobile" (Olessak 1981, p. 197). Architecturally, the automotive culture makes itself felt not only in residences fronted by mighty garage doors and the almost complete absence of people on the sidewalks (Nooteboom 2002 [1973, 1987]), but also in banks, supermarkets, cafés etc., which are often less impressive from the front than from the back, where they face onto the parking lot. Many such premises have no entrance from the street at all.

As an expression of technical-structural power, the joint automobile-freeway system has been infused with a concept of freedom as personal independence, as identity construct, and as social status handed down with authoritative power in the socialization of at least three generations. So far as the vehicle is concerned, age, type,

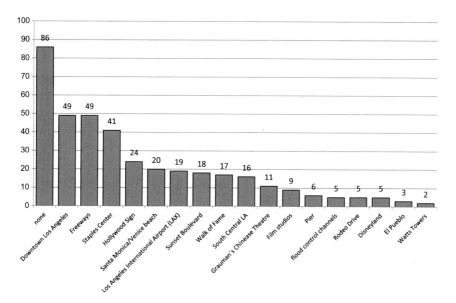

**Fig. 5.5** Significant symbolism in LA videos (absolute quotes, usually several objects per video: Kühne 2012)

and make are important in all three of these respects (see Johnston 1982; Matt 1998; Jakle and Sculle 2004; Kühne 2015). The social simplification of movement through the exclusion of pedestrians, streetcars, and cyclists is synonymous with liberation from 'the rest of society' (Ethington 2001), or as Baudrillard puts it: "Driving is a spectacular form of amnesia. Instant discovery. Instant deletion" Baudrillard (2004 [1986], p. 20). The freeways are often flanked, especially at intersections, where traffic is forced to slow down or comes to a standstill, by huge elevated advertisements known collectively as 'spamscape', sometimes animated by LED technology (see Fig. 5.6; McQuire 2009; Kühne 2012). Caught up in this system of technical power, the individual motorist signifies assent to the authoritative organization of society (Jakle and Sculle 2004): "Mobility is the lifestyle of the postmodern consumer nomad, always in pursuit of need or dream, an inveterate collector of pleasurable experiences" (Eickelpasch and Rademacher 2004, p. 47)—experiences, one might add, that often lie within the compass of suburban expectations (Kühne 2015).

In the wake of individually motorized mobility the aesthetic of landscape also changed, in particular transforming the wild "from a pioneer to a commercial mode, in which views of nature and of the city alternate" (Koshar 2008, p. 25). In Los Angeles this can be seen preeminently (also in movies, photographic volumes and guidebooks) in Mulholland Drive—an established aspect of the tourist program where one scenic observation point follows another. Even in its consumer aesthetics, the American urban landscape, of which Los Angeles is a prime example, has been physically transformed by the automobile more than by any other technical innovation (Jakle 2010). Thus Mulholland Drive has no facility at all for pedestrians, the nearest rail station to Disneyland is a half-hour away on foot, and a sidewalk in downtown LA can simply end in the middle of a bridge. In such a car-centered society, the contingent experience of landscape is difficult, if not impossible.

**Fig. 5.6** Freeway intersections as a preferred configuration for (partially animated) advertising boards due to the reduced speed, often also due to traffic jams. Because of their physical presence the messages of the advertising boards can hardly be ignored by drivers. From a landscape perspective, these arrangements can be described as 'spamscape' (Kühne 2012; *Photo* Olaf Kühne 2012)

Turning now to the vehicle whose technological and symbolic (Hanich 2007) force lies behind all this, the cultural and aesthetic assimilation of the automobile (and motorcycle) is recursively coupled with the development of Los Angeles as an urban-rural hybrid. Its individual customizing bears out in a particular way in Southern California what Hasse has observed for the socio-communicative function of the automobile as such, namely that "in certain subcultures the symbolic value of the vehicle, suitably enhanced and emotively charged, surpasses its utility value" (Hasse 2007, p. 182). Los Angeles has in this respect become a capital of customizing (Wolfe 2002 [1965]; Banham 2009 [1971]; Koshar 2008). Here the automobile is an integral part of popular culture (Davis 2004 [1998]), playing a major role not only in everyday activities (Steele 1997), but also in movies, novels, and songs (prototypically with the Beach Boys), often with an undertone of sports. And the Californian sport par excellence is surfing. But sport is lifestyle, and lifestyle soon becomes normative, so an endless loop of car, surf and sex is generated (Davis 2004 [1998]; Wolfe 2002 [1965]; Starr 2006; Banham 2009 [1971]), in which, oscillating recursively between physical object, social construct, and individual modality, Los Angeles becomes the incarnate symbol of a hedonistic and eccentric individualism (Molotch 1998; Hanich 2007; Kühne 2012, 2015).

However, the freedom symbolized by the automobile has its limits, formulated in the harsh imperative of signage—'Exit now', 'Right lane must turn right' etc. (instrumental power)—or the reversion of rush hour, when freeway becomes stoppage (Hanich 2007, p. 35). Even colloquial speech has its local expressions for such eventualities: "Where New Yorkers joke about traffic jams and Parisians dwell wistfully on their *embouteillage*, Angelinos are caught in a gridlock" (Hanich 2007, p. 35), which sounds considerably closer to violent crime. In LA, the street is cruel, cold and confusing (Sennett 1991). In gridlock the technological domestication of space (Stichweh 1997) meets its limits, and the 'mobile capsule of private space within the public sphere' that is the automobile (Kühne 2012) loses its sharply contoured differentiation from its surroundings. Kept at a distance by the sheer speed of travel, those surroundings now come dangerously close (Vanderbilt 2009), and the physical embarrassment of being within interactive distance to the next guy is routinely smothered by ferocious activity on the smartphone or dashboard computer, or simply by staring ahead into invisibility (see Kühne 2012). Even the density of the urban landscape seems wrong: arranged for an aesthetic of normal driving speed, it expands into over-simplicity when motion becomes stasis, and begins to contradict normal spatial expectations.

The dominant vehicle in Los Angeles is the—now ubiquitous—SUV (sport utility vehicle), which has largely swept limousines and station wagons from the streets (Schor 2004). And the SUV is even more of a "defense capsule" (Bauman 2008, p. 22) or "cocoon" (Jakle and Sculle 2004, p. 17) than those other vehicles: the physical expression of the residual invulnerability (*intégrité*) of the individual and his/her family (see Pareto 2006 [1916]; Kühne 2012) behind a few tenths of a millimeter of pressed steel. Here the martial superiority of form, the symbol of active power *in potentia*—compensates for physical reality. Elevated by the authoritative power of advertising, the SUV communicates immunity against the risks and contingencies of domestic life (Graham 2004; Vanderbilt 2009; Kühne 2012, 2015; on the social significance of the automobile in general see Katz 1999). Externally it stands for the opposite: the yearning for the greater potentialities of the wild, of life on the frontier, of freedom from the domestic round. Internally, on the other hand, it bespeaks comfort and safety, its leather upholstery echoing the furniture of home. In that sense the SUV is the urban-rural hybrid's co-evolutionary, mobile counterpart to the gated community (Kühne 2015).

## 5.2.3 Interim Summary

As in other metropolitan areas—especially but not only in the USA—the freeway (motorway, *autobahn* etc.) determines the development of settlement structures, on the one hand creating the infrastructural framework for suburbanization and desuburbanization, on the other hand destroying preexisting settlements by cutting them in half—or cutting them entirely off (Swift 2012). While deformation of this sort is a particularly striking example of the exercise of all four types of power described by Popitz, the formative function is a subtle example of the recursively reinforcing operations of technical power between the processes of public infrastructural development and private, market-directed population settlement. By simply suggesting that the logic of suburbanization and desuburbanization is self-perpetuating and endless, these processes reflect and reproduce the socially stabilizing imprint of (civil) authority. Southern Californian automobility is a social construct in (and into) which the physical artifacts of the urban landscape, as the product and expression of technical power, are increasingly appropriated. The workings of instrumental and authoritative power reinforce this process until it is simply taken for granted—a recursive development that again demonstrates the cohesive interface of the human with technology (Kühne 2015). This means, in turn, that machines, artifacts and technologies neither dominate nor are subordinate to social practices: their interrelations, too, are recursive (Urry 2000).

The symbolic connotations of infrastructure in particular are frequently the product of authoritative power processes and change with them—also determining social resistance beyond the common NIMBY phenomenon (Pietila 2010; Kühne 2012). An example is the shift in the discursive role of LA's 'Big Yellow' and 'Big Red Cars': initially a proud symbol of progress and of the availability of a rural dream within the urban agglomeration, they were rendered obsolete by the hegemonic advance of automobility, with its ideal of individual freedom, only to return in modern dress as a reincarnation of LA's golden age and a model for the future development of the city as a quintessentially urban-rural hybrid.

With the automobile defining large segments of private and social life, Vanderbilt can with some justification say that mobility has itself become a way of life (Vanderbilt 2009, p. 16), a spatial accommodation to the postmodern regime of transitory experience (Eickelpasch and Rademacher 2004, p. 39) which at the same time, however, conditions the yearning for escape into the greater stability of the wild (see the comments on the SUV at the end of the previous section). In both its dimensions (whether engaged or escaping from engagement) a transitory regime subjects daily life to the dictates of temporal economy, using all available means of spatiotemporal compression, from the automobile, through the airplane, to (most recently) the Internet (see Kühne 2015, p. 292)—which becomes a problem only when the machinery breaks down.

But the automobile, as the infrastructure of Los Angeles amply demonstrates, defines not only individual and social lifestyles, it (along with the co-evolutionary availability of cheap energy sources) has also profoundly shaped the physical space of the urban-rural hybrid, with its functionally rather than structurally (or architectonically) coherent 'edge' (Garreau 1991) and 'edgeless' (Lang 2003) cities. The authoritative and instrumental, as well as technical power embodied in this infrastructure (Engels 2010) allows its agents—from automotive industries to real estate developers—to "proceed largely undisturbed with their routines, [while] the options for new or alternative players" (ibid. 66) or systems—e.g. rail-based local public transportation—are severely restricted. After a century of unidirectional evolution, the scope for contingent interpretations or patterns of settlement and mobility (Koshar 2008) hardly exists any longer. The physical manifestations of technical-structural power have themselves become authoritative.

Its intense focus on automobility is perhaps the dominant reason why Los Angeles is regarded as a prototype of postmodern spatial development. Characterized by a democratic social context of untrammeled free market economics, it contrasts markedly with the city that will be the object of the second case study: Warsaw. Here the process of postmodernization has been forcefully imposed on a physical substrate indelibly inscribed by socialism.

## 5.3 Landscape and Power in the Development of Eastern Europe—The Example of Warsaw

After almost three decades of systemic transformation, East-Central and Eastern Europe has developed spatially in a way that can be perceived as specifically post-socialist, but whose historical roots lie partly in the socialist and partly in the pre-socialist era (see e.g. Andrusz et al. 2011; Czepczyński 2008; Gestwa 2003; Kühne 2016b; Marcińczak et al. 2015; Stanilov 2007). Two powerful driving factors in this development were first the post-WW2 wave of industrialization (especially in the 1950s), which catapulted the largely agricultural states of Eastern Europe into the industrial world (Popjaková 1998; on Stalinist spatial organization see Gestwa 2003), and secondly the establishment of a democratic free-market economic and social system after the fall of the Iron Curtain in 1989. This brought with it a post-Fordist logic whose spatial consequences in terms of fragmentation, polarization, and hybridization have been extensively analyzed above in relation to Southern California (Sect. 5.2; and see e.g. Scott 1988; Soja 1989, 1995, 1996;

Dear 2000; Davis 2004; Kühne 2012, 2015; Kühne and Schönwald 2015). The focus on Warsaw will illustrate the profound aesthetic impact of this twofold transformation, for the Polish capital has over the past eighty years experienced the physical manifestations of power and of aesthetic (re)interpretation with singular poignancy, making it an ideal object of study for the present argument. In the course of those decades, Warsaw has "undergone an intense change, from post-war ruin, through its function as a socialist industrial city, to its present role of metropolis" (Czesak et al. 2015, p. 167).

## 5.3.1 Warsaw—City of Socialist Modernism

The shaping of physical space during the socialist era of East-Central and Eastern Europe followed the aims and principles of communist-socialist thought (Czepczyński 2008), primary among which was "the creation of a classless, stateless social organization founded in the common ownership of the means of production" (Czepczyński 2008, p. 60). Urban planning was accorded a central role in this project (Domański 1997; Kühne 2001, 2003, 2016b; Czepczyński 2008), and although it rarely even approximately fulfilled this role—exceptions being new ventures like the Nowa Huta near Cracow (see Czesak et al. 2015)—it nevertheless left a defining mark on the cities of the region.

An example is Warsaw's Palace of Culture and Science (Fig. 5.7). Built in 1955 as a 'Gift of the Soviet Nation to the Polish Nation', this massive (234 m high) architectonic expression of Stalinist ideology remains the dominant image of the Polish capital (Koch 2012), intruding into and disrupting the scale of what had once been a densely populated area of the city. The Palace of Culture and Science symbolized the hegemony of the socialist idea and the Soviet Union as its prime representative (Czesak et al. 2015; Czepczyński 2008; Koch 2010; Kühne 2012). Its play on the Polish Renaissance architecture of Cracow and Zamość (Czepczyński 2008) reflected the "monumental, nationally tinted historicism" (Kadatz 1997, p. 15) that was also part of Stalin's legacy. However, the image of the socialist state was itself subject to change, and if the message of the Stalinist era to the residents of the new housing developments was that under socialism everyone could live like the bourgeois of previous centuries (Czepczyński 2008), this yielded after Stalin's death to a functionalism that itself gave way in the 1980s to a more liberal spirit, with a concomitant increase in private home ownership and stylistic pluralism (Czepczyński 2008).

Warsaw cannot, therefore, "be reduced to the status of a typical socialist city" (Koch 2010, p. 143, 2012), even if at the end of the socialist era it was spatially marked by the physical expressions of a superimposed industrialization as the symbol of socialist political doctrine and urban order (Domański 1997). Under socialism, city landscapes were considered a communicative medium that should remain "simple enough to be understood by the masses" (Czepczyński 2008, p. 63; Figs. 5.7 and 5.8). And other factors were also important: the historicist

## 5.3 Landscape and Power in the Development …

**Fig. 5.7** Palace of culture and science, Warsaw (left foreground: a post-socialist spamscape; *Photo* Olaf Kühne)

reconstruction of the 1950s, reminiscent of Hausmann's reshaping of Paris, made the erection of barricades virtually impossible and facilitated military access. The broad squares and magisterial avenues were ideal for parades and demonstrations of political power, and formed an appropriate backdrop for a socialist architecture that frequently cited national traditions. Here the assembled populace could feel itself part of something greater, something that stretched from past to future (Czepczyński 2008; Kühne 2016b).

The relation of the constructors of the socialist state to the pre-socialist urban heritage remained ambivalent. In Warsaw, the relatively few historical structures that had survived the war (Borodziej 2008) were on the one hand rejected as obsolete expressions of 19th century bourgeois capitalism, and—often still privately owned—left to deteriorate or fall victim to the demolition campaigns of the early socialist era (Domański 1997; Kühne 2001, 2003). On the other hand the Old Town and Royal Route were rebuilt on the basis of historical photographs, paintings, and plans, with the intention of demonstrating not only "national pride in the face of untold humiliation" (Koch 2010, p. 153), but also an awareness of the city's continuous European tradition (Czepczyński 2008, Fig. 5.9). It was, then, in a complex sense that "Warsaw rose again from its ruins to the dictates of an entirely new order" (Czesak et al. 2015, p. 178).

**Fig. 5.8** Piłsudki square, Warsaw (*Photo* Olaf Kühne)

**Fig. 5.9** Warsaw's historical Market Square, rebuilt after the Second World War, is today a major tourist attraction (*Photo* Olaf Kühne)

Alongside the display of political power in its various dimensions, a second major concern of socialist city planning was provision for the needs of the population. Mass housing, sports and cultural facilities, schools, stores, and administrative offices all belonged to the 'humanistic project of socialism' (Czepczyński 2008)—a project, however, that repeatedly experienced the entanglements of bad timing. Housing blocks would be finished before the subsidiary functions—even as basic as sewage disposal—were in place (Prawelska-Skrzypek 1988), which led on occasion to existential and ecological problems that were the very opposite of humanitarian (let alone humanistic) (Cierpiński 1993; Kühne 2001, 2003). The same is true of the transportation infrastructure required by any modern separation of supply and disposal functions: also in this respect Warsaw moved remarkably slowly (Czesak et al. 2015).

In line with modernist principles, the functionalist phase of reconstruction took advantage of economies of scale. Vast monofunctional housing complexes on the urban periphery fostered "a new social uniformity: the development of an egalitarian class of trade-union organized communist workers" (Lichtenberger 1995, p. 30; in more general terms Stakelbeck and Weber 2010; Weber 2013). A typical example is Warsaw's Ursynów complex, built in 1975 to house 130,000 people. The biggest of that era in Warsaw (Koch 2010; and see Gawryszewski 2010), it has been characterized by Juchnowicz (1990) as a prime example of 'pathological urbanization'. A second, in some ways similar project was the construction of the Huta Warszawa steelworks: an expression of the pure political will to establish an industrial working class in the Polish capital, the venture made little sense either economically or ecologically, as the Warsaw region has no substantial deposits of either coal or iron ore (see Kühne 2003).

In sum, socialist urban planning has been seen as a variant of "the rationalist drive" of modernism, evident also in the cities of West-Central and Western Europe, which "sought to erase the labyrinthine threat looming mythically within [...] the jungle of the European city" (Siebel 2004, p. 20).

### 5.3.2 Warsaw—City of Post-socialist Postmodernism

The aesthetic of 'form follows function' typical of the modernist age was followed in postmodernism by a polyvalence variously described as 'form follows fiction', 'form follows fear', 'form follows finesse', and 'form follows finance' (Ellin 1999). In post-socialist states this centrifugal phenomenon was recursively coupled with the change from a socialist to a democratic, free-market economic and social system which brought with it a transformation of the real-estate market, the privatization of industries, a free press, free and fair elections by secret ballot, the introduction of democratically elected regional and communal administrations, and the advent of new players like NGOs (see e.g. Kühne 2003; Czepczyński 2008).

In Poland, this radical systemic transformation also entailed a far-reaching withdrawal of the state at every level of spatial planning (for examples see

Jałowiecki 2012). This was particularly evident in Warsaw, whose numerous planning programs were for the most part "arbitrary and meaningless" (Koch 2010, p. 229) and had only marginal impact on the city's development. Moreover, until 2003 the capital suffered from extreme political and administrative fragmentation, with a total of 779 different bodies—ranging from *voivodeship* (provincial) council, through city council, ward and borough councils, municipal assembly, and central district councils—responsible for urban development (Koch 2010). As Piątek puts it: "Capitalist Warsaw, unlike its communist predecessor, had no overall development plan. Everything that happened after 1989 was the spontaneous expression of energy directed in particular to breaking the rules of the past forty years" (Piątek 2008, p. 30; Fig. 5.10).

The result of this spontaneity is apparent in the profound physical and symbolic contradictions of postmodern Warsaw, and the city's correspondingly fractal aesthetic and cognitive ascriptions (see Fig. 5.9). A pre-eminent example is the former central office of the Polska Zjednoczona Partia Robonticza (Polish United Workers Party)—the official title of the Communist Party that governed the country from 1948 through 1989. Known colloquially as the 'White House', the 1950s building was a dominant symbol of social and socialist power; today it houses Warsaw's financial center and serves the interests of global capitalism (Czepczyński 2008; see Fig. 5.11)—a more radical transformation in the denotations and connotations of a physical artifact is scarcely conceivable.

The rededication of some highly symbolic buildings along these lines was accompanied by the replacement of others, a striking example being the former Dziesieciolecia (Tenth Anniversary) Stadium. Opened in 1955, the capital's biggest sports venue staged its last soccer game (between Poland and Finland) in 1987, after which it was abandoned to inevitable dilapidation. Situated on the east bank of the Vistula near the Old Town—opposite the central finance and management district that has developed around the Central Station on the west bank of the river—it soon became the focal point of Warsaw's informal bazaar trade, with market stalls offering goods for every daily need, from food and clothing, through electrical appliances and automobile spares, to weapons, narcotics, prostitution, and illegal alcohol—mostly from Eastern Europe (Sulima 2012). Customers, generally from the poorer echelons of society (ibid.)—and especially traders—presented a multi-ethnic spectrum, and the old stadium assumed a correspondingly hybrid appearance (see Kühne 2012; Kühne and Schönwald 2015). Situated a mere three tram-stops from downtown Warsaw, it boasted a hub of nomadic trade that contributed significantly to the city's in any case fragmented aesthetic. So much so that before Dziesieciolecia was demolished to make way for the new National Stadium (Fig. 5.12), it found artistic expression in various media, especially film and photography (Sulima 2012). The successor stadium itself, built for the 2012 European Soccer Championship, is notable among other things for its vast parking lots (see Fig. 5.12 foreground) which—given that this is a postmodern project situated in the center of a capital city—illustrate the importance accorded to individual automobility by current Polish politics.

5.3 Landscape and Power in the Development …

**Fig. 5.10** Grzebowski Square strikingly embodies the patchwork aesthetic of postmodern Warsaw, with Pope John Paul II gesturing welcomingly toward the symbols of the city's embedment in the global economy (*Photo* Olaf Kühne)

**Fig. 5.11** Former central office of the Polish United Workers Party (*Photo* Olaf Kühne)

**Fig. 5.12** Poland's new National Stadium, built 2008–2011 (*Photo* Olaf Kühne)

## 5.3 Landscape and Power in the Development ...

The widespread absence of any political or administrative control of spatial development in Poland exposed the acquisition and shaping of physical space (as it had in Los Angeles, see Sect. 5.2) to unrestrained economic logic. Business towers erected with the resources of international capitalism (Fig. 5.7) challenge the aesthetic of the European city and contribute to the hybridization of urban and rural space. At the same time they mark Warsaw's unique economic situation in the country as a whole, with a GDP per inhabitant more than three times the Polish average (CIS: various years). Global patterns of consumption are reflected in physical structures: "Western-European and North American multinationals fight for space and (future) market share with greenfield hypermarkets and other big-box retailers strategically placed beside new arterial highways, boosting an automobile-dependent pattern of consumer behavior throughout the region" (Altrock et al. 2005, p. 9). Overloaded streets and highways point up an ostentatious consumption (Veblen 1899; Fig. 5.13) promoted by a veritable spamscape of advertising, ranging from the flyers pressed everywhere into one's hand to hoardings and placards sometimes covering an entire house façade (Kühne 2012; Fig. 5.7).

Situated immediately next to Warsaw Central Station and the striking business towers of the financial district, the shopping mall Złoty Terasy (Golden Terraces: Fig. 5.13) is just such an instance of postmodern consumer landscapes. Opened in 2007, it offers its customers some 65,000 m$^2$ of shopping facilities in a glossy

**Fig. 5.13** 'Złoty Terasy' ('Golden Terraces') shopping mall, Warsaw (*Photo* Olaf Kühne)

environment tinted with historicism. In overt contrast to the modernist aesthetic of 'form follows function', Złoty Terasy strikingly exemplifies the postmodern simulation of an urbanity divorced from the historical context of the city beyond its eclectic façade. That the investor is the Dutch 'ING Group' demonstrates the integration of the Polish capital in international financial and real estate operations.

The systemic transformation also affected the idea of house and home throughout Eastern-Central and Eastern Europe. Where "people had thought of their place of residence as an item of social service, [they] now considered it an economic good" (Sailer-Fliege 1999, p. 69; Radzimsky 2009), and in segments of the population furnished with medium or higher symbolic capital it became—above all with respect to location—a medium of representation and distinction. This took the form on the one hand of a belated suburbanization that made inroads into former agricultural land on the urban periphery (see e.g. Lisowski 2010; Staniszkis 2012) and brought with it an inevitable degree of segregation; on the other hand it fostered a different, markedly enhanced segregation in the form of gated communities (see Mierzejewska 2011; Kovács 2014; Brailich and Pütz 2014), of which Kusiak (2012) counted more than 400 in Warsaw alone. The motivation for this rapid development was, according to Kusiak, that "residents in a gated community could in the first place show off with their address, secondly feel safe, and thirdly imagine themselves living in an American movie" (Kusiak 2012, p. 48).

Marina Mokotów exemplifies these new landscapes of 'fear of the other' (see Kühne 2012), a syndrome rooted in a global media discourse of angst feeding on criminality, fear of sickness, and fear of fear itself (Gąsior-Niemiec et al. 2009). Gated communities counter the complexity of the world with a fenced-off space in which streets and parks, shopping and recreational facilities are tailored to the stereotypical aesthetic notions of potential customers (for further details on Warsaw see Gądecki 2014a, b; for Los Angeles see Kühne 2012, 2016b). Conveniently situated on a former industrial site between Okęcie Airport and downtown Warsaw, and with easy inner-city access via the main traffic artery of Żwirki i Wigury, Marina Mokotów accommodates some 6,000 residents (Czepczyński 2008; Johnsson 2012) in a ghetto for the rich that attracts angry as well as envious glances from those excluded from its gentrified precincts—after all it stands on what was once public space (see Gądecki 2014a, b; in more general terms Mills 1997). Fear of the other—fear of the strange and the stranger—produces a polarized aesthetic in which a stereotypical beauty erects bulwarks against a complex otherness (see Kühne 2012). In both size and style, Marina Mokotów is a prime example of Warsaw's gated communities; numerically, however, a different model dominates: the everyday gated community is often a former socialist apartment block surrounded with a fence and furnished with a securely locked gate (for details see Gąsior-Niemiec et al. 2009). Here too, however, the symbolic transformation is profound, for what was once an expression of equality now radiates social distinction.

## 5.3.3 Interim Summary

The development of the states of Eastern-Central and Eastern Europe has been deeply marked by the forced modernism of post-war socialism and the equally forced postmodernism that succeeded it upon the demise of 'real socialism' in 1989. The intensity of both processes is reflected in physical structures, which eminently symbolize the power interests behind them. This is more than ever true of places with a special function or significance, where the transformations in question took the form above all of fragmentation and polarization.

Due mainly to the ambivalent role of the state in urban planning, the transition from socialist modernism to post-socialist postmodernism in a city like Warsaw is even more marked than in democratic, free market societies. For the socialist state could and did shape society and its physical-spatial manifestations from above: hence the markedly more intense application of modernist rationalism (in socialist dress) in Eastern than in Western Europe. Moreover, the same is true, conversely, of the postmodern transition, where—in the almost entire absence of state control—the fragmentary aesthetics of economic interest dictated spatial development with an intensity largely foreign to Western European societies. In centers like Warsaw, which had experienced a significant accumulation of symbolic capital, this phenomenon was even more striking than elsewhere.

The result was what has been called a 'spatial pastiche' (Kühne 2012): the transformation of a symbolically intelligible landscape into a patchwork of variant property ownership, dedication, and use best described in terms of a meteoric fragmentation of the urban context. Forms that have little to do with actual function are stylistically peppered with quotations from international design idioms, visual landscape expectations remain largely unfulfilled, and the emotional ties to shapes and objects that foster a sense of home have been abandoned in pursuit of bigger ideas and more global challenges (Kühne and Hernik 2015).

The multi-faceted, largely uncontrolled jostling for position that characterizes the pastiche of contemporary Warsaw can be seen in the clamorous spamscapes of advertising, the commercial tower blocks and shopping malls, and the crass juxtapositions of the homeless squatting on wasteland in front of the headquarters of a global player (Fig. 5.14; see Kühne 2016b). Taken together, these phenomena make for what Czepczyński has called an "urban battlefield" (Czepczyński 2008, p. 109), not only on the level of physical space, but also and especially on that of aesthetic discourse with its implicit hegemonic potentiality. As in Los Angeles—the prototype of postmodern spatial development—the fragmentation of physical and symbolic space is at times intensified by factors ascribable to the modernist quest for order (e.g. gated communities, see Fayet 2003); but at other times this quest works in the opposite direction, suppressing the hybrid phenomena of postmodern fragmentation (e.g. Dziesieciolecia Stadium: see Kühne 2012; Kühne and Schönwald 2015; Sect. 5.2). Abruptly exposed to the logic and spatial requirements of global economics, the physical transformation of Warsaw has exceeded in speed and diversity even that of Los Angeles.

**Fig. 5.14** Spatial pastiche near Warsaw East Station—multinational office with wasteland and homeless people (*Photo* Olaf Kühne)

From the point of view of social equity, Warsaw's spatial development in recent years is open to criticism. In particular the proliferation of gated communities documents not only a gap in opportunity between 'insiders' and 'outsiders'; it also makes clear that the state has failed in its essential task of providing for the security of its inhabitants. When considerable segments of the population feel insecure, opt for segregation, privatize security, and in doing so perpetuate inequality, the state has a clear obligation to examine the reasons and seek remedies—not only in increased police presence but also, for example, in extending educational opportunities, especially for the underprivileged.

The trajectory of Warsaw's physical development after the fall of the Iron Curtain exemplifies that of many densely populated regions in Eastern-Central Europe and evinces with clarity the power structures behind the pastiche of its postmodern façade and post-socialist aesthetic. However, the specific path to postmodernism taken by these regions differs in essential respects from that of either Western Europe or North America (Kühne 2016b).

## 5.4 Power Conflicts—Landscape and the Impact of Renewable Energies

While the preceding case studies focused on densely populated areas, this section will be devoted to regions of sparser population whose physical structures have also been subjected in recent years to increasing change in the wake of Germany's transition from fossil-fueled to renewably sourced electrical energy (for repercussions in the landscape context see e.g. Pasqualetti 2001; Short 2002; Schultze 2006; Wolsink 2007; Krauss 2010; Selman 2010; Stremke 2010; Bosch and Peyke 2011; Dû-Blayo 2011; Kühne 2011a, 2013b; Rygg 2012; Schöbel 2012; Tobias 2012; Leibenath and Otto 2013; Kühne and Weber 2015). Forcefully propagated by German politics, this energy transition is seen by the population as a whole as necessary and even desirable: 93% of respondents in a TNS Emnid opinion poll of August 2015 considered the further expansion of the renewable energy sector as "important" or "extremely important" (German Renewable Energy Agency 2015 n.p.).

Despite this underlying consensus, the physical manifestations of the energy transition are often vehemently rejected, especially by local residents—a phenomenon by no means restricted to Germany (see e.g. Pasqualetti 2001; Wolsink 2007; Krauss 2010; Leibenath and Otto 2013). As well as health reasons, objectors cite issues specifically concerned with the social construct of landscape and its evaluative patterns. An interesting aspect in this context is that landscape is generally thought of as common property, in contrast (at least in capitalist societies) to the individual areas and objects—fields, trees, buildings, etc.—that make up its physical substrate. This double vision harbors immediate tension, for neither the private nature of the individual elements nor the communal nature of the panoptic vision of landscape is, as a rule, subjected to serious critique (see Apolinarski et al. 2006; Röhring 2008; Stotten 2015; Schneider 2016). Olwig (2002) sees this in light of the medieval origins of the concept of landscape as an area governed by traditional rules (e.g. common law) and exempt from feudal overlordship. Against this background, significant alteration of the physical structure of a landscape for individual profit is widely regarded as an illegitimate encroachment on established rights (Schneider 2016).

The rejection of the plant required for renewable energy generation (especially wind farms, see Fig. 5.15) also has a lot to do with the socialization of the concept of landscape as a conditioning factor of the environment people normally—and normatively—see as 'home' (see Sect. 4.1; Bosch and Peyke 2011; Kühne 2011b). It is against this background that change to the physical basis of the landscape is often construed as impairment or loss of an emotionally charged home environment —especially when that change is rapid and conspicuous and brings no immediate gain. The normative sense of home and its associated landscape values are, however, subject to generational change (Kühne 2008b, 2014a; Selman 2010; Forkel and Grimm 2014). Regenerative energy installations are for the most part fairly new and fall outside the range of the stereotypically beautiful, which is another reason— over and above their abrupt incursion into the local environment—for their

**Fig. 5.15** Wind farm on the borders of California's Mojave Desert. In contrast with the decentralized European model, several hundred wind turbines—along with their side-effects on the landscape—are packed here into a relatively small space (*Photo* Olaf Kühne)

widespread rejection. Wind farms, for example, are more readily associated with modernity, sustainability, and the awesome-sublime than with beauty in the received sense of that word.

In a survey of landscape preferences conducted in the Saarland in 2004 (n = 455) 39.8% of respondents presented with a photograph of an open landscape with a wind farm regarded this as 'modern', 33.4% as 'ugly', 9.9% as 'interesting', 7.0% as 'nondescript', only 0.4% as 'beautiful', and 0.2% as 'traditional'; the remainder used some other adjective or made no comment (Kühne 2006; and see Fig. 5.5). Consideration of the socio-demographic coordinates of the respondent group is also enlightening: among those less inclined to assess the photograph as 'modern' were women (highly significant variance), under 30 year-olds (very significant variance), and 31–45 year-olds (significant variance), as well as people with university entrance qualifications (significant variance), and those living in single households (highly significant variance); among those that characterized the photograph as 'ugly' the 46–65 year-old cohort was highly significant (Kühne 2006).

The German energy transition (or *Energiewende*, see Krause et al. 1980) has brought new relevance to the extended sensory impact of regenerative energy plant and processes: in the case of wind turbines this is visual and acoustic, in that of biomass visual, haptic, and olfactory. In both research and practice these factors are

## 5.4 Power Conflicts—Landscape and the Impact ...

relevant not only in the physical-spatial dimension (shaping and ordering of landscape change), but also in the social dimension (acceptance of such change). An example of the former is Schöbel's demand that wind farms "should be developed as a legible aspect of landscape that can be interpreted and filled with meaning by the observer" (Schöbel 2012, p. 76). Following Gipe (1995), he formulates proposals for the design of wind farms—e.g. that the turbines should look similar; that rotors should, in accordance with visual expectations, turn (unless there is no wind); that drives and generator should be suitably housed in a nacelle; and that the whole plant should be kept clean and not too widely fenced off from the public. Moreover, in order to avoid noise pollution and rotor shadow (Schöbel 2012), wind farms should not be too close to residential areas. A central requirement of Schöbel's is that they should blend into the physical structures of a landscape without dominating it. This could be achieved by planning the plant to follow the contours of the local geomorphology. What is striking about all this is that in Schöbel's list of demands, the social mechanisms of landscape construction and evaluation are left entirely aside. He is concerned only with the physical substrate of appropriated landscape. In that respect, he lays himself open to the accusation of *déformation professionnelle*—the occupational risk (in this case) of the designer.

Seen through the lens of the paradigmatic landscape discourses outlined in Sect. 4.2.5, the extent to which responses to the physical effects of the energy transition are conditioned by social constructs becomes clearer (Short 2002; Kühne 2011c, 2013a; Olwig 2011; Leibenath and Otto 2011, 2012). The four paradigms, and their bearing on the physical impact of renewables, are as follows:

- The paradigm of restoring the physical substrate of the appropriated (cultural) landscape often lies behind expert rejection of renewable energy plant: it is accused of being regionally 'untypical' and fostering (global) landscape uniformity. This position, however, harbors an inherent contradiction; for to vocally support the sustainability and preservation of a landscape on the basis of its postulated union with its inhabitants and at the same time to promote continued reliance on fossil fuels for electrical power is manifestly inconsistent. The contradiction can only be resolved by a strategy of self-sufficiency that dispenses with (imported) energy and accepts an inevitable lowering of living standards.
- The discourse of successional development generally tolerates renewable energy plant, albeit with reservations connected with the (overly) high subsidies it incurs and, in the case of biomass and photovoltaic plant, also with the very large areas these require.
- The discourse of reflective design allows renewable energy plant to be seen as an element of 'land art' (see Schöbel above).
- The discourse of reinterpretation implies the possibility of cognitive and, in particular, aesthetic revaluation of renewable energy plant as a coherent element of the physical landscape. This opens the way to enhanced acceptance of such plant. The following paragraphs are concerned with this process of reinterpretation (for greater detail see Selman 2010; Kühne 2011a).

Soyez notes the importance in this context "[...] of viewing not only historically grown but also currently growing developments within the framework of significant overall changes to the cultural-geographical landscape" (Soyez 2003, p. 37). And Schultze (2006) points out that the new energy landscapes, as a phenomenon representing current user-needs, are subject not only to past but also—and specifically—to current perspectives of appropriation. In this sense, the intergenerational shift in the construction of 'normative home landscape' will predictably absorb renewable energy plant as 'normal'. A similar process could be set in train for landscape stereotypes by systematically mediating alternative cognitive and aesthetic models and images through the socialization process—especially school-teaching, textbooks, and related literature (see Sect. 4.3.4). Models of this sort would offer alternatives to an anti-modernizing view of historical cultural landscape that in principle excludes renewable energy plant (see Kühne 2008a and Sect. 4.2.5).

Those renewable energy installations, especially wind farms, convey an impression of awe rather than beauty is largely a matter of their size—and it might be noted here that the postmodern era has a particular affinity to the awesome-sublime. Nevertheless, aesthetic attitudes to renewable energy plant do not stand alone; other important aspects enter the equation: "Local ownership guarantees local economic benefits and at least partial local involvement. Moreover, positive social impact can result from joint commitment to, and cooperation in, a wind scheme" (Breukers and Wolsink 2007, p. 2748; Table 5.1). And these other (economic, political and social) factors impinge on aesthetic ascriptions and change them (Hook 2008). Thus if no individual gain to the local population (apart perhaps from a single landowner) is forthcoming from an installation, because (for example) the company that owns it is headquartered overseas, aesthetic criteria will gain in importance (see the remarks above on landscape as common property). In the frequent case of an everyday small-town world, this—for well-known social reasons—often triggers the founding of a citizens' initiative opposing the projected building or extension of the plant in question. Opposition of this sort can be met by ensuring that installations are only erected on municipal (i.e. commonly owned and leased) property, and/or by a shareholding arrangement that allows financial participation by the local community (Kühne 2011a).

**Table 5.1** Impact of different renewable energy plant on the physical landscape (adapted from Dû-Blayo 2011)

|  | Visual impact | Impact on physical-spatial structure | Reversibility |
|---|---|---|---|
| Wind farm | Very high | Low | A few months |
| Agro biomass | Low | Very high | Decades |
| Forest biomass | Low | High | Decades |
| Large area photovoltaics | Medium | Medium | A few months |

The range and variety of arguments adduced by proponents as well as opponents of renewable energy installations is apparent from the Internet videos on the extension of the German power grid analyzed by Kühne and Weber (2015; on analysis of Internet videos in general see Sect. 5.6). Arguments for the grid extension are generally technical, abstract, and highly cognitive, presenting the need for power lines and pylons as the logical conclusion of a causal chain, a rational means to a 'good' end. A scattering of critical (lay) points of view occurs explicitly in these videos; more often, however, they appear as a subdiscourse contained (and hence neutralized) in an expert discourse voiced not infrequently in tones of didactic superiority—most notably in videos from the German Federal Network Agency. Lay reservations connected with health, aesthetics, or the home environment are either met with cognitive arguments or simply ignored. Overall, the relative paucity of references to landscape in these videos indicates its absence from expert discourse rather any intrinsic social irrelevance. Among others, Leibenath and Otto (2012) and Hübner and Hahn (2013) present the case for taking lay arguments and feelings about 'their' landscape seriously.

A growing self-confidence in citizens' organizations has further complicated the social perception of changes to the landscape—here exemplified in wind farms and grid extension measures. The situation can arise that, despite all the steps taken to clinch the planning process, ensure economic, scientific, and technological feasibility, confirm acceptability to the various associations involved, and secure press and media coverage, the implementation of major political decisions like the national energy transition which have immediate local repercussions can be effectively halted by (in most cases) local activists, with corresponding effects down the line. The manifest success of this kind of resistance, which de facto questions the ability of the political system to regulate public affairs, can be seen as a further instance of the crisis of democratic participation and legitimacy in contemporary politics (Walter et al. 2013). That the debate often reverts to the level of primary information long after relevant political decisions have been signed and sealed, heightens the impression that it questions the legitimacy of the entire process (see Walk 2008; Goldschmidt et al. 2012).

Far from merely defending the value of private property, citizens' initiatives set out to "consciously preserve and shape the quality of life in their home environment" (Walter et al. 2013, p. 106). They consider landscape a common value (Apolinarski et al. 2006; Olwig 2008, 2009; Schneider 2016) with emotional and aesthetic connotations (Ipsen 2006), and regard technological intrusions in the form of renewable energy plant as rationalizations in the Weberian sense (Weber 1976) and physical expressions of systemic political economics in Habermas' sense (Habermas 1981)—both interpretive models deriving from the 1970s and 80s. At that time, "technology was seen as driving society toward ever higher efficiency, functionality, and profit: goals whose legitimacy many doubted" (Renn 2005, p. 30). But the fascination of technology has diminished over the centuries—after all, the 17th century Dutch masters painted windmills not for their romantic value (an altogether later development), but because they were the high-tech of the day (Berr 2014). "What is new is the organized nature of the shaping hand, which points function in a new direction" (Kornwachs 2013, p. 311). New, too, in this context, is the type of social capital accrued by protest, which—especially for people at a biographical turning point (e.g. retirement)—endows life with new meaning (Van der Horst and Toke 2010; Walter et al. 2013; Kühne 2014b).

## 5.5 Model Railroad Landscapes—Power Versus Contingency

The previous case study highlighted the reactions of stereotypical and/or familiar landscape constructs in the face of change to the physical substrate of landscape. Model railroading landscapes, on the other hand, are per se prime instances of contingency, reflecting what at least potentially could be a highly individualized actualization of appropriated social landscape. For landscape is a key element in any model railroad configuration: more than a mere background, it serves—according to relevant discourse conventions among enthusiasts—as the determining reason for the entire enterprise (for greater detail see Kühne 2008a; Kühne and Schmitt 2012a, b). Modeling a ten-track station in a village, for example, is as absurd as building an arbitrarily twisting track across a plain.

Table 5.2 illustrates landscape references (by specific element and frequency) in the discourse of model railroaders (Swales 1990). Like all discourse communities, the modelers also exhibit a complex system of (generally unconscious) spoken and written conventions (Weber 2013, 2015). The table below is based on two empirical surveys: a quantitative and qualitative analysis of 17 publications (in German) citing a total of 95 model railroads (n = 95; Kühne 2008a), and a qualitative study

**Table 5.2** Selected references to landscape elements in model railroad discourse (Kühne 2006, 2008a)

|  | Interviews (n = 455) | Publications (n = 95) (in %) |
|---|---|---|
| Forests | 96.26 | 93.68 |
| Fields and meadows | 95.16 | 82.11 |
| Lakes and watercourses | 91.21 | 69.47 |
| Villages | 83.08 | 75.79 |
| Farms | 73.63 | 44.21 |
| Mountains | 59.12 | 87.37 |
| Country roads | 44.84 | 83.16 |
| Small towns | 32.09 | 18.95 |
| People | 40.88 | 95.79 |
| Industrial plant | 14.07 | 18.95 |
| Wind turbines | 10.99 | 2.11 |
| Cities | 8.79 | 13.68 |
| Highways/autobahns | 8.79 | 2.11 |
| Automobiles | 6.37 | 87.37 |

## 5.5 Model Railroad Landscapes—Power Versus Contingency

of 25 guideline-based interviews with members of four model railroad clubs in the Saarland (n = 455; Kühne and Schmitt 2012a, b). There is marked congruence in the landscapes most favored in these surveys (Kühne 2008a), with a clear preference for rural, hilly, and semi-open country. A comparison of the preferences expressed for specific landscape elements indicates not only a high correlation coefficient (0.615) but also a high exponential coefficient of determination (0.35; Kühne and Schmitt 2012a). Statistical deviance is due on the one hand to the technological bias of model railroaders (hence the higher incidence of roads and automobiles), and on the other to the difficulty of modeling lakes and watercourses. The incidence of human figures reflects one of the primary purposes of railroads: personal mobility. The specific landscape preferences for the type of hill country found in the central German massif can be explained in terms of modeling requirements—a true-to-scale representation of Alpine mountains would, as a rule, disrupt the dimensions of the model. Tunnels, on the other hand, are not only 'realistic' but allow trains to be hidden from view (as if in sidings), enhancing differentiation in the overall model profile without overloading it.

Model railroad discourse reveals as a core aim of its participants the achievement of a credible symbiosis between track and landscape, with the track fitting the landscape, not vice versa. Credibility of this kind is a distinguishing mark of the 'serious' as opposed to 'playful' model railroader (Kühne 2008a; Kühne and Schmitt 2012a, b). Another indication of seriousness is that the trains do not just 'run around in circles' but as far as possible (appear to) return from the direction in which they last traveled; or if the model is too small to allow this, then at least they should take as long as possible to complete their circuit, which masks the problem. Finally, verisimilitude demands consistency in the historical epoch presented in the model—i.e. rolling stock, buildings and figures should all appear, as they would have done in the period in question, which, as the literature reveals, is most often that of the 1950s to early 1960s.

As well as the 'playful' group, German model railroader discourse documents two other deviant tendencies: 'rivet counters' and 'leaf counters'. Both are obsessive about accuracy of detail: the former with regard to rolling stock—the number of rivets on a locomotive must exactly correspond with that on its real counterpart—the latter with regard to landscape elements. The discursive equation of obsession with deviance already reveals the imposition of social norms setting the limits of both interpretation and practice. The discourse in question is on the one hand face to face, on the other via specialist publications, which are particularly interesting in this context for their frequently imperative language and dichotomous presentation of the 'rights' and 'wrongs' of railroad modeling.

Interviews as well as publications show a preference among model railroaders for historical cultural landscapes, with a sometimes explicit, sometimes implicit yearning for a romantically idealized rural, communality (see Vicencotti 2005) in a premodern world free from the insecurity that plagues our century. Hence the predilection for the post-WW2 era (so-called Epoch III), a time of hope in a burgeoning future, with a stable class society and a still moderate pace of life (and transportation), symbolized by the still actively deployed steam engine; 68.4% of the models in the survey depict this period. However, most interview respondents were aware of the discrepancy between idyllic landscapes of this sort and the postulate of verisimilitude; only a minority insisted on modeling a purely ideal world (Kühne and Schmitt 2012a, transcript 3, l. 101). Accordingly, 'ugly' elements like forest fires, industrial installations, or dams under construction also occur as guarantees of normative realism, even if they are few and far between.

It is evident from the data presented here that the discourse conventions of model railroaders represent a synthesis of general landscape preferences and a special group focus on railroads. The norms formulated in this discourse determine social recognition and/or withdrawal within the group (Kühne and Schmitt 2012a). Landscape thereby becomes a medium of social distinction: as in the expert discourse of landscape architects, planners, geographers etc., a hierarchy soon develops (Bourdieu 1979), with a ruling class determining—especially via relevant publications—what should and should not be part of the model. Most modelers belong to the 'middle class', which on the whole simply follows these directives. It must be noted, however, that a model railroad is almost always in a state of change: transformation is also of the essence. And although the normative discourse sets the boundaries of acceptability, within these limits there is still freedom without loss of social standing. As one respondent said: "There's nothing worse than being told 'Oh, I put that house there, too'" (Kühne and Schmitt 2012a, transcript 4, ll. 54–55). So although the potential contingency of model railroad landscapes is severely reduced—often to matters of detail—it is not entirely abolished.

In this context, the definition of deviance is manifestly a powerful instrument for ensuring the maintenance of conventions. This is apparent in its stigmatization of 'players', 'rivet counters', and leaf counters'—the former not serious enough, the latter overly serious. Evidently, therefore, the key discursive negotiation in model railroading is concerned with the legitimate scope of compromise in a scale model whose original is known in detail. A scale model of fictive landscapes could in principle explore many avenues of contingency, form new syntheses, express individual conceptions, and create miniaturized physical alternatives to the expressions of economic, social and political logic that mold our existent landscapes. But the ideology of 'realistic' presentation of a world as it was in a specific time and place discards this possibility in favor of a stereotypical vision which—for the most part uncritically—perpetuates the normative transformatory mechanisms of unequally distributed symbolic capital (Kühne and Schmitt 2012a).

## 5.6 Landscape and Power in the Media—The Reproduction of Social Landscape Stereotypes in Internet Videos of Southern California

While the last case study, along with earlier references, showed the significance of specialist media for the social construction of landscapes, the present section is devoted to the impact of media in their own right. Focusing on Southern California,

## 5.6 Landscape and Power in the Media …

it summarizes chapters of Kühne 2012 (with analyses of 216 videos on Los Angeles) and Kühne and Schönwald 2015 (with analyses of 120 videos each on San Diego and Tijuana).

With the spread first of radio, then of television, and finally of the Internet—the latter especially in the context of Web 2.0—the social significance of media in shaping our encounter with the world has surged, with profound effects on the development and character of common notions of landscape. Following Faulstich, media can be defined as "institutionalized systems employing organized channels of communication of varied outreach and social dominance" (Faulstich 2002, p. 26). But different media appeal to different age groups: younger people visit the Internet for knowledge of landscape more readily than their elders (Fig. 5.10), for whom personal observation, guided tours, or print media still play a larger role (Kühne 2014b).

The following analysis of media presentations of Los Angeles, San Diego, and Tijuana concentrates on Internet videos for a number of reasons:

(a) "the emergence of digital publications has remorselessly and successfully supplanted" the dominance of print journalism and radio (Münker 2009, p. 34);
(b) Internet videos provide a broader perspective on the cities in question than film or television, as they include a wide spectrum of nonconformist creative and documentary work from semi-professionals and amateurs that would scarcely be granted air time, as well as the work of professional journalists, directors, and producers with commercial interests;
(c) Web 2.0 in particular is recursively linked to the differentiation and individualization of society—a process already taken into account in the program offers and timing of television channels (Münker 2009);
(d) young people spend more time online than they do watching television, while 'YouTube', for example, also has an increasing following in older groups (Schmidt 2011);
(e) as the presentation of landscape in film has already been widely researched, new insights in that direction are strictly limited.

Moreover, unidirectional media like newspapers, television, radio and early Internet (Web 1.0) were (and are) characterized by uniform modernist patterns of hierarchical division between author and recipient: on the one hand informed (expert) writer-broadcasters, on the other uninformed (lay) reader-listeners. This division reflects a modernism "whose enlightened emancipatory impulses were long borne aloft on the sense of mission of a handful of great prophets and governed by the ideals of their grand narratives and the implicit promise of 'one true society'" (Münker 2009, p. 47; see also Leggewie 1998). As late as 1997 Gupta and Ferguson could describe the impact of contemporary media on concepts of space in vivid terms: "The mass media violate the notion that places are containers of integrated cultures: the words and images of mass media travel to you rather than you having to travel to them; they are commonly understood to be alienated discourse, not the expression of the consciousness and world of a collective 'people' seen as originating authors; and they are mediated by the market, not rooted in place, tradition or locality" (Gupta and Ferguson 1997, p. 9).

Web 2.0, on the other hand, can be seen as an aspect of the postmodern social development that has largely abolished the author-recipient hierarchy. Characteristically decentralized, it replaces a single grand narrative with many small narratives, which are, however, open to anti-emancipatory tendencies and may rapidly sink into racist or sexist discourse, or be seen to glorify violence (Münker 2009; and see Kühne 2012).

Landscape is a key feature of visual media; for Sergei Eisenstein it was as central to film as music. He called it "the freest element in film, but the one that must transport the greatest narrative burden of changing moods, emotional states, and spiritual experiences" (Eisenstein 1987 [1921], p. 217). Documentaries and videos —whether private or posted on the Internet—as well as feature films in the cinema and on television are a major resource for the social construction of landscape; for "while they tell stories and recount events, they at the same time picture cities, countries and regions" (Escher and Zimmermann 2001, p. 228; Ford 1994; Bollhöfer 2003; Lukinbeal 2005). And doing so, they set in train a process of branding, "promoting one or two dominant ways of 'reading' a city" (Löw 2010, p. 84)—or, one might add, a landscape—and thereby systematically reducing its potentiality for contingency: "visitors and residents alike now gain the impression that they know it" (ibid.). Far from being unidirectional, however, this process is highly recursive: the medially constructed landscape and the appropriated physical landscape intimately intermesh, or in Dear's words, "as cities are represented, so representations become cities" (Dear 2000, p. 167).

Films in general and Internet videos in particular are, as Harvey has observed, singularly adapted to investigate "the conditions of postmodernity, especially the confusing and conflict-ridden experience of space and time" (Harvey 1989, p. 322). Not that they are simply "a pre-eminent reflection of reality: they are a language speaking a pre-existent language—both languages influencing each other in their systems of conventions" (Eco 1972, p. 254). They are the product of acts of "selection, evaluation, and interpretation of social events and processes" (Werlen 1997, p. 383; Kühne 2013a), and their message, according to Liessmann (following Anders 1980), is constitutive of what is socially accepted as good, true, and beautiful. In this (recursive) sense "the measure of reality becomes whether or not it was 'in the picture'" (Liessmann 1999, p. 113). Münker meanwhile makes the rather more basic point that "there are many reasons for calling our society a 'media society': anything that transcends our immediate experience comes to us *exclusively* via the media—and we know we are dealing with mediated, medialized information" (Münker 2009, p. 33, original emphasis). In fact Internet videos, like their earlier counterpart in films and series, present "a double medialization: from social role [...] to theatrical medium [...], and from theatrical to electronic medium [...]" (Faulstich 2002, p. 254; and see Graham 2005).

The fading, and final disappearance of the border between representing and represented is a core feature of postmodern social development. Kubsch has observed that "postmodernism was not founded in the cinema, but the cinema and the worlds of TV and Internet are where it is most concentratedly manifest" (Kubsch 2007, p. 49). Moreover, these media, more than anything else, accelerate the processes of which they are an integral part—not only via social networks, but also in the constant production and distribution of goods, views, and patterns of interpretation (many of them very short-lived). Television films and series, feature films (now predominantly streamed and viewed at home), and self-posted presentations (e.g. on YouTube) form an increasingly significant resource for the "patchwork identity" (Keupp 1992, p. 176) and fragmented world-view of *homo medialis* (Hall 1980; Benton 1995; Bourdieu 1996; Hepp 1998)—a phenomenon that filters down, more or less intensely, into everyday "reflections on one's own life, clothing, and hairstyle, [in which] popular films, TV productions, and video clips assume the function of expert systems" (Winter 1997, p. 68).

The same applies to landscape: patterns of interpretation are presented, received, and cemented in social landscape, from whose compartmentalized structures they recursively draw their forms: "Deposits of cultural memory, narratives, psycho-geographical visions: all this is absorbed in the spatial practice of filmic perception" (Bruno 2008, p. 74)—a practice caught up in the continuous invocation and recreation of stereotypes and 'simulacra' (Baudrillard 1976, 1994), which often have little to do with the way local people see their city or region. In fact "the actual inhabitants [...] are frequently reduced to the role of observers" of the medial staging of their lives (Dimendberg 2008, p. 42); but however remote from their reality this may be, its expectations press on them with such insistence that they find themselves in the end conforming to the stereotype.

And what is true of social contexts also holds good for the material basis of appropriated physical landscape. While films of every type and genre contain iconographic, indexical and symbolic references to common landscapes, these frequently undergo stereotyping and idealization (Ford 1994; Lukinbeal 2005; Lefebvre 2006; Kühne 2012; Kühne and Schönwald 2015). Vernacular landscapes occur with far less regularity than items selected by the hegemonic masters of interpretation to bear their message on the media channels of the day—currently Internet videos (Kühne 2012). In 1923 a different medium was used by the real estate developer who imposed his will so demonstratively on a piece of land just outside Los Angeles by erecting what has become a unique historic icon: the big white letters of HOLLYWOODLAND (Braudy 2011). Even when part of the material structure of the sign (the four letters of LAND) later collapsed under the impact of latent physical forces (and the rest has needed restoration), its symbolic power has—even among other great American icons—remained unparalleled. For, as Braudy points out, unlike the Statue of Liberty, Mount Rushmore, or the Golden Gate Bridge, the 'Hollywood' sign evokes neither patriotism nor historical awareness: all it does, for better or worse, is focus our inner life and dreams (ibid.).

With the dawn of Web 2.0—a designation based on the "ongoing functional leaps" (Schmidt 2011, p. 13) of software—the representation of physical space in the media grew remarkably and today many scientific inquiries into the social construction of space (e.g. urban-rural hybrids: see Kühne 2012; Kühne and Schönwald 2015; Sect. 4.2.3 above) focus on this area. Largely responsible for this is what Münker sees as "the general trend that has made user participation *a core feature* of Internet design" (Münker 2009, p. 15, original emphasis). This ranges from commentaries and evaluations (e.g. at Amazon.com) to full production of content by the user (e.g. at Wikipedia.org). Between these two poles, according to Münker (ibid.), lie video, photo, and music portals like the source used here, YouTube. While the Internet before Web 2.0 was characteristically unidirectional (from writer to reader), it is now "a medium for a culture of reading *and* writing that is at present, so to speak, in training" (ibid. 16–17, original emphasis). What Münker says of the read/write culture can be extended in the case of videos to a do-it-yourself watch/listen/film culture. Along with social networks, the new information and entertainment platforms are recording the highest Internet growth rates; for many users they represent a "conscious alternative to the more conventional offers of traditional mass media" (ibid. 18).

As an essential element of Web 2.0, social sharing means making one's own films and photos available to others, "creating joint knowledge bases via links" (Ebersbach et al. 2011, p. 117). The content of videos etc. can often be commented, tagged, and evaluated by means of a rating function. Social sharing platforms are among the oldest and most widely used of Web 2.0 applications (Ebersbach et al. 2011), and the recognition—and with it distinction potential—derived from their use grows in direct relation to the level of activity and knowledge of the user (see Zillien and Hargittai 2009). Three distinct roles are apparent: users' activity protocols are exploited to create preference lists that provide a level of order in platform contents; evaluators classify and comment on contents, actively enhancing that order; producers upload existing or newly created material onto the platform (Ebersbach et al. 2011).

The landmarks of San Diego (Kühne and Schönwald 2015) and Los Angeles (Kühne 2012; see Fig. 5.5) are subject to a continuous circular process of updating in the media—here specifically Internet videos—and physical interface with tourists and the guide books, brochures, flyers etc. they bring with them. Landmarks like the Hotel Del Coronado, or the logos of San Diego Zoo, Sea World, Balboa Park, or Old Town function as sites of meaning recursively actualized by the presence of tourists, as well as by their own presence in other media (see e.g. Lukinbeal 2000; Ford 2004, 2005; Lüke 2008; Fine 2000). Taken together, these presences create San Diego's image of leisure-leaning port, with its impressive central business district and waterfront development (34). In comparison with the Los Angeles videos in the same category (Kühne 2012), San Diego's symbols are less specific. Where the LA videos present a mere 15 iconic sites, the San Diego set counts 27, but their distribution is far less focused than in LA, where 85.2% show the downtown skyline—despite the fact that, for scientific geographers, Los Angeles is *the* typical horizontal city, which incidentally reveals something of the gap between expert and lay perspectives. A further 74.1% of the LA videos show the city's freeways, while the Hollywood sign, for all its global fame, only occurs in 48.1% of the videos in the survey. None of San Diego's iconic symbols occurs in more than 75% of the videos, but four achieve between 50 and 75%: the harbor with 72%, and the CBD, Waterfront, and beaches with 68% each. Tijuana is represented by the

Avenida Revolución with 76.2%, the city's CBD (50%), and the arch at the top of the Avenida Revolución (45.2%). A further significant qualitative difference between San Diego's and Tijuana's images is that the Californian city excludes anything from the videos that might contradict its image as 'America's finest city', whereas the Tijuana videos in the same category (travel reports and municipal presentation) also articulate topical problems and challenges—especially those connected with the Mexico–U.S. border (Kühne and Schönwald 2015) (Fig. 5.16).

In many Internet videos the "desire for authenticity replaces the demand for objectivity" (Münker 2009, p. 100). Subjective experiences take center stage: a youth group visiting San Diego Zoo, driving through Tijuana with 'spontaneous' commentaries, a TV personality giving 'personal' tips for a weekend in San Diego —individual responses of this kind make no attempt to describe the cities 'as they are'; what they show is what can be said about a social landscape without loss of social recognition. Interpretations that diverge too much from the established line (as in a video praising the high quality of life in Tijuana) are met by a counterblast of uninhibited comment—this is, after all, the anonymous Internet.

Another aspect of these videos is the differences they reveal between lay and expert filmmakers. Videos by experts (e.g. journalists) are the exception in one direction; films by amateurs with neither technical nor rhetorical skill in the other. The vastly improved user-friendliness of webcams and editing software, the new ease of private product distribution, and the drastically reduced cost of equipment have brought about a hybridization of lay and expert spheres boosted by the expansion of education and the availability of Internet information. Three further modes of hybridization are also evident in these videos:

(a) Hybridization of public and private spheres—private activities like a visit to San Diego Zoo become globally available, usually confirming established patterns of social discourse about landscape, sometimes relativizing and extending it (e.g. Tijuana for Dummies No. 207).
(b) Videos are an integral part of the interactive social network of "identity, personal relations, information, and communication management," but as "reproductions of structures [from other contexts], they at the same time go beyond the Internet," revoking any "clear division between the virtual and real worlds" (Schmidt 2011, p. 75). Content posted on the Internet is taken as personal to the one who posts it.
(c) A hybridization of person and role evolves directly from the blurring of the public/private and virtual/real distinctions. In Internet videos, people often want to present themselves 'authentically' rather than as role-players (e.g. as experts seeking interpretive sovereignty)—but the 'authentic self' is also an image, albeit contextually a bit more individual than an official role.

There is also some hybridity in the main content of both batches of videos; less so, however, in the San Diego (40%) than the Tijuana (72.5%) group, where hybridity is often treated implicitly—e.g. by presenting the Avenida Revolución (generally in an affirmative sense) as a Mexican–U.S. amusement strip. As the San

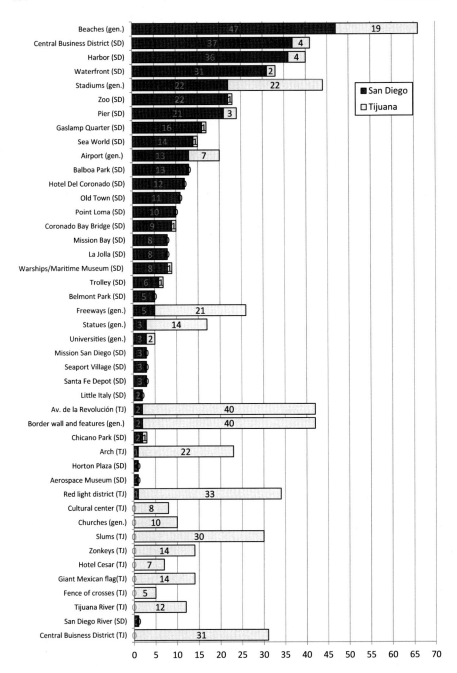

**Fig. 5.16** Iconic items in Internet travel and local advertising videos for San Diego (n = 25) and Tijuana (n = 42; Kühne and Schönwald 2015)

Diego videos mostly see the city as purely U.S. American, the real consequences of proximity to the Mexican border and the presence of a Hispanic population are not addressed. The question of hybridity therefore scarcely arises.

The presentation of Tijuana strongly reflects U.S. American ideas about Mexico, and the physical structures of the city have also been adapted to this perspective, which is even present in Mexican videos, often with voyeuristic or deconstructive overtones—as in *Tijuana for Dummies*, a video notable for its humorous approach to common clichés about Tijuana, and the surreal techniques it uses to comment on the culture of corruption and denunciation, but also on the economic success of the border community. The video starts with the highly symbolic *Mona de Tijuana* (official name *Tijuana III Millennium*, Fig. 5.17), a large statue in a quarter whose inhabitants have little symbolic capital, and which regularly features in videos critical of the city's development trajectory (generally in comparison with the USA). Created in 1989 by art-school student Armando Muñoz Garcia, this millennium commemoration work raises an arm aloft in a gesture echoing the Statue of Liberty. It was initially intended for a position in the center of Tijuana, but the city leaders decided otherwise and put a stainless steel arch at the head of the Avenida Revolución instead. Undeterred, the sculptor installed the work in his own backyard and initially lived in it; in fact, it still houses his studio. Today, while the arch in the city center—reminiscent of the Gateway Arch in St. Louis, Missouri—possesses little symbolic force (it occurs frequently in videos, but with low intensity), the *Mona* remains an icon of the irrepressible will for freedom and independence.

Under the *Mona* in *Tijuana for Dummies* a young blonde woman rebuffs a Hispanic-looking man who evidently thinks she is an easy prey. This is the beginning of a series of role swaps (especially by the woman) and caricature-like exaggerations, for example, when two men in Mexican uniforms are led like dogs on a leash by a man in a suit and then let loose to harass the blonde woman. In sequences like this, the video parodies common Tijuana stereotypes, from crime (sharpshooters) and corruption (the men in uniform) to the sale of candy at the border (by the blonde woman to a Hispanic-looking man in a Ford Mustang).

A characteristic way of dealing with the undesirable products of society and its actions, or (what amounts to the same thing) with landscape materials that run counter to social stereotypes and expectations, is to desensualize them or—given the dominance of the visual in Western constructions of landscape—to make them invisible (for greater detail see Marcus 1995; Kühne 2012). Hence, in our 'real' worlds, physical objects regarded as socially unacceptable (e.g. trash heaps, sewage) are subjected to technical processing that largely removes them from the sphere of common social activity; likewise, citizens with manifestly low symbolic capital (e.g. the homeless) are refused access to sites of intense consumerism (e.g. shopping malls). In addition, what happens in physical space happens all the more readily in its virtual presentation: again, the two dimensions stand in a recursive relation. The videos examined in the survey express discursive limits very clearly, revealing what can and what cannot be shown/made visible/sensualized of San Diego and Tijuana.

**Fig. 5.17** The Mona de Tijuana (official title Tijuana III Millennium; TIJUANALANDIA 2010; Atlas Obscura 2013; Kühne and Schönwald 2015) (Photo Olaf Kühne)

In the Tijuana videos, the implicit limits of the social landscape are set remarkably wide. Some of the things shown—dirt, deserts, prostitution, slums— directly contradict social stereotypes of beauty and morality. In comparison, the producers of the San Diego videos keep a much firmer grip on discursive freedoms:

the landscape of 'America's finest city' must not be allowed to diverge from the normative image. Whatever that might break that image—border systems and defenses, Hispanic cultural structures etc.—is only shown in connection with deviant behavior like criminality or illegal entry into the U.S. The city's abundant homeless, socially, morally and aesthetically stigmatized, do not appear at all: they are radically desensualized. Nor is the technical infrastructure of San Diego or Los Angeles shown: e.g. power supply installations, the ecological side effects of long distance water supply, or the high attrition rate of San Diego's technical infrastructure. Rather than depicting the physical and spatial problems of cities set in a desert (or semi-desert) climate, these videos present an image of San Diego—even more than Los Angeles—as a green paradise under an everlasting sun.

## References

Agentur für Erneuerbare Energien [German Renewable Energy Agency](2015). Die deutsche Bevölkerung will mehr Erneuerbare Energien: Repräsentative Akzeptanzumfrage zeigt hohe Zustimmung für weiteren Ausbau. http://www.unendlich-viel-energie.de/die-deutsche-bevoelkerung-will-mehr-erneuerbare-energien (last access 07.03.2016)

Altrock, U., Güntner, S., Huning, S., & Peters, D. (2005). Zwischen Anpassung und Neuerfindung. In U. Altrock, S. Güntner, S. Huning & D. Peters (Eds.), Zwischen Anpassung und Neuerfindung. Raumplanung und Stadtentwicklung in den Staaten der EU-Osterweiterung (pp. 9–24). Berlin: Springer VS.

Anders, G. (1980). Die Antiquiertheit des Menschen. Volume 1. Über die Seele im Zeitalter der zweiten industriellen Revolution. München: Verlag C. H. Beck.

Andrusz, G., Harloe, M., & Szelényi, I. (Eds.). (2011). *Cities after socialism: Urban and regional change and conflict in post-socialist societies*. New York: Wiley-Blackwell.

Apolinarski, I., Gailing, L., & Röhring, A. (2006). Kulturlandschaft als regionales Gemeinschaftsgut. Vom Kulturlandschaftsdilemma zum Kulturlandschaftsmanagement. In U. Matthiesen, R. Danielzyk, St. Heiland & S. Tzschaschel (Eds.), Kulturlandschaften als Herausforderung für die Raumplanung. Verständnisse – Erfahrungen – Perspektiven (pp. 81–98). Hannover: Verlag der ARL.

Atlas Obscura. (2013). La Mona. Giant nude woman of Tijuana is also the artist's home. http://www.atlasobscura.com/places/la-mona-by-armando-garcia. Accessed September 30, 2013.

Baker, G. (2007). *San Diego. Another HarborTown History*. Santa Barbara: Harbortown Histories.

Banham, R. (2009 [1971]). Los Angeles. The architecture of four ecologies. Berkeley, Los Angeles, London: University of California Press.

Baudrillard, J. (1976). *L'échange symbolique et la mort*. Paris: Gallimard.

Baudrillard, J. (1994). *Simulacra and simulation*. Ann Arbor: University of Michigan Press.

Baudrillard, J. (2004 [1986]). America. London, New York: Routledge.

Bauman, Z. (2008). *Flüchtige Zeiten. Leben in der Ungewissheit*. Hamburg: Hamburger Edition.

Benton, L. M. (1995). Will the reel/real Los Angeles please stand up? *Urban Geography, 16*(2), 144–164.

Berr, K. (2014). Zum ethischen Gehalt des Gebauten und Gestalteten. *Ausdruck und Gebrauch, 12*, 30–56.

Bierling, St. (2006). *Kleine Geschichte Kaliforniens*. München: C. H. Beck.

Bollhöfer, B. (2003). Stadt und Film – neue Herausforderungen für die Kulturgeographie. *Petermanns Geographische Mitteilungen, 147*, 54–59.

Borodziej, W. (2008). Warschau von Warschau aus betrachten. *Bauwelt, 12*, 10–19.

Bosch, S., & Peyke, G. (2011). Gegenwind für die Erneuerbaren – Räumliche Neuorientierung der Wind-, Solar- und Bioenergie vor dem Hintergrund einer verringerten Akzeptanz sowie zunehmender Flächennutzungskonflikte im ländlichen Raum. *Raumforschung und Raumordnung, 69*, 105–118.

Bottles, S. (1987). *Los Angeles and the Automobile. The Making of the Modern City*. Berkeley, Los Angeles, London: University of California Press.

Bourdieu, P. (1979). *La distinction: critique sociale du jugement*. Paris: Les Editions de Minuit.

Bourdieu, P. (1996). *Sur la télévision*. Paris: Raisons d'agir.

Brailich, A., & Pütz, R. (2014). Gated Community vs. Großwohnsiedlung. Identitätskonstruktionen in städtebaulichen Auseinandersetzungen einer Transformationsgesellschaft am Beispiel von Warschau-Ursynów. *Europa Regional, 20*(2–3), 76–88.

Bratzel, S. (1995). *Extreme der Mobilität. Entwicklungen und Folgen der Verkehrspolitik in Los Angeles*. Basel et al.: Birkhäuser.

Braudy, L. (2011). *The Hollywood sign. Fantasy and reality of an American Icon*. New Haven, London: Yale University Press.

Breukers, S., & Wolsink, M. (2007). Wind power implementation in changing institutional landscapes: an international comparison. *Energy Policy, 35*(5), 2737–2750.

Brodsly, D. (1981). *L. A. Freeway. An appreciative essay*. Berkeley et al.: University of California Press.

Bruno, G. (2008). Bildwissenschaft. Spatial Turns in vier Einstellungen. In J. Döring & T. Thielmann (Eds.), Spatial turn. Das Raumparadigma in den Kultur- und Sozialwissenschaften (pp. 71–74). Bielefeld: transcript.

Bruns, A., & Gee, K. (2009). From state-centered decision-making to participatory governance-planning for offshore wind farms and implementation of the water framework directive in Northern Germany. *GAIA-Ecological Perspectives for Science and Society, 18*(2), 150–157.

Buntin, J. (2009). *L.A. Noir. The struggle for the soul of America's most seductive city*. New York: Broadway Books.

Campbell, N. (2000). *The cultures of the American new west*. Edinbourgh: Psychology Press.

Cierpiński, J. (1993). Wstęp. In Warszawa Rozwój Przestrzenny. Warszawa, 5–8.

Clarke, G. (1993). Introduction. A critical and historical overview. In G. Clarke (Eds.), The American landscape. Literary sources and documents (2 Vols., pp. 3–51). The Banks, Mountfield: Helm Information.

Conzen, M. (2010). Introduction. In M. Conzen (Ed.), *The making of American landscape* (pp. 1–10). New York, London: Routledge.

Corner, J. (1992). Representation and Landscape. In: Swaffield, S. (Ed.), Theory in Landscape Architecture. A Reader (pp. 144–165). Philadelphia: Penn Studies in Landscape Architecture.

Corner, J. (2002a). The hermeneutic landscape (1991). In S. Swaffield (Ed.), Theory in landscape architecture. A reader (pp. 130–131). Philadelphia: University of Pennsylvania Press.

Corner, J. (2002b). Representation and landscape (1992). In S. Swaffield (Ed.), Theory in landscape architecture. A reader (pp. 144–165). Philadelphia: University of Pennsylvania Press.

Cosgrove, D. E. (1984). *Social formation and symbolic landscape*. London, Sydney: University of Wisconsin Press.

Cronon, W. (1996b). The trouble with wilderness; or, getting back to the wrong nature. In W. Cronon (Ed.), Uncommon ground. Rethinking the human place in nature (pp. 69–90). New York, London: W. W. Norton.

Culver, L. (2010). *The frontier of leisure*. Oxford et al.: Oxford University Press.

Czepczyński, M. (2008). *Cultural landscapes of post-socialist cities: Representation of powers and needs*. London: Ashgate Publishing Ltd.

Czesak, B., Pazdan, M., & Różycka-Czas, R. (2015). Die städtische Landschaft in der Transformation: Krakau und Warschau. In O. Kühne, K. Gawroński, & J. Hernik (Eds.), *Transformation und Landschaft – landschaftliche Folgen gesellschaftlicher Veränderungsprozesse* (pp. 165–182). Wiesbaden: Springer VS.

Davis, M. (2004 [1998]). Ökologie der Angst. Das Leben mit der Katastrophe. München, Zürich: Antje Kunstmann.

Dear, M. (1998). In the city, time becomes visible: Internationality and urbanism in Los Angeles, 1781–1991. In A. Scott & E. Soja (Eds.), *The city: Los Angeles and urban theory at the end of the twentieth century* (pp. 76–105). Berkeley: University of California Press.

Dear, M. (2000). *The postmodern urban condition*. Oxford: Blackwell.

Dimendberg, E. (2008). Das Kino und die Entstehung einer modernen Stadt. In A. Ofner & C. Siefen (Eds.), Los Angeles. Eine Stadt im Film (pp. 42–48). Wien: Lit.

Domanski, B. (1997). *Industrial control over socialist towns. Benevolence or exploitation?* Westport, London: Praeger Publishers.

Dû-Blayo, Le L. (2011). How do we accommodate new land uses in traditional landscapes? Remanence of landscapes, resilience of areas. Resistance of people. *Landscape Research, 36* (4), 417–434.

Durkheim, É. (2013[1912]). Les formes élémentaires de la vie religieuse. Paris: PUF.

Ebersbach, A., Glaser, M., & Heigl, R. (2011). Social Web. Konstanz: utb.

Eco, U. (1972). *Einführung in die Semiotik*. München: Fink.

Egner, H. (2006). Mythos ‚West'. Die Canyon Country (USA) als ‚Freizeitpark'. In M. Flintner & J. Lossau (Eds.), Themenorte (pp. 59–74). Münster: Lit.

Eickelpasch, R., & Rademacher, C. (2004). Identität. Bielefeld: Transcript.

Eisel, U. (2009). *Landschaft und Gesellschaft. Räumliches Denken im Visier*. Münster: Westfälisches Dampfboot.

Eisenstein, S. (1987). *Nonindifferent Nature*. Cambridge: Cambridge University Press.

Ellin, N. (1999). *Postmodern Urbanism*. New York: Princeton archctektural Press.

Engels, J. (2010). Machtfragen. Aktuelle Entwicklungen und Perspektiven der Infrastrukturgeschichte. *Neue Politische Literatur, 55*, 51–70.

Escher, A., & Zimmermann, S. (2001). Geography meets Hollywood. Die Rolle der Landschaft im Film. *Geographische Zeitschrift, 89*, 227–236.

Ethington, P. (2001). Ghost neighborhoods: Space, time and alienation in Los Angeles. In C. Salas & A. Roth (Eds.), *Looking for Los Angeles* (pp. 29–56). Los Angeles: Getty Research Institute.

Faulstich, W. (2002). Einführung in die Medienwissenschaft. Probleme – Methoden – Domänen. München: Fink.

Fayet, R. (2003). *Reinigungen. Vom Abfall der Moderne zum Kompost der Nachmoderne*. Wien: Passagen-Verlag.

Fehl, G. (2004). Stadt im ‚National Grid'. Zu einigen historischen Grundlagen der US-amerikanischen Stadtproduktion. In U. v. Petz (Ed.), Going West. Stadtplanung in den USA – gestern und heute (pp. 42–68). Dortmund: IRPUD.

Fine, D. (2000). *Imagining Los Angeles. A city in fiction*. Reno, Las Vegas: University of New Mexico Press.

Fogelson, R. (1993 [1967]). The Fragmented Metropolis: Los Angeles 1850–1930. Cambridge, Massachusetts: Harvard University Press.

Ford, L. (1994). Sunshine and shadow: Lighting and color in the depiction of cities on film. In S. Aitken & L. Zonn (Eds.), *Place, power, situation and spectacle: A geography of film* (pp. 119–136). Lanham: Rowman & Littlefield Publishers.

Ford, L. (2004). The visions of the builders. The historical evolution of the San Diego cityscape. In P. Pryde (Ed.), San Diego. An introduction to the region (pp. 195–212). San Diego: Kendall/Hunt Publishing Company.

Ford, L. (2005). *Metropolitan San Diego. How geography and lifestyle shape a new urban environment*. Philadelphia: University of Pennsylvania Press.

Forkel, J., & Grimm, M. (2014). Die Emotionalisierung durch Landschaft oder das Glück in der Natur. *Sozialwissenschaften & Berufspraxis, 37*(2), 251–266.

Fröhlich, H. (2003). *Learning from Los Angeles. Zur Rolle von Los Angeles in der Diskussion um die postmoderne Stadt*. Bayreuth: Univ. Bayreuth, Abt. Raumplanung.

Gądecki, J. (2014a). The maketers of dreams vs. romantic gentrifiers—Reflections on consumption in the Polish housing market. *Europa Regional, 20*(2), 30–41.
Gądecki, J. (2014). Gating Warsaw. Enclosed housing estates and the aesthetics of luxury. In M. Grubbauer & J. Kusiak (Eds.), *Chasing Warsaw. Socio-material dynamics of urban change since 1990* (pp. 109–132). Frankfurt a. M.: Campus.
Garreau, J. (1991). *Edge city. Life on the new frontier.* New York: Doubleday.
Gąsior-Niemiec, A., Glasze, G., Lippok, D., & Pütz, R. (2009). Gating a city: The case of Warsaw. *Regional and Local Studies, Special Issue, 2009,* 78–101.
Gawryszewski, A. (2010). Ludność Warszawy w XX wieku. Warszawa: Instytut Geografii i Przestrzennego Zagospodarowania PAN im. Stanisława Leszczyckiego: Urząd Pracy.
Gestwa, K. (2003). Sowjetische Landschaften als Panorama von Macht und Ohnmacht. *Historische Anthropologie, 11*(1), 72–100.
Gipe, P. B. (1995). Design as if People Matter: Aesthetic Guidelines for the Wind Industry. http://www.wind-works.org/articles/design.html. Accessed 29 April 2012.
Goldschmidt, R., Scheel, O., & Renn, O. (2012). Zur Wirkung und Effektivität von Dialog- und Beteiligungsformaten. Stuttgarter Beiträge zur Risiko- und Nachhaltigkeitsforschung 23. Stuttgart: Universität Stuttgart.
Graham, S. (2004) Postmodern city. Towards an urban geopolitics. Regional Studies Association, City. *Analysis of Urban Trends, Culture, Theory, Policy, Action, 8*(2), 165–196.
Graham, G. (2005). *Philosophy of the arts: An introduction to aesthetics.* London, New York: Routledge.
Graham, S., & Marvin, S. (2001). *Splintering urbanism: Networked infrastructures, technological mobilities and the urban condition.* New York, London: Routledge Psychology Press.
Gupta, A., & Ferguson, J. (1997). Culture, power, place: Ethnography at the end of an era. In A. Gupta & J. Ferguson (Eds.), *Culture, power, place: Explorations in critical anthropology* (pp. 1–29). Durham: Duke University press.
Habermas, J. (1981). Theorie des kommunikativen Handelns. 2 Volumes. Frankfurt a. M.: Suhrkamp.
Hall, S. (1980). Encoding, decoding. In S. Hall, D. Hobson, A. Lowe & P. Willis (Eds.), Culture, media, language. Working papers in cultural studies 1972–1979 (pp. 139-156). London, New York: Hutchinson.
Halle, D. (2003a). The New York and Los Angeles schools. In D. Halle (Ed.), New York and Los Angeles. *Politics, Society, and Culture* (pp. 1–48). Chicago, London: University of Chicago Press.
Halle, D. (2003b). Conclusion. In D. Halle (Ed.), New York and Los Angeles. Politics, Society, and Culture (pp. 449–459). Chicago, London: University of Chicago Press.
Hanich, J. (2007). Automobile Einsamkeit. Freeways im Dauerstau – in Los Angeles wird total mobil gemacht. *Lettre International, 76,* 34–37.
Hardinghaus, M. (2004). Zur amerikanischen Entwicklung der Stadt. Ein Beitrag zur Kulturgenese des City-Suburb-Phänomens unter besonderer Berücksichtigung protestantisch-calvinistischer Leitbilder. Frankfurt a. M. et al.: Peter Lang.
Harvey, D. (1989). *The condition of postmodernity.* Oxford: Wiley.
Hasse, J. (2007). Übersehene Räume. Zur Kulturgeschichte und Heterotologie des Parkhauses. Bielefeld: Transkript.
Hepp, A. (1998). *Fernsehaneignung und Alltagsgespräche. Fernsehnutzung aus der Perspektive der Cultural Studies.* Opladen, Wiesbaden: Westdeutscher Verlag.
Holzner, L. (1996). *Stadtland USA: Die Kulturlandschaft des American Way of Life.* Gotha: Klett/SVK.
Hook, S. (2008). Landschaftsveränderungen im südlichen Oberrheingebiet und Schwarzwald. Wahrnehmung kulturtechnischer Maßnahmen seit Beginn der 19. Jahrhunderts. Saarbrücken: VDM publishing company Dr. Müller.
Hübner, G., & Hahn, C. (2013). *Akzeptanz des Stromnetzausbaus in Schleswig-Holstein.* Halle: University Halle-Wittenberg.
Ipsen, D. (2006). *Ort und Landschaft.* Wiesbaden: Springer VS.

# References

Jackson, J. (1970). Jefferson, Thoreau and After. In E. H. Zube (Ed.), Landscapes, 15. Selected Writings of J. Jackson (pp. 1–9). New York: University of Massachusetts Press.

Jacobs, J. (1961). *The death and life of Great American cities.* New York: Vintage Books.

Jakle, J. (2010). Paving America for the automobile. In M. Conzen (Ed.), *The making of American landscape* (pp. 403–422). New York, London: Routledge Chapman & Hall.

Jakle, J., & Sculle, K. (2004). *Lots of parking. Land use in a car culture.* Charlottesville, London: University of Virginia Press.

Jałowiecki, B. (2012). Aktuelle Tendenzen der Metropolisierung von Warschau/Aktualny tendenje w metropolizacji Warszawy. In J. Sulzer (Ed.), *Stadtheimaten/Miekie ojczyzny* (pp. 211–225). Berlin: Jovis.

Jarvis, B. (1998). *Postmodern cartographies. The geographical imagination in contemporary American culture.* London: St. Martin's Press.

Johler, R. (2001). „Wir müssen Landschaft produzieren". Die Europäischen Union und ihre ‚Politics of Landscape and Nature'. In R. Brednich, A. Schneider & U. Werner (Eds.), Natur – Kultur. Volkskundliche Perspektiven auf Mensch und Umwelt (pp. 77–90). Münster, New York, München, Berlin: Waxmann.

Johnson, H. (2010 [1990]). Gridding a national landscape. In M. Conzen (Ed.), The making of American landscape (pp. 142–161). New York, London: Spon Press.

Johnsson, P. (2012). Gated communities. Poland holds the European record in housing for the distrustful. *Baltic Worlds, 3–4,* 26–32.

Johnston, R. (1982). *The American urban system. A geographical perspective.* New York: St. Martins Press.

Juchnowicz, S. (1990). Zródla patologicznej urbanizacji i kryzysu ekologicznego obszaru Krakowa. - Przyczyny, Terazniejszość, Perspektywy Ekorozwoju Miasta. In Polski Klub Ekologiczny (Ed.), Klęska Ekologiczna Krakowa. Przyczyny, Terażniejszość, Perspektywy Ekorozwoju Miasta Kraków (pp. 248–255). Kraków: PKE.

Kadatz, H. (1997). Städtebauliche Entwicklungslinien in Mittel- und Osteuropa. DDR, Tschechoslowakei und Ungarn. Erkner: IRS.

Karrasch, H. (2000). Los Angeles – Traumstadt mit Problemen. In H. Karrasch, W. Gamerith, T. Schwan, K. Sachs & U. Krause (Eds.), Megastädte – Weltstädte (Global Cities). Heidelberger Geographische Gesellschaft Journal 15 (pp. 33–58). Heidelberg: Springer.

Katz, J. (1999). *How emotions work.* Chicago et al.: University of Chicago Press.

Kaufmann, S. (2005). *Soziologie der Landschaft.* Wiesbaden: Springer VS.

Keil, R. (1993). *Weltstadt-Stadt der Welt.* Internationalisierung und lokale Politik in Los Angeles, Münster: Westfälisches Dampfboot.

Keupp, H. (1992). Verunsicherungen. Risiken und Chancen des Subjekts in der Postmoderne. In Th. Rauschenbach & H. Gängler (Eds.), Soziale Arbeit und Erziehung in der Risikogesellschaft (pp. 165–183). Neuwied: Hermann Luchterhand.

Knox, P., Bartels, E., Holcomb, B., Bohland, J., & Johnston, R. (1988). *The United States. A contemporary human geography.* Harlow: Longman Scientific & Technical.

Koch, F. (2010). *Die Europäische Stadt in Transformation. Stadtplanung und Stadtentwicklungspolitik im postsozialistischen Warschau.* Wiesbaden: Springer VS.

Koch, F. (2012). Anspruch und Realität europäischer Stadtentwicklung: Das Beispiel Warschau/ Aspiracje i rzeczywistość w europejskim rozwoju miast: przykład Warszawy. In J. Sulzer (Ed.), *Stadtheimaten/Miekie ojczyzny* (pp. 147–168). Berlin: Jovis.

Kocks, D. (2000). *Dream a little. Land and social justice in modern America.* Berkeley, Los Angeles, London: University of California Press.

Körner, S. (2010). *Amerikanische Landschaften. J. B. Jackson in der deutschen Rezeption.* Stuttgart: Steiner.

Kornwachs, K. (2013). Von der Faszination technischer Oberflächen. In H. Friesen & M. Wolf (Eds.), Kunst, Ästhetik, Philosophie. Im Spannungsfeld der Disziplinen (pp. 291–312). Münster: mentis.

Koshar, R. (2008). Driving cultures and the meanings of roads. Some comparative examples. In C. Mauch & T. Zeller (Eds.), *The world beyond the windshield. Roads and landscapes in the United States and Europe*. Athens, Stuttgart: Ohio University Press.

Kotkin, J. (2006). *The city. A global history*. New York: Modern Library.

Kovács, Z. (2014). New post-socialist urban landscapes: The emergence of gated communities in East Central Europe. *Cities, 36*, 179–181.

Krause, F., Bossel, H., & Müller-Reißmann, K. (1980). Energiewende – Wachstum und Wohlstand ohne Erdöl und Uran. Frankfurt a. M.: Fischer S. Verlag.

Krauss, W. (2010). The 'Dingpolitik' of wind energy in northern German landscapes: An ethnographic case study. *Landscape Research, 35*(2), 195–208.

Kubsch, R. (2007). *Die Postmoderne. Abschied von der Eindeutigkeit*. Holzgerlingen: SCM Hänssler.

Kühne, O. (2001). The interaction of industry and town in Central Eastern Europe—An intertemporary comparison based on systems theory and exemplified by Poland. *Die Erde, 132* (2), 161–185.

Kühne, O. (2003). Transformation und Umwelt in Polen. Eine kybernetisch-systemtheoretische Analyse. Mainzer Geographische Studien 51. Mainz: Geographisches Institut Johannes-Gutenberg-Universität.

Kühne, O. (2006). *Landschaft in der Postmoderne. Das Beispiel des Saarlandes*. Wiesbaden: Deutscher Universitätsverlag.

Kühne, O. (2008a). *Distinktion, Macht, Landschaft*. Springer VS: Zur sozialen Definition von Landschaft. Wiesbaden.

Kühne, O. (2008b). Die Sozialisation von Landschaft – sozialkonstruktivistische Überlegungen, empirische Befunde und Konsequenzen für den Umgang mit dem Thema Landschaft in Geographie und räumlicher Planung. *Geographische Zeitschrift, 96*(4), 189–206.

Kühne, O. (2011a). Akzeptanz von regenerativen Energien—Überlegungen zur sozialen Definition von Landschaft und Ästhetik. Stadt+Grün, 60(8), 9–13.

Kühne, O. (2011b). Akzeptanz von regenerativen Energien – Überlegungen zur sozialen Definition von Landschaft und Ästhetik. *Stadt+Grün, 60*(8), 9–13.

Kühne, O. (2011c). Akzeptanz von regenerativen Energien – Überlegungen zur sozialen Definition von Landschaft und Ästhetik. Stadt + Grün, 60(8), 9-13.

Kühne, O. (2012). *Stadt – Landschaft – Hybridität. Ästhetische Bezüge im postmodernen Los Angeles mit seinen modernen Persistenzen*. Wiesbaden: Springer VS.

Kühne, O. (2013a). *Landschaftstheorie und Landschaftspraxis. Eine Einführung aus sozialkonstruktivistischer Perspektive*. Wiesbaden: Springer VS.

Kühne, O. (2013b). Landschaftsästhetik und regenerative Energien – Grundüberlegungen zu De- und Re- Sensualisierungen und inversen Landschaften. In L. Gailing & M. Leibenath (Eds.), *Neue Energielandschaften – Neue Perspektiven der Landschaftsforschung* (pp. 101–120). Wiesbaden: Springer VS.

Kühne, O. (2014a). Die intergenerationell differenzierte Konstruktion von Landschaft. *Naturschutz und Landschaftsplanung, 46*(10), 297–302.

Kühne, O. (2014b). Zur sozialen Akzeptanz des Abbaus mineralischer Rohstoffe in Zeiten von Governance – Überlegungen auf Grundlage der sozialkonstruktivistischen Landschaftstheorie. *Akademie für Geowissenschaften und Geotechnologien, Veröffentlichungen, 30*, 81–88.

Kühne, O. (2015). The Streets of Los Angeles—About the integration of infrastructure and power. *Landscape Research, 40*(2), 139–153.

Kühne, O. (2016a). Los Angeles – machtspezifische Implikationen einer Verkehrsinfrastruktur. In S. Hofmeister & O. Kühne (Eds.), StadtLandschaften. Die neue Hybridität von Stadt und Land (pp. 253–281). Wiesbaden: Springer VS.

Kühne, O. (2016b). Räume, Grenzen und Ränder – Aspekte gesellschaftlicher Raumorganisation. In A. Schaffer, E. Lang, & S. Hartard (Eds.), *An und in Grenzen – Entfaltungsräume für eine nachhaltige Entwicklung* (pp. 303–326). Marburg: Metropolis.

Kühne, O., & Hernik, J. (2015). Zur Bedeutung materieller Objekte bei der Konstitution von Heimat – unter besonderer Berücksichtigung von deutschstämmigen Objekten deutschen

Ursprungs aus der Teilungsära Polens. In O. Kühne, K. Gawroński, & J. Hernik (Eds.), *Transformation und Landschaft – landschaftliche Folgen gesellschaftlicher Veränderungsprozesse* (pp. 221–230). Wiesbaden: Springer VS.

Kühne, O., & Schmitt, J. (2012a). Spiel mit Landschaft – Logiken der Konstruktion von Landschaft in der Diskursgemeinschaft der Modelleisenbahner. *Berichte zur Deutschen Landeskunde, 86*(2), 175–194.

Kühne, O., & Schmitt, J. (2012b). Landschaft und Modelleisenbahn – zur Erzeugung von Landschaft im Spannungsfeld von ästhetischen Präferenzen, Anpassung an das Vorbild und Technikbezogenheit. Zoll+. *Österreichische Schriftenreihe für Landschaft und Freiraum, 21,* 108–111.

Kühne, O., & Schönwald, A. (2015). *San Diego – Eigenlogiken, Widersprüche und Entwicklungen in und von ‚America's finest city'*. Wiesbaden: Springer VS.

Kühne, O., & Weber, F. (2015). Der Energienetzausbau in Internetvideos – eine quantitativ ausgerichtete diskurstheoretisch orientierte Analyse. In S. Kost & A. Schönwald (Eds.), *Landschaftswandel – Wandel von Machtstrukturen* (pp. 113–126). Wiesbaden: Springer VS.

Kusiak, J. (2012). Die Gated-Community-Stadt Warschau. *StadtBauwelt, 196,* 41–45.

Lang, R. (2003). *Edgeless cities. Exploring the elusive metropolis.* Washington: Brookings Institution Press.

Lefebvre, M. (2006). Between setting and landscape in the cinema. In M. Lefebvre (Ed.), *Landscape and film* (pp. 19–60). New York, London: Routledge.

Leggewie, C. (1998). *Internet & Politik: von der Zuschauer- zur Beteiligungsdemokratie?* Köln: Bollmann.

Leibenath, M., & Otto, A. (2011). Diskursive Konstituierung von Kulturlandschaft am Beispiel politischer Windenergiediskurse. Lecture at the Kulakon-final conference on 13th May 2011 in Hannover.

Leibenath, M., & Otto, A. (2012). Diskursive Konstituierung von Kulturlandschaft am Beispiel politischer Windenergiediskurse in Deutschland. *Raumforschung und Raumordnung, 70*(2), 119–131.

Leibenath, M., & Otto, A. (2013). Windräder in Wolfhagen – eine Fallstudie zur diskursiven Konstituierung von Landschaften. In M. Leibenath, S. Heiland, H. Kilper & S. Tzschaschel (Eds.), Wie werden Landschaften gemacht? Sozialwissenschaftliche Perspektiven auf die Konstituierung von Kulturlandschaften (pp. 205–236). Bielefeld: transcript.

Lichtenberger, E. (1995). Vorsozialistische Siedlungsmuster, Effekte der sozialistischen Planwirtschaft und Segmentierung der Märkte. In H. Fassmann & E. Lichtenberger (Eds.), Märkte in Bewegung. Metropolen und Regionen in Ostmitteleuropa (pp. 27–35). Wien: Böhlau.

Liessmann, K. (1999). *Philosophie der modernen Kunst. Eine Einführung.* Wien: WUV.

Light, I. (1988). Los Angeles. In M. Dogan & J. Kasarda (Eds.), *The metropolis era, Volume 2: Mega cities* (pp. 56–96). Newbury Park et al.: Sage.

Lisowski, A. (2010). Suburbanizacja w obszarze metropolitalnym Warszawy. In S. Ciok & P. Migoń (Eds.), Przekształcenia struktur regionalnych. Aspekty społeczne, ekonomiczne i przyrodnicze, Uniwersytet Wrocławski, Instytut Geografii i Rozwoju Regionalnego (pp. 93–108). Wrocław: Instytut Geografii i Rozwoju Regionalnego.

Löw, M. (2010). *Soziologie der Städte.* Suhrkamp: Frankfurt a. M.

Lüke, B. (2008). Los Angeles im Film. Das Hollywoodkino zwischen Traum und Alptraum. Saarbrücken: VDM, Verlag Dr. Müller.

Lukinbeal, C. (2000). *‚On location' in San Diego. Film, television and urban thirdspace.* Santa Barbara: San Diego State University.

Lukinbeal, C. (2005). Cinematic landscapes. *Journal of Cultural Geography, 23*(1), 3–22.

Madsen, D. (1998). *American excepionalism.* Edinburg: Edinburgh University Press.

Marcińczak, S., et al. (2015). Patterns of socioeconomic segregation in the capital cities of fast-track reforming postsocialist countries. *Annals of the Association of American Geographers, 105*(1), 183–202.

Marcus, M. (1995). Geography as visual ideology: Landscape, knowledge, and power in Neo-Assyrian art. In M. Liverani (Ed.), *Neo-Assyrian geography* (pp. 193–202). Rome: Università di Roma.

Massey, D. (2006). *For space*. London: Sage.

Matless, D. (2005). *Landscape and englishness*. London: Reaktion Books.

Matt, S. (1998). Frocks, finery, and feelings. Rural and urban woman's envy, 1890–1930. In P. Stearns & J. Lewis (Eds.), *An Emotional History of the United States* (pp. 377-395). New York, London: New York University Press.

McQuire, S. (2009). Public screens, civic architecture and the transnational public sphere. In J. Döring & T. Thielmann (Eds.), *Mediengeographie. Theorie – Analyse – Diskussion* (pp. 565–586). Bielefeld: transcript.

Merchant, C. (1996). Reinventing Eden. Western culture as a recovery narrative. In W. Cronon (Ed.), *Uncommon ground. Rethinking the human place in nature* (pp. 132–170). New York, London: WW Norton & Company.

Mierzejewska, L. (2011). Appropriation of public urban space as an effect of privatisation and globalisation. *Quaestiones Geographicae, 30*(4), 39–46.

Mills, C. (1997a). Myths and meanings of gentrification. In J. S. Duncan & D. Ley (Eds.), *Place, culture, representation* (pp. 149–170). London: Routledge.

Mills, S. (1997b). *The American landscape*. Edinburgh: Keele.

Moïsi, D. (2009). Kampf der Emotionen. Wie Kulturen Angst, Demütigung und Hoffnung die Weltpolitik bestimmen. Darmstadt: Deutsche Verlags-Anstalt.

Molotch, H. (1998). L.A. as design product. How art works in a regional economy. In A. J. Scott & E. Soja (Eds.), *The city. Los Angeles and urban theory at the ende of the twentieth century* (pp. 225–275). Berkeley, Los Angeles, London: University of California Press.

Muller, E. (2010). Building American cityscape. In M. Conzen (Ed.), *The making of American landscape* (pp. 303–328). New York, London: Routledge.

Münker, S. (2009). Emergenz digitaler Öffentlichkeiten. Die Sozialen Medien im Web 2.0. Frankfurt a. M.: Suhrkamp.

Newman, D., & Paasi, A. (1998). Fences and neighbours in the postmodern world: Boundary narratives in political geography. *Progress in Human Geography, 22*(2), 186–207.

Nooteboom, C. (2002[1973 and 1987]). "Autopia" (1973) and Passages from "The Language of Images" (1987). In Ch. G. Salas & A. S. Roth (Eds.), *Looking for Los Angeles* (pp. 13–28). Los Angeles: Getty Research Institute.

Olessak, E. (1981). *Kalifornien*. München: Prestel.

Olmsted, F., Bartholomew, H., & Cheney, C. (1924). *A major traffic street plan for Los Angeles*. Los Angeles: City of Los Angeles.

Olwig, K. (2002). *Landscape, nature, and the body politic. From Britain's renaissance to America's new world*. London: University of Wisconsin Press.

Olwig, K. (2008). The Jutland Ciper: Unlocking the meaning and power of a contested landscape. In M. Jones & K. Olwig (Eds.), *Nordic Landscapes. Region and Belonging on the Northern Edge of Europe* (pp. 12–52). Minneapolis, London: University of Minnesota Press.

Olwig, K. (2009). The landscape of 'customary' law versus that 'natural' law. In K. Olwig & D. Mitchell (Eds.), *Justice, power and the political landscape* (pp. 11–32). London, New York: Routledge.

Olwig, K. (2011). The earth is not a globe: Landscape versus the 'globalist' agenda. *Landscape Research, 36*(4), 401–415.

Pareto, V. (2006 [1916]). Allgemeine Soziologie. München: J. C. B. Mohr.

Pasqualetti, M. (2001). Wind energy landscapes: Society and technology in the California desert. *Society & Natural Resources, 14*(8), 689–699.

Piątek, G. (2008). In Warschau wäre Gleichheit undenkbar. *Bauwelt, 12*, 28–37.

Pietila, A. (2010). *Not in my neighborhood. How bigotry shaped a Great American city*. Chicago: Rowman & Littlefield.

Popitz, H. (1992). *Phänomene der Macht*. Tübingen: Mohr Siebeck.

Popjaková, D. (1998). *Socioekonomická transformácia. Folia geografica, 1*, 317–339.

# References

Prawelska-Skrzypek, G. (1988). Social differentiation in old central city neighbourhoods in Poland. *Area, 20*(3), 221–232.

Pregill, P., & Volkman, N. (1999). *Landscapes in history. Design and planning in the eastern and western traditions*. New York et al.: Wiley.

Radding, C. (2005). *Landscapes of power and identity: Comparative histories in the Sonoran desert and the forests of Amazonia from colony to republic*. Durham, London: Duke University Press.

Radzimsky, A. (2009). Die Liberalisierung der Wohnungspolitik in Polen und ihre sozialräumlichen Auswirkungen. *Das Beispiel Poznań. Europa regional, 17*(3), 157–168.

Renn, O. (2005). Technikakzeptanz: Lehren und Rückschlüsse der Akzeptanzforschung für die Bewältigung des technischen Wandels. *Technikfolgenabschätzung – Theorie und Praxis, 14* (3), 29–38.

Röhring, A. (2008). Gemeinschaftsgut Kulturlandschaft. Dilemma und Chancen der Kulturlandschaftsforschung. In D. Fürst, L. Gailing, K. Pollermann & A. Röhring (Eds.), Kulturlandschaft als Handlungsraum. Institutionen und Governance im Umgang mit dem regionalen Gemeinschaftsgut Kulturlandschaft (pp. 35–48). Dortmund: Rohn-Verlag.

Ruchala, F. (2009). Crude city. Oil. In K. Varnelis (Ed.), The infrastructural city. Networked ecologies in Los Angeles (pp. 54–67). Barcelona, New York: Scribner.

Rygg, B. (2012). Wind power—An assault on local landscapes or an opportunity for modernization? *Energy Policy, 48*, 167–175.

Sachs, W. (1989). Die auto-mobile Gesellschaft. Vom Aufstieg und Niedergang einer Utopie. In F.-J. Brüggemeier & T. Rommelspacher (Eds.), Besiegte Natur. Geschichte der Umwelt im 19. und 20. Jahrhundert (pp. 106–123). München: C. H. Beck.

Sailer-Fliege, U. (1999). Wohnungsmärkte in der Transformation: Das Beispiel Osteuropa. In R. Pütz (Ed.), Ostmitteleuropa im Umbruch. Wirtschafts- und sozialgeographische Aspekte der Transformation (pp. 69–84). Mainz: Geographisches Institut des Johannes Gutenberg-Universität.

Schäfer, P. (1998). *Alltag in den Vereinigten Staaten. Von der Kolonialzeit bis zur Gegenwart*. Graz, Wien: Styria.

Schmidt, J. (2011). *Das neue Netz. Merkmale, Praktiken und Folgen des Web 2.0*. Konstanz: UVK Verlagsgesellschaft mbH.

Schneider, M. (2016). Der Raum – ein Gemeingut? Die Grenzen einer artorientierten Raumverteilung. In F. Weber & O. Kühne (Eds.), Fraktale Metropolen. Stadtentwicklung zwischen Devianz, Polarisierung und Hybridisierung (pp. 179–214). Wiesbaden: Springer VS.

Schneider-Sliwa, R. (2005). *USA*. Darmstadt: Wissenschaftliche Buchgesellschaft.

Schöbel, S. (2012). *Windenergie und Landschaftsästhetik. Zur landschaftlichen Anordnung von Windfarmen*. Berlin: Jovis.

Schor, J. (2004). *Born to buy*. New York, London, Toronto, Sydney: Scribner.

Schultze, C. (2006). Landnutzungsdynamik. Vom Zeitfenster für Energielandschaften. In Institut für Landschaftsarchitektur und Umweltplanung – Technische Universität Berlin (Ed.), *Perspektive Landschaft* (pp. 307–320). Berlin: Wissenschaftlicher Verlag Berlin.

Schwentker, W. (2006). Die Megastadt als Problem der Geschichte. In W. Schwentker (Ed.), Megastädte im 20. Jahrhundert (pp. 7–26). Göttingen: Vandenhoeck & Ruprecht.

Scott, A. J. (1988). *Metropolis. From the division of labor to urban form*. Berkeley: University of California Press.

Selman, P. (2010). Learning to love the landscapes of carbon-neutrality. *Landscape Research, 35*, 157–171.

Sennett, R. (1977). *The fall of public man*. Cambridge et al.: Cambridge University Press.

Sennett, R. (1991). *Civitas*. Suhrkamp: Die Großstadt und die Kultur des Unterschieds. Frankfurt a. M.

Short, L. (2002). Wind power and English landscape identity. In M. Pasqualetti, P. Gipe, & P. Righter (Eds.), *Wind power in view: Energy landscapes in a crowded world* (pp. 43–58). San Diego et al.: Academic Press.

Siebel, W. (2004). Einleitung: Die europäische Stadt. In W. Siebel (Ed.), *Die europäische Stadt* (pp. 11–50). Frankfurt a. M.: Suhrkamp.

Sloane, D. (2003). Medicine in the (Mini) Mall: An American Health Care Landscape. In Ch. Wilson & P. Groth (Eds.), Everyday America. Cultural landscape studies after J. B. Jackson (pp. 293–308). Berkeley, Los Angeles, London: University of California Press.

Slotkin, R. (1973). *Regeneration through violence: The mythology of the American frontier, 1600–1860*. Middletown: University of Oklahoma Press.

Soja, E. (1989). *Postmodern geographies. The reassertion of space in critical social theory*. London, New York: Verso.

Soja, E. (1995). Postmodern urbanization: The six restructurings of Los Angeles. In S. Watson & K. Gibson (Eds.), *Postmodern cities and spaces* (pp. 125–137). Oxford: Blackwell.

Soja, E. (1996). *Thirdspace. Journeys to Los Angeles and other real and imagined places*. Cambridge: Blackwell.

Soja, E., & Scott, A. (2006). Los Angeles 1870–1990. Historische Geographie einer amerikanischen Megastadt. In W. Schwentker (Ed.), Megastädte im 20. Jahrhundert (pp. 283–304). Göttingen: Vandenhoeck & Ruprecht.

Soyez, D. (2003). Kulturlandschaftspflege: Wessen Kultur? Welche Landschaft? Was für eine Pflege? *Petermanns Geographische Mitteilungen, 147*, 30–39.

Stakelbeck, F., & Weber, F. (2010). Heidelberg – Mannheim – Ludwigshafen: Stadtentwicklung zwischen Idealstadtmodellen, Leitbildern und historischem Einfluss. *Mitteilungen der Fränkischen Geographischen Gesellschaft, 57*, 51–86.

Stanilov, K. (Ed.). (2007). *The post-socialist city: Urban form and space transformations in Central and Eastern Europe after socialism. GeoJournal Library* (Vol. 92). Dordrecht: Springer.

Staniszkis, M. (2012). Continuity of Change vs. Change of Continuity: A Diagnosis and Evaluation of Warsaws's Urban Transformation. In M. Grubbauer & J. Kusiak (Eds.), Chasing Warsaw. Socio-material dynamics of urban change since 1990 (pp. 81–108). Frankfurt a. M.: Campus.

Starr, R. G. (1986). *San Diego. A pictorial history*. Norfolk: The Donning Company.

Starr, K. (2006). *Coast of dreams. California on the edge, 1990–2003*. New York: Modern Library.

Starr, K. (2007). *California. A history*. New York: Modern Library.

Starr, K. (2009). Los Angeles. The culture of imagination. In J. Norwich (Ed.), The great cities in history (pp. 259–263). New York: Thames & Hudson.

Steele, J. (1997). *Los Angeles architecture. The contemporary condition*. London: Phaidon Press.

Stichweh, R. (1997). Professions in modern society. *International Review of Sociology, 7*, 95–102.

Stotten, R. (2015). Landscape as a common good: The agrarian view. In M. Kohe, A. Koutsouris, R. B. Larsen, D. Maye, E. Noe, T. Oedl-Wieser, L. Philip, P. Pospěch, E. Rasch, M. Rivera, M. Schermer, S. Shortall, P. Starosta, S. Sumane, R. Wilkie & M. Woods. (Eds.), *Places of possibility? Rural societies in a Neoliberal World. On-line Proceedings of the XXVI European Society for Rural Sociology Congress* (pp. 172–173). Craigiebuttler Aberdeen: The James Hutton Institute.

Stremke, S. (2010). *Designing sustainable energy landscapes. concepts, principles and procedures*. Wageningen: Wageningen University.

Sulima, R. (2012). The laboratory of polish postmodernity. An ethnographic report from Stadium-Bazaar. In M. Grubbauer & J. Kusiak (Eds.), *Chasing Warsaw. Socio-material dynamics of urban change since 1990* (pp. 241–268). Frankfurt a. M.: Campus.

Swales, J. (1990). *Genre analysis: English in academic and research settings*. Cambridge: Cambridge University Press.

Swift, E. (2012). *The big roads. The untold story of the engineers, visionairies, and trailblazers who created the American Superhighways*. Boston, New York: Mariner Books.

Swyngedouw, E. (1997). Power, nature, and the city. The conquest of water and the political ecology of urbanization in Guayaquil, Ecuador: 1880–1990. *Environment and Planning, 29*, 311–332.

Thieme, G., & Laux, H. (1996). Los Angeles. Prototyp einer Weltstadt an der Schwelle zum 21. Jahrhundert. *Geographische Rundschau, 48*(2), 82–88.

Tijuanalandia (2010). La Mona. http://www.tijuanalandia.com/tag/la-mona/. Accessed September 30, 2013.

Tobias, K. (2012). Zukunftslandschaften. In D. Bruns & O. Kühne (Eds.), *Landschaften: Theorie, Praxis und internationale Bezüge* (pp. 323–334). Schwerin: Oceano.

Tuan, Y.-F. (1979a). *Landscapes of fear*. New York: Pantheon Books.

Tuan, Y.-F. (1979b). Sight and pictures. *Geographical Revue, 69*(4), 413–422.

Urry, J. (2000). *Sociology beyond societies*. London: Routledge.

U.S. Department of Transportation. (1921). Federal highway act of 1921. Date: November 9, 1921. Number: 42 Stat. 22.

U.S. Department of Transportation. (1956). Federal-aid highway act of 1956. Date: June 29, 1956. *Public Law*, 84–627.

Van der Horst, D., & Toke, D. (2010). Exploring the landscape of wind farm developments, local area characteristics and planning process outcomes in rural England. *Land Use Policy, 27*(2), 214–221.

Vanderbilt, T. (2009). *Traffic. Why we drive the way we do (and What It Says About Us)*. New York: Vintage.

Varnelis, K. (2009a). Introduction. Networked ecologies. In K. Varnelis (Ed.), *The infrastructural city. Networked ecologies in Los Angeles* (pp. 6–17). Barcelona, New York: ACTAR.

Varnelis, K. (2009b). Invisible city. Telecommunication. In K.Varnelis (Ed.), *Networked ecologies in Los Angeles* (pp. 120–131). Barcelona, New York: ACTAR.

Varnelis, K. (2009c). *The infrastructural city. Networked ecologies in Los Angeles*. Barcelona, New York: ACTAR.

Veblen, T. (1899). *The theory of the leisure class. An economic study of the evolution of institutions*. New York: B.W. Huebsch.

Vicenzotti, V. (2005). Kulturlandschaft und Stadt-Wildnis. In I. Kazal, A. Voigt, A. Weil & A. Zutz (Eds.), *Kulturen der Landschaft. Ideen von Kulturlandschaft zwischen Tradition und Modernisierung* (pp. 221–236). Berlin: Universitätsverlag der TU Berlin.

Wachs, M. (1984). Autos, transit, and sprawl of Los Angeles: The 1920s. *Journal of the American Planning Association, 50*, 297–310.

Wachs, M. (1998). The evolution of transportation policy in Los Angeles. Images of past policies and future prospects. In A. J. Scott & E. Soja (Eds.), *The city. Los Angeles and urban theory at the end of the twentieth century* (pp. 106–159). Berkeley, Los Angeles, London: University of California Press.

Waldie, D. (2005). *Holy land. A suburban memoir*. New York: W. W. Norton.

Walk, H. (2008). *Partizipative Governance. Beteiligungsrechte und Beteiligungsformen im Mehrebenensystem der Klimapolitik*. Wiesbaden: Springer VS.

Walter, F., et al. (2013). *Die neue Macht der Bürger. Was motiviert die Protestbewegungen?* Reinbek bei Hamburg: Rowohlt.

Weber, F. (2013). Soziale Stadt – Politique de la Ville – Politische Logiken. (Re-)Produktion kultureller Differenzierungen in quartiersbezogenen Stadtpolitiken in Deutschland und Frankreich. Wiesbaden: Springer VS.

Weber, F. (2015). Diskurs – Macht – Landschaft. Potenziale der Diskurs- und Hegemonietheorie von Ernesto Laclau und Chantal Mouffe für die Landschaftsforschung. In S. Kost & A. Schönwald (Eds.), *Landschaftswandel – Wandel von Machtstrukturen* (pp. 97– 112). Wiesbaden: Springer VS.

Weber, M. (1976 [1922]). *Wirtschaft und Gesellschaft. Grundriß der verstehenden Soziologie*. Tübingen: Mohr Siebeck.

Weinstein, R. (1996). The first American City. In A. J. Scott & E. Soja (Eds.), *The city. Los Angeles and Urban Theory at the Ende of the Twentieth Century* (pp. 22–46). Berkeley, Los Angeles, London: University of California Press.

Werlen, B. (1997). *Gesellschaft, Handlung und Raum – Grundlagen handlungstheoretischer Sozialgeographie*. Stuttgart: Franz-Steiner Verlag.

Wescoat, J. (2010). Watering the Deserts. In M. Conzen (Ed.), *The Making of American Landscape* (pp. 207–228). New York, London: Routledge.

Williams, M. (2010). Clearing the forests. In M. Conzen (Ed.), *The making of American Landscape* (pp. 162–187). New York, London: Routledge.

Winter, R. (1997). Vom Widerstand zur kulturellen Reflexivität. Die Jugendstudien der British Cultural Studies. In M. Charlton & S. Schneider (Eds.), *Rezeptionsforschung* (pp. 59–72). Opladen: Westdeutscher Verlag.

Wolfe, T. (2002 [1965]). The Kandy-Kolored Tangerine-Flake Streamline Baby. In D. L. Ulin (Ed.), Writing Los Angeles. A literary anthology (pp. 438–464). New York: Library of America.

Wolsink, M. (2007). Planning of renewables schemes: Deliberative and fair decision-making on landscape issues instead of reproachful accusations of non-cooperation. *Energy Policy, 35*(5), 2692–2704.

Zelinsky, W. (2010). Asserting central authority. In M. P. Conzen (Ed.), *The making of American Landscape* (pp. 329–356). New York, London: Routledge Chapman & Hall.

Zillien, N., & Hargittai, E. (2009). Digital distinction: Status-specific types of internet usage. *Social Science Quarterly, 90*(2), 274–291.

Zukin, S. (1993). *Landscapes of power. From detroit to disney world.* Berkeley, Los Angeles, London: University of California Press.

Zukin, S., Lash, S., & Friedman, J. (1992). Postmodern urban landscapes: Mapping culture and power. In S. Lash & J. Friedman (Eds.), *Modernity and Identity* (pp. 221–247). Oxford et al.: Blackwell Publishers.

# Chapter 6
# Conclusion

The present book has focused less on the physical substrate of what we know as landscape, and its origins and development through inscriptions of power, than on the genesis, change and, above all, perpetuation of landscape-related ascriptions as the expression of social forces. This perspective broadens existing research on landscape and power. It also embraces the shift in the current research focus on globalization and 'neoliberalism', which—often implicitly and sometimes even explicitly—calls for greater political influence on society and its physical structures. Given the logic at work here—above all in planning authorities, but also in scientific advice to politicians—this particular path may well be seen as supporting an administrative approach to landscape.

From the point of view of social constructivism, landscape is neither a value-free physical entity, of the sort postulated by the positivist school, nor one that possesses its own inherent value, as typically maintained by conservative and essentialist scholars. Landscape, from a social-constructivist perspective, is defined by social processes that frequently serve the interests of the powers that be; hence the importance of maintaining symbolic dimensions of material structures (see Greider and Garkovich 1994); hence, too, the tendency of landscape to become a medium for the perpetuation (or even enhancement) of unequal social opportunities. The processes at work in the genesis of landscape create a recursive relation between the physical and social substrates of what we appropriate as landscape. On the level of social landscape, interpretations and evaluations of landscape are generated and assimilated in the individual process of socialization. On this basis, the objects in a specific physical space are synthesized into a 'landscape' which is then subjected to continuous comparison with existing social constructs such as stereotypical landscapes or the normative landscape of the home environment.

In this process of synthesis, comparison, and adjustment, landscape in its various dimensions (including the aesthetic) evolves into an expression of power structures and relations and an instrument for their continued development and preservation (particularly through the transmission of taste). A key discursive aspect of this process is the production of paradigms of landscape, especially by experts (see Hannigan 2014), for whom they serve as hegemonic instruments. This inevitably involves inequality of opportunity for others, be they discursive underdogs or laypeople: neither group gains a hearing when it comes to expressing their spatial ideas and requirements adequately.

However, constructivist perspectives have made such inroads into sociological landscape research that this is now an essential avenue of approach. In addition—largely as a result of this development—lay participation in planning processes is noticeably increasing (see also Mitchell 2003; Collins 2004; Jones 2007; Kühne 2013). Today it is widely accepted that

- laypeople are experts in their own environment and know their needs in relation to landscape better than experts;
- the expansion of education and the availability of information on the Internet have undermined the lay/expert dichotomy;
- planners should no longer seek to modify the physical basis of landscape in the interests of creating a 'good' society, but content themselves with supporting local people in the development of a landscape that best fulfills their actual requirements.

The rise of a social-constructivist mindset has brought a meta-level of reflection to considerations of landscape, inasmuch as the scope and limits of other kinds of knowledge—whether implicit, conceptual, or systematic—are often critically scrutinized. This extends to the ontological and reifying tendencies found in some approaches informed by contemporary ideas of governance (see Gailing 2012, 2015). Thus, by laying bare the constructed character of spatial entities, socio-spatial exclusion processes—especially that of 'home' versus 'other' environments—may be avoided. After all, both aesthetic and cognitive interpretations are reversible, as are the emotional components of landscape awareness. The normative landscape of the home environment is a symbol of continuity and tranquility in a breakneck world; but it, too, is subject to intergenerational change—for in the social evaluation of landscape, aesthetic ascriptions, cognitive interpretations, and emotional forces form a circle of mutually conditioning bonds (see Hook 2008).

The integration of aesthetics in this recursive interface underlines the fact that landscape can never become a space of contemplative flight from the self-perpetuating workings of social and governmental powers. The sacralization of landscape as a refuge from everyday systems of this kind fails to recognize the systemic interests that colonize landscape in all its dimensions—social, individual, and physical (see Warnke 1992). All these contextual modalities are the product of a specific history, the "result of a situation determined by power and dominance" (Hauser and Kamleithner 2006, p. 168). It follows that the concept of 'cultural landscape', if taken at face value, is a euphemism concealing rather than expressing the forces at work in the genesis of such landscapes (Kühne 2006, 2008). These, in Popitz's terms, amount to technical power, and they can be seen to operate on two levels: on the physical landscape as power to shape the forces of nature and on the human environment as power to communicate that shaping (see Popitz 1992).

# 6 Conclusion

The sacralization of these processes as 'culture' reveals "a yearning for harmony between man and nature" (Hauser 2001) that obscures social power structures, thereby implicitly reinforcing them (Kühne 2008).

As such, it contradicts the principle of equality of opportunity (Kühne 2011), whose intergenerational connotations (see e.g. Tremmel 2009) with reference to landscape immediately bring up the question of sustainable deployment of both physical (e.g. conservation of soil fertility and species diversity) and financial resources. This question involves not only whether long-term technical infrastructure, for example, will meet future needs, but whether it will burden coming generations with a mortgage for structures they may find useless. Seen in this light, equality of opportunity reflects not so much the paradigm of preserving and restoring the appropriated physical landscape—except in issues like maintenance of species diversity—as the paradigms of successional development, reflective design, and reinterpretation (see Kühne 2011).

Applied to questions of a sustainable future, the social-constructivist perspective of landscape research has a number of implications (see Kühne 2012). Avoiding hegemonic inscriptions of aesthetic value and cognitive meaning, appropriated physical landscapes may foreseeably become expressions of an aesthetic of intergenerational, inter-gender, and intercultural tolerance and diversity that is open to multiple authorships and interpretations, as well as to physical change (Riley 1994; Potteiger and Purinton 2002 [1998]; Kühne 2013). As such, they will combine emotional accessibility with cognitive intelligibility; and they will take history seriously, albeit not exclusively. The principle of equality of opportunity requires that the development of physical space provide for all segments of the population in comparable measure, whether in education, recreation (parks and green belts), or safety and security. Moreover, ecological sustainability requires that the results of environmental research also be taken seriously and that its philosophical roots should be expressed in the governance and administration of landscape (see Fainstein 2010). This, in turn, must be so organized that no new inequalities of opportunity arise or old ones be perpetuated—for example by staffing the offices of governance exclusively with people endowed with higher symbolic capital, to the systematic detriment of those with less (Kühne and Meyer 2015).

The processes of governance should constitutionally engage all those concerned with landscape, and should do so at all levels—social, individual, and physical. This means lay participation in every step of planning procedures, from the selection of elements for analysis, through the development of a synthesis, to the physical realization of plans, as well as projects for the extension of a social understanding of landscape (e.g. landscape biographies). Postmodern appropriated physical landscapes that arise under these conditions can be understood as expressions of social contingency that also extend to and embrace 'unsuccessful' socialization. Their processes will not be entirely free from the workings of power, but integrated levels of reflection will operate to reduce systemic inequalities (such as between experts and lay participants). People can, after all, be relied upon to know their needs better than any expert. As Sofsky remarks, paternalism, whether conservative or socialist, "is a product of state pessimism, spawned by the erroneous notion that ordinary people are incapable of fending for themselves and recognizing what is good for them" (Sofsky 2007, p. 42).

The postmodern deconstruction of science and planning, with its outreach in the practical awareness that scientific concepts are radically embedded in the everyday world, brings with it not only a reinforcement of the need for social participation in processes concerned with landscape, but above all a rehabilitation of the concept of landscape in scientific discourse. If science, even in the 1970s (e.g. Paffen 1973), could still look modernistically askance at a concept with such an inexact 'semantic train', landscape today has been rehabilitated in the context of the realignment of scientific and everyday models of discourse. It is, in fact, regarded as contributing to the social legitimation of science (see Latour 1999; Weingart 2003). In such a perspective, appropriated physical landscape is accepted as a manifestation of a pluralist society socialized along inclusive rather than exclusive lines, wary of the fundamentalist dichotomies of 'good versus 'bad' and 'urban' versus 'rural', and open to feeling (in the sense advocated by Lipps 1891, 1902) and intuition (as advocated by Croce 1930) as alternative modes of perception alongside the (Western) cultural dominance of the visual. Against this background—especially in the wake of today's 'citizen science' movement—claims to discursive sovereignty can be deconstructed and many different landscape discourses judged on equal terms (see e.g. Finke 2016).

The growing significance of landscape in scientific as well as general contexts suggests two further questions: first, whether it is possible to synthesize the various perspectives of sociologists, cultural and natural scientists, and planners into a single comprehensive science of landscape—which might relativize the rampant semantic competition between these specialties (see Küster 2009; Kühne 2013)—and second, whether landscape should not take a more prominent place in the center of spatial concerns; for 'landscape' is anchored in everyday contexts, whereas 'space' is a more abstract, expert-weighted term.

The examination undertaken in this book of the meaning of landscape in discursive configurations of power and processes of social distinction has shown how profoundly activities related to landscape can influence the development of vital human opportunities—especially when these employ aesthetic categories and values. This is not only a matter of the obvious material foundations of the appropriated physical landscape, but also of the less obvious social dimensions of these phenomena. An approach to landscape that seeks to enhance the equality of opportunity of all concerned will place at the forefront of its considerations the question whether, and to what extent, human opportunities are affected by its actions.

# References

Collins, T. (2004). Konturen einer Ästhetik der Vielfalt. In H. Strelow (Ed.), *Ökologische Ästhetik. Theorie und Praxis künstlerischer Umweltgestaltung* (pp. 170–179). Basel, Berlin, Boston: Birkhäuser.

Croce, B. (1930). *Aesthetik als Wissenschaft vom Ausdruck und allgemeine Sprachwissenschaft*. Tübingen: J.C.B. Mohr.

Fainstein, S. (2010). *The just city*. Ithacta: Cornell University Press.

Finke, P. (2016). Was ist eine Grenze? Wider die digitale Weltsicht oder: Was wir von den Fröschen lernen können. In A. Schaffer, E. Lang, & S. Hartard (Eds.), *An und in Grenzen—Entfaltungsräume für eine nachhaltige Entwicklung* (pp. 19–47). Marburg: Metropolis.

Gailing, L. (2012). Sektorale Institutionensysteme und die Governance kulturlandschaftlicher Handlungsräume. Eine institutionen- und steuerungstheoretische Perspektive auf die Konstruktion von Kulturlandschaft. *Raumforschung und Raumordnung, 70*(2), 147–160.

Gailing, L. (2015). Landschaft und productive Macht. Auf dem Weg zur Analyse landschaftlicher Gouvernementalität. In S. Kost & A. Schönwald (Eds.), *Landschaftswandel—Wandel von Machtstrukturen* (pp. 37–41). Wiesbaden: Springer VS.

Greider, T., & Garkovich, L. (1994). Landscapes: The social construction of nature and the environment. *Rural Sociology, 59*(1), 1–24.

Hannigan, J. (2014). *Environmental sociology*. London, New York: Taylor & Francis Group.

Hauser, S. (2001). *Metamorphosen des Abfalls. Konzepte für alte Industrieareale*. Campus: Frankfurt a. M.

Hauser, S., & Kamleithner, C. (2006). *Ästhetik der Agglomeration*. Wuppertal: Müller und Busmann.

Hook, S. (2008). *Landschaftsveränderungen im südlichen Oberrheingebiet und Schwarzwald. Wahrnehmung kulturtechnischer Maßnahmen seit Beginn der 19. Jahrhunderts*. Saarbrücken: VDM publishing company Dr. Müller.

Jones, M. (2007). The European landscape convention and the question of public participation. *Landscape Research, 32*(5), 613–633.

Kühne, O. (2006). Landschaft, Geschmack, soziale Distinktion und Macht—von der romantischen Landschaft zur Industriekultur. Eine Betrachtung auf Grundlage der Soziologie Pierre Bourdieus. Beiträge zur Kritischen Geographie 6. Wien: Verein kritische Geographie.

Kühne, O. (2008). *Distinktion, Macht, Landschaft. Zur sozialen Definition von Landschaft*. Springer VS: Wiesbaden.

Kühne, O. (2011). Die Konstruktion von Landschaft aus Perspektive des politischen Liberalismus. Zusammenhänge zwischen politischen Theorien und Umgang mit Landschaft. *Naturschutz und Landschaftsplanung, 43*(6), 171–176.

Kühne, O. (2012). *Stadt—Landschaft—Hybridität. Ästhetische Bezüge im postmodernen Los Angeles mit seinen modernen Persistenzen*. Wiesbaden: Springer.

Kühne, O. (2013). *Landschaftstheorie und Landschaftspraxis. Eine Einführung aus sozialkonstruktivistischer Perspektive*. Wiesbaden: Springer.

Kühne, O., & Meyer, W. (2015). Gerechte Grenzen? Zur territorialen Steuerung von Nachhaltigkeit. In O. Kühne & F. Weber, F. (Eds.), *Bausteine der Regionalentwicklung* (pp. 25–40). Wiesbaden: Springer.

Küster, H. (2009). *Schöne Aussichten. Kleine Geschichte der Landschaft*. München: C.H. Beck.

Latour, B. (1999). *Pandora's hope. Essays on the reality of science studies*. Cambridge, Massachusetts: Harvard University Press.

Lipps, T. (1891). *Ästhetische Faktoren der Raumanschauung*. Hamburg: L. Voss.

Lipps, T. (1902). *Vom Fühlen, Wollen und Denken*. Hamburg: JA Barth.

Mitchell, D. (2003). Cultural landscapes: Just landscapes or landscapes of justice? *Progress in Human Geography, 27*(6), 787–796.

Paffen, K. (1973). Einleitung. In K. Paffen (Ed.), *Das Wesen der Landschaft* (pp. 9–37). Darmstadt: Wissenschaftliche Buchgesellschaft.

Popitz, H. (1992). *Phänomene der Macht*. Tübingen: Mohr Siebeck.
Potteiger, M., & Purinton, J. (2002 [1998]). Landscape narratives. In S. Swaffield (Ed.), *Theory in Landscape Architecture. A Reader* (pp. 136–144). Philadelphia: University of Pennsylvania Press.
Riley, R. (1994). Gender, landscape, culture. Sorting out some questions. *Landscape Journal, 13*, 153–163.
Sofsky, W. (2007). *Verteidigung des Privaten. Eine Streitschrift*. München: C.H. Beck.
Tremmel, J. (2009). *A theory of intergenerational justice*. London: Earthscan.
Warnke, M. (1992). *Politische Landschaft. Zur Kunstgeschichte der Natur*. München, Wien: Carl Hanser Verlag.
Weingart, P. (2003). Wissenschaftssoziologie. Bielefeld: transcript.